Complex Analysis

An Invitation

Complex Analysis
An Invitation

2nd Edition

Murali Rao
University of Florida, USA

Henrik Stetkær
Aarhus University, Denmark

Søren Fournais
Aarhus University, Denmark

Jacob Schach Møller
Aarhus University, Denmark

 **World Scientific**

NEW JERSEY · LONDON · SINGAPORE · BEIJING · SHANGHAI · HONG KONG · TAIPEI · CHENNAI

Published by

World Scientific Publishing Co. Pte. Ltd.

5 Toh Tuck Link, Singapore 596224

USA office: 27 Warren Street, Suite 401-402, Hackensack, NJ 07601

UK office: 57 Shelton Street, Covent Garden, London WC2H 9HE

Library of Congress Cataloging-in-Publication Data
Complex analysis : an invitation. -- 2nd edition / Murali Rao (University of Florida, USA),
Henrik Stetkaer (Aarhus University, Denmark), Soren Fournais (Aarhus University, Denmark),
Jacob Schach Moller (Aarhus University, Denmark).
    pages cm
  ISBN 978-9814579582 -- ISBN 978-9814579599 (pbk)
  1. Functions of complex variables. I. Rao, Murali. II. Stetkær, Henrik. III. Fournais, Søren
IV. Møller, Jacob S.
  QA331.7.R36 2015
  515'.9--dc23
                            2014044826

British Library Cataloguing-in-Publication Data
A catalogue record for this book is available from the British Library.

Printed in Singapore

Preface

This textbook is a rigorous introduction to the theory of functions of one complex variable. Much of the material has been used in both graduate and undergraduate courses at Aarhus University. It is our impression that courses in complex function theory these days at many places are postponed to leave space for other topics like point set topology and measure theory. Thus we have felt free to assume that the students as background have some point set topology and calculus of several variables, and that they understand $\varepsilon\delta$ arguments. From Chapter 10 on we even use Lebesgue's dominated and monotone convergence theorems freely. But the proofs are meant for the students and so they are fairly detailed.

We have made an effort to whenever possible to give references to literature that is accessible for the students and/or puts the theory in perspective, both mathematically and otherwise. E.g., to the Chauvenet prize winning paper Zalcman (1974) and to the fascinating gossip on Bloch's life in Campbell (1985) and its sequel Cartan and Ferrand (1988). Our goal is not to compete with the existing excellent textbooks for historical notes and remarks or for wealth of material. For that we refer the interested reader to, say the monumental work Burckel (1979) and the classic Sansone and Gerretsen (1960, 1969, I–II).

We have tried to reach some of the deeper and more interesting results (Picard's theorems, Riemann's mapping theorem, Runge's approximation theorems) rather early, and nevertheless to give the very basic theory an adequate treatment.

Standard notation is enforced throughout. A possible exception is that $B[a, r]$ is the closed ball with center a and radius r in analogy with the notation for a closed interval.

An important part of any course is the set of exercises. We have exercises after each chapter. They are meant to be doable for the students, so we have quite often provided hints about how to proceed.

A couple of times we have succumbed to the temptation of making a digression to an interesting topic, that will not be pursued, e.g., Tauber's theorem, Hadamard's gap theorem and the Prime Number Theorem. We hope that the reader will be irresistibly tempted as well.

The authors welcome correspondence with criticism and suggestions. In particular about literature on the level of students that are about to start their graduate studies in mathematics.

Preface to the second edition

The known typographical errors in the first edition have been corrected.

Furthermore, two of the authors, Fournais and Møller, have written six chapters with new material. These chapters provide a concise and accessible treatment of recent work on capacities and Loewner theory, not just in the disc, but also in the half-plane. Parts of the new material are described in research papers only or appear here for the first time. The enlarged edition bridges the gap between the classical material of the first edition and recent advances on Loewner's equation, thus providing the complex analysis backbone of Loewner theory.

Like the first edition the new material contains a number of worked-out examples and many exercises to illuminate the theory.

We are indebted to our colleague Jørgen Tornehave for sharing his deep insight in the topology and geometry of the complex plane with us.

We thank LATEX expert Lars Madsen for his invaluable help with practical details and his insistence on quality and consistency in the typesetting of our book.

Contents

Chapter 1

Power Series

1.1 Elementary facts

This section contains basic results about convergence of power series. The simplest and most important example is the *geometric series*

$$\sum_{k=0}^{\infty} z^k = \frac{1}{1-z}$$

which converges absolutely for any z in the open unit disc. It emerges in so many other contexts than complex function theory that it may be considered one of the fundamental elements of mathematics (see the thought-provoking paper Halmos, 1980/81).

Definition 1.1. *A **power series** around $z_0 \in \mathbb{C}$ is a formal series of the form*

$$\sum_{k=0}^{\infty} a_k(z - z_0)^k \tag{1.1}$$

where the coefficients $a_k \in \mathbb{C}$ are fixed and where $z \in \mathbb{C}$.

We shall very often only consider the case $z_0 = 0$, i.e.,

$$\sum_{k=0}^{\infty} a_k z^k \tag{1.2}$$

because it will be obvious how to derive the general case from this more handy special case.

We shall in Section 6.1 encounter Laurent series, i.e., "power" series in which the summation ranges over $\mathbb{Z}$, not just $\mathbb{N}$. A Laurent series will be treated as a sum of a power series in z and another in the variable $1/z$.

Proposition 1.2. *Consider the power series (1.2), and let us assume that the set $\{a_k\zeta^k \mid k = 0, 1, 2, \ldots\}$ is bounded for some $\zeta \in \mathbb{C}$.*

Then for any $\rho < |\zeta|$ the power series (1.2) converges absolutely and uniformly in the closed disc $B[0, \rho] = \{z \in \mathbb{C} \mid |z| \leq \rho\}$.

Proof. Because of boundedness $|a_k||\zeta^k| \leq M$ for some M and all k. Thus if $|z| \leq \rho < |\zeta|$, then

$$|a_k z^k| \leq |a_k||\zeta|^k \rho^k |\zeta|^{-k} \leq M \left(\frac{\rho}{|\zeta|}\right)^k$$

So the series is dominated by a convergent series (the geometric) and the proposition follows from Weierstrass' M-test. $\square$

Proposition 1.2 says that the power series (1.2) either converges for all complex z or there is a number m such that $\sup_k |a_k m^k| = \infty$. We define the *radius of convergence*, $\rho \in [0, \infty]$, of the power series (1.1) by

$$\rho := \inf\{m > 0 \mid \sup_k |a_k m^k| = \infty\} \tag{1.3}$$

and the *circle of convergence* as $\{z \in \mathbb{C} \mid |z - z_0| = \rho\}$, provided $\rho < \infty$. The reason for this terminology is that (1.1) converges at each interior point of the circle of convergence, and diverges at each exterior point.

A side remark: If z is not a complex number but, say a matrix, then the series may converge even if the norm of z is bigger than ρ; that can happen in exponentiating a matrix.

Theorem 1.3. *Let ρ denote the radius of convergence of the power series (1.1).*

(a) *(The Cauchy-Hadamard Radius of Convergence Formula)*

$$\rho = \frac{1}{\limsup_{k \to \infty} |a_k|^{\frac{1}{k}}}. \tag{1.4}$$

(b) *(The Quotient Formula). If $a_n \neq 0$ for all n, then*

$$\rho = \lim_{n \to \infty} \left|\frac{a_n}{a_{n+1}}\right|$$

provided that the limit exists in $[0, \infty]$.

(c) *(1.1) converges for any $R < \rho$ uniformly on the disc $B[z_0, R]$, and it converges absolutely on $B(z_0, \rho)$.*

Proof. Item (a). We prove that

$$\limsup_{k\to\infty} |a_k|^{\frac{1}{k}} \geq \tfrac{1}{\rho}$$

by contradiction: If

$$\limsup_{k\to\infty} |a_k|^{\frac{1}{k}} < \tfrac{1}{\rho}$$

then there exists $m > \rho$ such that

$$\limsup_{k\to\infty} |a_k|^{\frac{1}{k}} < \tfrac{1}{m}$$

So except for finitely many k we have $m|a_k|^{\frac{1}{k}} < 1$, i.e., $|a_k m^k|$ is bounded, contradicting the definition of ρ.

Next we shall prove that

$$\limsup_{k\to\infty} |a_k|^{\frac{1}{k}} \leq \tfrac{1}{\rho}$$

or equivalently

$$\limsup_{k\to\infty} |a_k|^{\frac{1}{k}} \leq \tfrac{1}{b}$$

for any $b < \rho$. In this case we note that there exists an $M > 0$ such that $|a_k|b^k \leq M$ for all k. Hence

$$\limsup_{k\to\infty} |a_k|^{\frac{1}{k}} b \leq \lim_{k\to\infty} M^{\frac{1}{k}} = 1$$

Item (b) is left to the reader as Exercise 1.2.

Item (c) is part of Proposition 1.2. □

As examples we note that the geometric series has radius of convergence 1, and that the series

$$\sum_{n=0}^{\infty} \frac{z^n}{n!}$$

has radius of convergence $\rho = \infty$.

Corollary 1.4. *If the power series $\sum_{k=0}^{\infty} a_k z^k$ has radius of convergence ρ, then so does the formally differentiated series $\sum_{k=0}^{\infty} a_{k+1}(k+1)z^k$.*

Proof. Note that $\sum a_{k+1}(k+1)z^k$ converges iff $\sum a_k k z^k$ does. Then use Cauchy-Hadamard and that $\lim_{k\to\infty} k^{\frac{1}{k}} = 1$. □

Recalling that a uniformly convergent series of continuous functions has continuous sum, we get

Proposition 1.5. *The sum of a power series is continuous inside its circle of convergence.*

Much more is true inside, as we will discover in the next chapter, in which we discuss differentiability. The behavior of the power series at the convergence circle is more delicate, and aspects of it will be treated in Section 1.2.

An N'th order polynomial which is 0 at $N + 1$ points is identically 0. A primitive extension of this to power series, viewed as polynomials of infinite order, is the following:

Proposition 1.6. *Assume that the power series* (1.1) *has positive radius of convergence, and let f denote its sum.*

(a) *If $f(z) = 0$ for z in a set which has z_0 as an accumulation point, then the coefficients a_k are all 0.*

(b) *In particular, f determines the coefficients uniquely.*

Our final version of Proposition 1.6 will be the Unique Continuation Theorem (Theorem 4.11).

Proof of the proposition. Item (b) is an immediate consequence of (a), so we will only do (a). We also take $z_0 = 0$.

Let $z_1, z_2, \ldots$ be a sequence from $\mathbb{C}$, such that $z_k \to 0$ as $k \to \infty$, $z_k \neq 0$ and $f(z_k) = 0$ for $k = 1, 2, \ldots$ By the continuity of f we get

$$a_0 = f(0) = \lim_{k \to \infty} f(z_k) = 0$$

But then

$$\frac{f(z)}{z} = \sum_{k=0}^{\infty} a_{k+1} z^k$$

so replacing $f(z)$ by $f(z)/z$ and by repeating the argument we find that $a_1 = 0$. Proceeding in this way we get successively that $a_n = 0$ for $n = 0, 1, 2, \ldots$ $\square$

For more detailed information on power series we refer the reader to Knopp (1951).

1.2 The theorems of Abel and Tauber

In this paragraph we will in special cases study the behavior of the power series (1.2) on its circle of convergence. Unfortunately, in general almost anything can happen. One can here mention a famous example, due to L. Fejér, of a power series that converges uniformly, but not absolutely, on the closed unit disc. (See Zalcman, 1974, p. 125 ff or Hille, 1959, p. 122.) Of course, absolute convergence on the circle of convergence holds either everywhere or nowhere. In case of absolute convergence the power series even converges uniformly on the closed disc $B[0, \rho] = \{z \in \mathbb{C} \mid |z| \leq \rho\}$, where ρ is the radius of convergence. We are thus left with the case of nonabsolute convergence which is technically unpleasant.

Theorem 1.7 (Abel's Theorem). *Assume that the power series*

$$f(z) = \sum_{k=0}^{\infty} a_k z^k$$

converges for $z = 1$ and hence for each $|z| < 1$.
  Then $\sum_{k=0}^{\infty} a_k x^k$ converges uniformly to $f(x)$ on the closed interval $[0,1]$. In particular, f is continuous on $[0,1]$, so

$$f(x) \to \sum_{k=0}^{\infty} a_k \qquad \text{as } x \to 1 \text{ in } [0,1].$$

Proof. Note that the sequence

$$r_k := - \sum_{m>k}^{\infty} a_m, k = 0, 1, 2, \ldots$$

converges to 0 as $k \to \infty$. In particular it is bounded so that the series $\sum_{k=0}^{\infty} r_k z^k$ converges whenever $|z| < 1$. Now, for any z such that $|z| < 1$ and any natural number N we have

$$f(z) - \sum_{k=0}^{N} a_k z^k = \sum_{k=N+1}^{\infty} a_k z^k = \sum_{k=N+1}^{\infty} (r_k - r_{k-1}) z^k$$

$$= \sum_{k=N+1}^{\infty} r_k z^k - \sum_{k=N+1}^{\infty} r_{k-1} z^k$$

$$= \sum_{k=N+1}^{\infty} r_k z^k (1 - z) - r_N z^{N+1}$$

from which we for $z = x \in [0, 1[$ get the estimate

$$\left| f(x) - \sum_{k=0}^{N} a_k x^k \right|$$

$$\leq \sup_{k>N}\{|r_k|\} \sum_{k=N+1}^{\infty} x^k(1-x) + |r_N| \leq \sup_{k>N}\{|r_k|\} + |r_N|,$$

the case $x = 1$ being trivial. Thus we have for any $x \in [0, 1]$ that

$$\left| f(x) - \sum_{k=0}^{N} a_k z^k \right| \leq 2 \sup_{k \geq N}\{|r_k|\}.$$

$\square$

Example 1.8. A combination of Abel's theorem with the identities

$$\text{Log}(1 + z) = \sum_{k=1}^{\infty} \frac{(-1)^{k-1}}{k} z^k \qquad \text{for } |z| < 1$$

$$\arctan z = \sum_{k=0}^{\infty} \frac{(-1)^k}{2k+1} z^{2k+1} \qquad \text{for } |z| < 1$$

yields the following pretty formulas

$$\text{Log } 2 = 1 - \tfrac{1}{2} + \tfrac{1}{3} - \tfrac{1}{4} + \cdots$$

$$\frac{\pi}{4} = 1 - \tfrac{1}{3} + \tfrac{1}{5} - \tfrac{1}{7} + \cdots$$

which are called respectively *Brouncker's series* and *Leibniz' series*.

Our next theorem provides a partial inverse to Abel's theorem.

Theorem 1.9 (Tauber's Theorem). *If $f(z) = \sum_{k=0}^{\infty} a_k z^k$ converges on the unit disc, $k a_k \to 0$ as $k \to \infty$ and $\lim f(x)$ exists for $x \to 1_-$, then $\sum_{k=0}^{\infty} a_k$ converges.*

Proof. Introducing

$$s_n := \sum_{k=0}^{n} a_k \quad \text{and} \quad \omega(n) := \sup_{k \geq n}\{k|a_k|\}$$

we find since

$$1 - x^k \leq k(1-x) \quad \text{and} \quad \sum_{k=n+1}^{\infty} \frac{x^k}{k} \leq \frac{1}{n+1} \sum_{k=n+1}^{\infty} x^k \leq \frac{1}{(1-x)(n+1)}$$

that

$$|f(x) - s_n| \leq \sum_{k=1}^{n} |a_k|(1 - x^k) + \omega(n+1) \sum_{k=n+1}^{\infty} \frac{x^k}{k} \qquad (1.5)$$

$$\leq \omega(0)(n+1)(1-x) + \frac{\omega(n+1)}{(1-x)(n+1)}$$

The expression on the right hand side of (1.5) is of the form $At + Bt^{-1}$ and when $t^2 = B/A$ it equals $2\sqrt{AB}$. So when x_n satisfies

$$(1 - x_n)^2 = \frac{\omega(n+1)}{(n+1)^2 \omega(0)}$$

we have from (1.5) that

$$|f(x_n) - s_n| \leq 2\sqrt{\omega(0)\omega(n+1)}$$

which tends to 0 as $n \to \infty$ because $\omega(n) \to 0$. $\qquad \square$

Remark. More difficult is Littlewood's Tauberian theorem which requires only that $\{ka_k\}$ is bounded. For an accessible proof (due to Karamata and Wieland) see p. 129 ff of Zalcman (1974).

Even though the following identity is easy to state and prove it is often useful in manipulations with explicitly given series. We used it, without actually saying so, in the proof of Abel's theorem.

Proposition 1.10 (Abel's Partial Summation Formula). *Let c_k and d_k, $k = 0, 1, \ldots, n$ be complex numbers, and put $s_k := \sum_{j=0}^{k} d_j$ for $k = 0, 1, \ldots, n$.*
Then

$$\sum_{k=0}^{n} c_k d_k = \sum_{k=0}^{n-1} (c_k - c_{k+1}) s_k + c_n s_n. \qquad (1.6)$$

1.3 Liouville's theorem

Liouville's theorem (that a bounded entire function is constant), is not only interesting in itself, but it is also useful in many unexpected situations. It has a generalization which says that a power series that grows at most polynomially at ∞ is a polynomial. We present the generalization here:

Theorem 1.11 (Liouville's Theorem). *Let the power series*

$$f(z) = \sum_{k=0}^{\infty} a_k z^k$$

have infinite radius of convergence.

(a) *If f is bounded then f is constant.*

(b) *More generally, if f for some nonnegative constants a, b and m satisfies the estimate*

$$|f(z)| \le a|z|^m \qquad \text{for all } |z| > b$$

then f is a polynomial of degree $\le m$.

Of course Liouville's theorem does not hold for smooth functions of one or more real variables. For example, the function $f(x) = \sin x$ is a bounded, smooth function of one real variable, but it is certainly not constant.

Liouville's theorem can be generalized to harmonic functions (see Chapter 12. Exercise 12.10).

Proof. It suffices to prove (b). The coefficient a_n may for any $n = 0, 1, 2, \ldots$ be found by the formula

$$a_n = \frac{1}{2\pi R^n} \int_0^{2\pi} e^{-in\theta} f(Re^{i\theta}) \, d\theta \qquad \text{for any } R > 0$$

which crops up when we introduce the power series expansion of f on the right hand side. In particular we get the estimate

$$|a_n| \le \frac{1}{2\pi R^n} \int_0^{2\pi} |f(Re^{i\theta})| \, d\theta$$

Substituting the estimate for f from (b) into this we find that

$$|a_n| \le aR^{m-n} \qquad \text{for all } R > b.$$

Letting $R \to \infty$ we see that $a_n = 0$ whenever $n > m$, so the power series reduces to the polynomial $f(z) = \sum_{n=0}^{N} a_n z^n$, where N is the integer part of m. $\square$

For proof without calculus(!) see Lenard (1986).

1.4 Important power series

In this paragraph we collect the most important power series expansions. Some of them will be derived later. In particular we will discuss the exponential function in the next chapter.

$$\frac{1}{1-z} = \sum_{n=0}^{\infty} z^n \qquad\qquad \text{for } |z| < 1,$$

$$e^z = \sum_{n=0}^{\infty} \frac{z^n}{n!} \qquad\qquad \text{for } z \in \mathbb{C},$$

$$\cos z = \frac{e^{iz} + e^{-iz}}{2} = \sum_{n=0}^{\infty} \frac{(-1)^n}{(2n)!} z^{2n} \qquad \text{for } z \in \mathbb{C},$$

$$\sin z = \frac{e^{iz} - e^{-iz}}{2i} = \sum_{n=0}^{\infty} \frac{(-1)^n}{(2n+1)!} z^{2n+1} \quad \text{for } z \in \mathbb{C},$$

$$\cosh z = \frac{e^z + e^{-z}}{2} = \sum_{n=0}^{\infty} \frac{1}{(2n)!} z^{2n} \qquad \text{for } z \in \mathbb{C},$$

$$\sinh z = \frac{e^z - e^{-z}}{2} = \sum_{n=0}^{\infty} \frac{1}{(2n+1)!} z^{2n+1} \quad \text{for } z \in \mathbb{C},$$

$$\text{Log}(1+z) = \sum_{n=1}^{\infty} \frac{(-1)^{n-1}}{n} z^n \qquad\qquad \text{for } |z| < 1,$$

$$(1+z)^\alpha = \sum_{n=0}^{\infty} \binom{\alpha}{n} z^n \qquad\qquad \text{for } |z| < 1,$$

$$\arctan z = \sum_{n=0}^{\infty} \frac{(-1)^n}{2n+1} z^{2n+1} \qquad\qquad \text{for } |z| < 1,$$

$$\arcsin z = \sum_{n=0}^{\infty} \frac{1 \cdot 3 \cdot 5 \cdots (2n-1)}{2 \cdot 4 \cdot 6 \cdots 2n} z^{2n+1} \qquad \text{for } |z| < 1,$$

$$J_\nu(z) = \frac{1}{2\pi} \int_0^{2\pi} e^{i(z \sin t - \nu t)} \, dt$$

$$= \sum_{n=0}^{\infty} \frac{(-1)^n (\frac{z}{2})^{\nu+2n}}{n! \Gamma(\nu + n + 1)} \qquad\qquad \text{for } z \in \mathbb{C}.$$

Above the power series expansion for $(1+z)^\alpha$ is of the principal branch, and J_ν is a Bessel function.

Exercises

1.1. Determine the radius of convergence and study the behavior on the circle of convergence of the following four power series:

$$\sum_{k=0}^{\infty} z^k, \quad \sum_{k=1}^{\infty} \frac{z^k}{k^2}, \quad \sum_{k=0}^{\infty} \frac{z^k}{k+1}, \quad \sum_{k=1}^{\infty} \frac{z^{kn}}{k}$$

where n is a positive integer.

1.2. Let $a_n > 0$ and $b_n \in \mathbb{C} \setminus \{0\}$ for $n = 0, 1, \ldots$

(α) Show that

$$\liminf_{n \to \infty} \frac{a_{n+1}}{a_n} \leq \liminf_{n \to \infty} a_n^{\frac{1}{n}} \leq \limsup_{n \to \infty} a_n^{\frac{1}{n}} \leq \limsup_{n \to \infty} \frac{a_{n+1}}{a_n}.$$

(β) Show that the radius of convergence of $\sum_{n=0}^{\infty} b_n z^n$ is $\lim_{n \to \infty} |b_n/b_{n+1}|$, provided that the limit exists in $[0, \infty]$.

1.3. Let $a \neq 0$. Show that the radius of convergence of the power series

$$1 + az + a(a - 2b)\frac{z^2}{2!} + a(a - 3b)^2 \frac{z^3}{3!} + \cdots$$

is $1/(e|b|)$.

1.4. Find an example of a power series $\sum_{n=0}^{\infty} a_n z^n$ that converges only at $z = 0$.

1.5.
(α) Let $\sum_{n=0}^{\infty} a_n z^n$ and $\sum_{n=0}^{\infty} b_n z^n$ be power series with radii of convergence ρ and σ respectively. Consider the product series $\sum_{n=0}^{\infty} c_n z^n$, defined by

$$c_n := \sum_{k=0}^{n} a_k b_{n-k} \qquad \text{for } n = 0, 1, \ldots$$

Show that its radius of convergence τ satisfies $\tau \geq \min(\rho, \sigma)$, and that

$$\sum_{n=0}^{\infty} c_n z^n = \left(\sum_{n=0}^{\infty} a_n z^n \right) \left(\sum_{n=0}^{\infty} b_n z^n \right) \qquad \text{whenever } |z| < \min(\rho, \sigma).$$

(β) Let $\sum_{n=0}^{\infty} a_n$ and $\sum_{n=0}^{\infty} b_n$ be convergent series of complex numbers, and let $\sum_{n=0}^{\infty} c_n$ denote the series given by

$$c_n = \sum_{k=0}^{n} a_k b_{n-k} \qquad \text{for } n = 0, 1, \ldots$$

Show that

$$\sum_{n=0}^{\infty} c_n = \left(\sum_{n=0}^{\infty} a_n \right) \left(\sum_{n=0}^{\infty} b_n \right)$$

provided that $\sum_{n=0}^{\infty} c_n$ is convergent.

1.6 (Eneström-Kakeya).

(α) If $a_n \geq a_{n-1} \geq \cdots \geq a_1 \geq a_0 > 0$, then

$$\sum_{k=0}^{n} a_k z^k \neq 0 \qquad \text{for all } |z| > 1.$$

Hint: Abel's partial summation formula.

(β) Deduce from (α), that $a_n > a_{n-1} > \cdots > a_1 > a_0 > 0$ implies that $\sum_{k=0}^{n} a_k z^k$ has all its roots in the open disc $|z| < 1$.

For extensions of the Eneström-Kakeya result consult e.g., Dewan and Govil (1984). The result has applications in digital signal analysis, where it is used in examinations of stability of filters.

1.7. In this problem we generalize the Alternating Series Theorem from the real domain to the complex domain (point (α)). We consider a power series $\sum_{k=0}^{\infty} a_k z^k$ where $\{a_k\}$ is a sequence of positive numbers decreasing to 0. We let $f(z)$ denote its sum.

(α) Prove Abel's test (due to E. Picard): The series converges everywhere on the unit circle except possibly at $z = 1$.

Hint: Abel's partial summation formula. Or use the geometrical proof in the short note Beck (1986).

(β) Give an example in which the series diverges at $z = 1$ and another in which it converges at $z = 1$.

(γ) Show that $(1 - z)f(z) \to 0$ as $z \to 1_-$.

1.8. Let $f(z) := \sum_{k=0}^{\infty} a_k z^k$ have radius of convergence $\rho > 0$. Show for each ξ such that $|\xi| = 1$ and such that ξ is not a root of unity (i.e., $\xi^n \neq 1$ for all $n \geq 1$) that

$$\lim_{n \to \infty} \frac{1}{n+1} \sum_{k=0}^{n} f(\xi^k z) = 0$$

uniformly in z on compact subsets of $B(0, \rho)$.

Deduce the *Cauchy's inequality*

$$|a_0| \leq \sup_{|z|=r} |f(z)| \qquad \text{for each } r \in \,]0, \rho[.$$

1.9 (A Generalization of Abel's Theorem). Let us assume that the power series $f(z) := \sum_{k=0}^{\infty} a_k z^k$ converges for $z = 1$ and hence for each $|z| < 1$.

(α) Let $C > 0$. Show that $\sum_{k=0}^{\infty} a_k z^k$ converges to $f(z)$ uniformly on the set

$$\left\{ z \in B(0,1) \,\middle|\, \frac{|1-z|}{1-|z|} \leq C \right\} \cup \{1\}.$$

(β) Let $\theta \in [0, \frac{\pi}{2}[$. Show that $\sum_{k=0}^{\infty} a_k z^k$ converges to $f(z)$ uniformly on the set (sketch it!)

$$\{ z \in B(0,1) \,|\, |1-z| \leq \cos\theta, \; |\mathrm{Arg}|(1-z) \leq \theta \} \cup \{1\}.$$

1.10. Devise an example, which shows that the assumption $ka_k \to 0$ as $k \to \infty$ in Tauber's theorem cannot be deleted.

Chapter 2

Holomorphic and Analytic Functions

2.1 Basics of complex calculus

Definition 2.1. *Let f be a complex valued function defined on an open set Ω in the complex plane. f is said to be **complex differentiable** at $z_0 \in \mathbb{C}$ if the limit*

$$\lim_{h \to 0} \frac{f(z_0 + h) - f(z_0)}{h} \tag{2.1}$$

exists in $\mathbb{C}$.

*The limit is called the **complex derivative of f at** z_0 and is denoted $f'(z_0)$ or $\frac{df}{dz}(z_0)$. We say that f is **holomorphic on** Ω if f is complex differentiable at each point of Ω. The function $f' \colon \Omega \to \mathbb{C}$ is then called the **complex derivative of f** or just the **derivative**. If there exists a holomorphic function F defined on Ω such that $F' = f$, we say that F is a **primitive** of f. If f is holomorphic in all of $\mathbb{C}$ then f is said to be **entire**.*

Like in real variable theory we find that f is continuous on an open set Ω if it is holomorphic on Ω.

By routine calculations the usual rules for differentiation also hold in the complex case:

Theorem 2.2.

(a) *If f and g are holomorphic on Ω then so are $f + g$ and fg, and $(f + g)' = f' + g'$, $(fg)' = f'g + fg'$.*

(b) *Let f and g be holomorphic on Ω. If g is not identically 0 on Ω then f/g is holomorphic on the open set $\{z \in \Omega \mid g(z) \neq 0\}$, and on this set*

$$\left(\frac{f}{g} \right)' = \frac{f'g - fg'}{g^2}$$

(c) *The chain rule: If f is holomorphic on Ω, g on Ω_1 and $f(\Omega)$ is contained in Ω_1, then the composition $g \circ f$ is holomorphic on Ω, and*

$$(g \circ f)'(z) = g'(f(z))f'(z) \qquad \text{for all } z \in \Omega$$

As is easy to see from the theorem, the set of functions which are holomorphic on Ω forms an algebra over the complex numbers. We denote this algebra $\mathrm{Hol}(\Omega)$.

Examples.

(i) Constants are holomorphic and their derivatives are 0.

(ii) The function $z \to z$ is holomorphic and its derivative is 1.

(iii) The function $z \to z^n$ is holomorphic on $\mathbb{C}$ for $n = 0, 1, 2, \ldots$, and on $\mathbb{C} \backslash \{0\}$ for $n = -1, -2, \ldots$ In both cases with derivative $(z^n)' = nz^{n-1}$.

(iv) A polynomial $p(z) = \sum_{k=0}^{n} a_k z^k$ is entire and $p'(z) = \sum_{k=1}^{n} k a_k z^{k-1}$.

Theorem 2.3. *Let $f \colon \Omega \to \mathbb{C}$, where Ω is an open subset of the complex plane, and write $f = u + iv$, where u and v are real valued functions.*

(a) *If f is holomorphic on Ω then all the first order partial derivatives of u and v exist in Ω and the **Cauchy-Riemann equations** are satisfied there, i.e.,*

$$\frac{\partial u}{\partial x} = \frac{\partial v}{\partial y} \quad \text{and} \quad \frac{\partial u}{\partial y} = -\frac{\partial v}{\partial x} \qquad (2.2)$$

or in a more compact notation

$$\frac{\partial f}{\partial \bar{z}} = 0, \qquad \text{where} \quad \frac{\partial}{\partial \bar{z}} := \frac{1}{2}\left(\frac{\partial}{\partial x} + i\frac{\partial}{\partial y}\right).$$

(b) *If $f \in C^1(\Omega)$ satisfies the Cauchy-Riemann equations in Ω then f is holomorphic in Ω.*

Proof. Ad (a): This is seen from (2.1) by letting h tend to 0 along the real and the imaginary axes respectively.

Ad (b): Left to the reader as Exercise 2.4. $\qquad\qquad\qquad\qquad\qquad$ □

Remark 2.4 (Cauchy-Riemann Equations). The differentiability conditions in point (b) of the theorem can be relaxed considerably. One generalization is the *Looman-Menchoff Theorem:*

Let $f \in \mathbb{C}(\Omega)$. Then $f \in \text{Hol}(\Omega)$, if $\partial f / \partial x$ and $\partial f / \partial y$ exist in all of Ω and satisfy the Cauchy-Riemann equations there (see Narasimhan, 1985, Theorem 1.6.1 p. 48).

Another generalization: It suffices that the Cauchy-Riemann equations are satisfied in the sense of distributions (see Trèves, 1975, Theorem 5.1 p. 36). For more information see Gray and Morris (1978).

The Cauchy-Riemann equations say roughly speaking that f does not depend on $\bar{z}$, and so it is a function of z only.

If f is holomorphic on an open set Ω, then (use the Cauchy-Riemann equations) $|f'|^2 = |\text{grad } u|^2 = |\text{grad } v|^2$, and $|\text{grad } u|^2 = (\partial u / \partial x)^2 + (\partial u / \partial y)^2$. In particular, $f' = 0$ throughout Ω implies that f is constant on each connected component of Ω.

Not all functions are holomorphic: E.g., $z \to \bar{z}$ is not.

As we saw in an example above, the functions $f_0 = 1$ and $f_1(z) = z$ are entire. Hence so is each polynomial in z. And each rational function, i.e., each function of the form $f = P/Q$ where P and Q are polynomials, is holomorphic off the zeros of the denominator Q. It turns out (Theorem 3.5 and Proposition 3.21 of the next chapter) that continuous roots and logarithms are holomorphic, too. So are power series; we give a direct proof here. A shorter, but more sophisticated proof is given below following Lemma 2.13.

Proposition 2.5. *The sum of a power series is holomorphic inside its circle of convergence, and its derivative can be got by term-by-term differentiation: If*

$$f(z) = \sum_{n=0}^{\infty} a_n (z - z_0)^n$$

has radius of convergence $\rho > 0$, then f is holomorphic on $B(z_0, \rho)$ and

$$f'(z) = \sum_{n=1}^{\infty} a_n n (z - z_0)^{n-1} \qquad \text{for } z \in B(z_0, \rho)$$

In particular all the complex derivatives $f', f'', \ldots$ exist in $B(z_0, \rho)$.

Proof. Observe that the formula for f' makes sense by Corollary 1.4. Let us for convenience of writing assume that $z_0 = 0$.

The technical key to the proof is the following inequality:

$$|(z + h)^n - z^n - h n z^{n-1}| \le \left|\frac{h}{\delta}\right|^2 (|z| + \delta)^n$$

valid for $n \in \mathbb{N}$, $h, z \in \mathbb{C}$ such that $0 < |h| \le \delta$.

Indeed, using the inequality we find with $\delta < \rho - |z|$ that

$$
\left| \frac{f(z+h) - f(z)}{h} - \sum_{n=1}^{\infty} a_n n z^{n-1} \right|
$$

$$
= \left| \frac{1}{h} \sum_{n=1}^{\infty} a_n \{ (z+h)^n - z^n - h n z^{n-1} \} \right|
$$

$$
\leq \frac{1}{|h|} \sum_{n=1}^{\infty} |a_n| \left| \frac{h}{\delta} \right|^2 (|z| + \delta)^n
$$

$$
= \frac{|h|}{\delta^2} \sum_{n=1}^{\infty} |a_n| (|z| + \delta)^n \to 0 \qquad \text{as } h \to 0.
$$

The inequality follows from the binomial formula:

$$
(z+h)^n - z^n - h n z^{n-1} = \sum_{k=0}^{n} \binom{n}{k} z^{n-k} h^k - z^n - h n z^{n-1}
$$

$$
= \sum_{k=2}^{n} \binom{n}{k} z^{n-k} h^k = \left\{ \sum_{k=2}^{n} \binom{n}{k} z^{n-k} h^{k-2} \right\} h^2
$$

so

$$
|(z+h)^n - z^n - h n z^{n-1}| \leq \sum_{k=2}^{n} \binom{n}{k} |z|^{n-k} |h|^{k-2} |h|^2
$$

$$
\leq \sum_{k=2}^{n} \binom{n}{k} |z|^{n-k} \delta^{k-2} |h|^2 = \sum_{k=2}^{n} \binom{n}{k} |z|^{n-k} \delta^k \left| \frac{h}{\delta} \right|^2
$$

$$
\leq \sum_{k=0}^{n} \binom{n}{k} |z|^{n-k} \delta^k \left| \frac{h}{\delta} \right|^2 = (|z| + \delta)^n \left| \frac{h}{\delta} \right|^2
$$

$$\square$$

Definition 2.6. *Let Ω be an open subset of $\mathbb{C}$. A function $f \colon \Omega \to \mathbb{C}$ is said to be **analytic** on Ω if for each point $z_0 \in \mathbb{C}$ there exists an open disc $B(z_0, \rho)$ in Ω in which f can be written as the sum of a power series centered at z_0, i.e., f can be written*

$$
f(z) = \sum_{n=0}^{\infty} a_n (z - z_0)^n \qquad \text{for } z \in B(z_0, \rho).
$$

We have already observed that the coefficients $\{a_n\}$ in the power series around z_0 are uniquely determined by f (Proposition 1.6). By help of Proposition 2.5 we can even say what the coefficients are:

$$a_n = \frac{1}{n!} f^{(n)}(z_0) \qquad \text{for } n = 0, 1, 2, \ldots$$

So, restricting to the real line, we see that the power series expansion of an analytic function coincides with its Taylor series expansion. Given any sequence $\{a_n\}$ of complex numbers the power series $\sum_{n=0}^{\infty} a_n z^n$ may or may not converge around 0. There always exists a C^∞ function on the real line with $\sum_{n=0}^{\infty} a_n z^n$ as its Taylor series. Many in fact (see e.g., Meyerson, 1981). But such functions will in general not be restrictions of power series to the real line.

A consequence of Proposition 2.5 is that an analytic function is holomorphic, a result which is nice, but certainly not surprising. What is surprising and remarkable is that the converse is true (as will be proved in Chapter 4). Under the (apparently) humble assumption that f is complex differentiable, we shall infer that f is infinitely often differentiable and even that f around each point in its domain of definition can be expanded in a power series.

The situation is vastly different in the case of functions on the real line. There the derivative of a differentiable function need not even be continuous, let alone expandable in a power series.

But for functions in the complex plane the word "holomorphic" is synonymous with the word "analytic". However, until we have proved that result, we must distinguish between the two concepts.

2.2   Line integrals

In this paragraph we introduce the line integrals which are important tools in complex function theory.

We shall until further notice not require any finer theory of integration and convergence theorems. All we need is to consider piecewise continuous functions on the real line and uniform convergence.

Definition 2.7. *Let $\gamma\colon [a,b] \to \mathbb{C}$, where $-\infty < a < b < \infty$, be a curve, that is a continuous map.*

(a) *γ is said to be a **path** if it is piecewise differentiable. This means that there are points*

$$a = t_0 < t_1 < t_2 < \cdots < t_{n-1} < t_n = b$$

such that the derivative γ' exists in the open subinterval $]t_j, t_{j+1}[$ and extends to a continuous function on the closed subinterval $[t_j, t_{j+1}]$ for each $j = 0, 1, 2, \ldots$

(b) *The **length** $l(\gamma)$ of the path γ is the number*

$$l(\gamma) := \int_a^b |\gamma'(t)|\, dt \in [0, \infty[.$$

(c) *We will say that a complex number z belongs (or does not belong) to the curve γ if $z \in \gamma^*$ (resp. $z \notin \gamma^*$). We express the same thing by saying that γ passes through z (resp. does not pass through z).*

(d) *The curve γ is said to be **closed** if $\gamma(a) = \gamma(b)$.*

(e) *If γ^* is contained in a subset Ω of $\mathbb{C}$, then we will say that γ is a curve in Ω.*

Thus we distinguish between the curve γ, which is a map, and its image $\gamma^* := \gamma([a, b])$ which is a compact subset of $\mathbb{C}$.

Examples 2.8.

(i) The positively oriented circle γ around $z_0 \in \mathbb{C}$ with radius $R > 0$, i.e., the curve

$$\gamma(t) = z_0 + \mathrm{R}e^{it} \qquad \text{for } t \in [0, 2\pi]$$

is a closed path with length $2\pi R$. We will often write $|z - z_0| = R$ instead of γ.

(ii) The line segment $[A, B]$, where A and B are complex numbers, i.e., the curve

$$[A, B] := (1 - t)A + tB \qquad \text{for } t \in [0, 1]$$

is a path of length $l([A, B]) = |B - A|$.

(iii) Let $A, B, C \in \mathbb{C}$. By the boundary $\partial\Delta$ of the triangle Δ with vertices A, B and C (the order is important!) we understand the closed curve defined by

$$\partial\Delta\,(t) := \begin{cases} [A,B](t) & \text{for } t \in [0,1], \\ [B,C](t-1) & \text{for } t \in [1,2], \\ [C,A](t-2) & \text{for } t \in [2,3]. \end{cases}$$

It is a path of length $l(\partial\Delta) = |B - A| + |C - B| + |A - C|$.

The length of a path does not depend on its parametrization. More formally:

Proposition 2.9. *Let* $\gamma\colon [a,b] \to \mathbb{C}$ *be a path. Let* $\phi\colon [c,d] \to [a,b]$ *be a continuously differentiable monotone map of* $[c,d]$ *onto* $[a,b]$. *Then* $l(\gamma) = l(\gamma \circ \phi)$.

Proof. The change of variables theorem. $\square$

Definition 2.10. *Let* $\gamma\colon [a,b] \to \mathbb{C}$ *be a path, and let* $f\colon \gamma^* = \gamma([a,b]) \to \mathbb{C}$ *be a continuous function.* **The line integral of** f **along** γ *is the complex number*

$$\int_\gamma f = \int_\gamma f(z)\,dz := \int_{[a,b]} f(\gamma(t))\gamma'(t)\,dt.$$

To emphasize what the variable is, we sometimes write $\int_\gamma f(w)\,dw$ or the like instead of $\int_\gamma f(z)\,dz$.

The next proposition collects properties of the line integral which will be useful for us in the sequel.

Proposition 2.11. *Let* $\gamma\colon [a,b] \to \mathbb{C}$ *be a path, and let* f *be a continuous, complex-valued function on* γ^*. *Then*

(a)

$$\left| \int_\gamma f(z)\,dz \right| \le \sup_{z \in \gamma^*} \{|f(z)|\} l(\gamma).$$

(b) *(A primitive version of the fact that the line integral is invariant under orientation-preserving changes of parameter).*

   Let $-\infty < c < d < \infty$. *If* $\phi\colon [c,d] \to [a,b]$ *has the form*

$$\phi(s) = ks + l, s \in [c,d], \qquad \text{where } k,l \in \mathbb{R}$$

and if ϕ maps $[c, d]$ onto $[a, b]$ then

$$\int_\gamma f = \text{sign}(k) \int_{\gamma \circ \phi} f.$$

For two points A and B in $\mathbb{C}$ we have in particular

$$\int_{[A,B]} f = -\int_{[B,A]} f.$$

(c) *Let $c \in]a, b[$ and define paths γ_1 and γ_2 by*

$$\gamma_1(t) := \gamma(t) \qquad \text{for } t \in [a, c],$$
$$\gamma_2(t) := \gamma(t) \qquad \text{for } t \in [c, b].$$

Then

$$\int_\gamma f = \int_{\gamma_1} f + \int_{\gamma_2} f.$$

If $C \in [A, B]^*$, where A and B are points of $\mathbb{C}$, then in particular

$$\int_{[A,B]} f = \int_{[A,C]} f + \int_{[C,B]} f.$$

(d) *If $\{f_n\}$ is a sequence of continuous functions on γ^* that converges uniformly on γ^* to f, then*

$$\int_\gamma f_n \to \int_\gamma f \qquad \text{as } n \to \infty.$$

(e) *If f is defined on a neighborhood of γ^* and has a primitive, say F, there, then*

$$\int_\gamma f = F(\gamma(b)) - F(\gamma(a)).$$

If furthermore γ is closed then $\int_\gamma f = 0$.

Proof. Left to the reader.   □

We conclude this chapter by some applications of line integrals. They are investments for the future.

Definition 2.12. *A subset A of $\mathbb{R}^n$ is said to be **starshaped**, if there exists a point $a \in A$ such that all the segments*

$$\{tx + (1-t)a \mid t \in [0,1]\}, \qquad x \in A$$

*are contained in A; we say that A is **starshaped with respect to** a.*

Clearly any convex set is starshaped. An interval is a simple, but important special case.

Lemma 2.13. *Let Ω be an open subset of $\mathbb{C}$ which is starshaped with respect to a point $a \in \Omega$. Let $g \in C(\Omega)$ have the property that $\int_{\partial\triangle} g = 0$, whenever $\triangle$ is a triangle in Ω with a as one of its vertices. Then the function*

$$G(z) := \int_{[a,z]} g, \qquad z \in \Omega$$

is holomorphic on Ω and $G' = g$. In particular g has a primitive.

Proof. Let $z \in \Omega$. If $h \in \mathbb{C}$ is so small that $[z, z+h]$ is contained in Ω then we get from our assumption that

$$G(z+h) - G(z) = \int_{[a,z+h]} g - \int_{[a,z]} g = \int_{[z,z+h]} g = h \int_0^1 g(z+th)\,dt$$

Division through by h and use of the continuity of g at z gives us the lemma. $\qquad\square$

We can now give the promised simpler proof of Proposition 2.5:

Proof of Proposition 2.5. Let us, as in the earlier proof, for convenience take $z_0 = 0$. The power series

$$g(z) := \sum_{n=1}^{\infty} a_n n z^{n-1}$$

converges according to Corollary 1.4 uniformly on compact subsets of $B(0, \rho)$, so term by term integration is permitted and yields

$$f(z) = \int_{[0,z]} g(w)\,dw + a_0 \qquad \text{for } z \in B(0, \rho).$$

By Lemma 2.13 it now suffices to prove that the integral of g along any triangle $\triangle$ in $B(0, \rho)$ vanishes. But that is easy: Indeed,

$$\int_{\partial\triangle} g(z)\,dz = \sum_{n=1}^{\infty} a_n n \int_{\partial\triangle} z^{n-1}\,dz = 0$$

because the integral of z^{n-1} vanishes (z^{n-1} has a primitive, viz. z^n/n). $\quad\square$

Lemma 2.14. *Let $\phi\colon \gamma^* \to \mathbb{C}$ be a continuous function on a path γ. Then the function*

$$f(z) := \frac{1}{2\pi i} \int_\gamma \frac{\phi(w)}{w - z}\, dw, \qquad z \in \mathbb{C} \setminus \gamma^*$$

is analytic, hence holomorphic and

$$f^{(n)}(z) = \frac{n!}{2\pi i} \int_\gamma \frac{\phi(w)}{(w - z)^{n+1}}\, dw \qquad for\ n = 0, 1, 2, \ldots$$

If $z_0 \in \mathbb{C} \setminus \gamma^*$ then the power series expansion of f around z_0 converges everywhere in the open disc around z_0 with radius $\mathrm{dist}(z_0, \gamma^*)$.

As support for our memory we note that the formula for $f^{(n)}$ appears when we differentiate f under the integral sign.

Proof. If $|z - z_0| < \mathrm{dist}(z_0, \gamma^*)$, then the geometric series

$$\sum_{n=0}^{\infty} \left(\frac{z - z_0}{w - z_0} \right)^n$$

converges uniformly for $w \in \gamma^*$ to

$$\frac{1}{1 - \frac{z - z_0}{w - z_0}}$$

so writing

$$w - z = \left\{ 1 - \frac{z - z_0}{w - z_0} \right\}(w - z_0)$$

we find

$$2\pi i f(z) = \int_\gamma \frac{\phi(w)}{w - z}\, dw = \int_\gamma \sum_{n=0}^{\infty} \left(\frac{z - z_0}{w - z_0} \right)^n \frac{\phi(w)}{w - z_0}\, dw$$

$$= \sum_{n=0}^{\infty} \int_\gamma \left(\frac{z - z_0}{w - z_0} \right)^n \frac{\phi(w)}{w - z_0}\, dw$$

$$= \sum_{n=0}^{\infty} \left\{ \int_\gamma \frac{\phi(w)}{(w - z_0)^{n+1}}\, dw \right\}(z - z_0)^n$$

which exhibits the power series expansion of f around z_0. $\square$

Exercises

2.1. Let f be holomorphic on an open subset Ω of the complex plane. Define a function f^* on $\Omega^* := \{z \in \mathbb{C} \mid \bar{z} \in \Omega\}$ by $f^*(z) := \overline{f(\bar{z})}$ for $z \in \Omega^*$.

(α) Show that f^* is holomorphic on Ω^*.

Suppose that f has the power series expansion $f(z) = \sum_{n=0}^{\infty} a_n z^n$ on Ω.

(β) Find the power series expansion of f^* on Ω^*.

2.2. Show that the function $f(z) := |z|^2$ has a complex derivative at $z = 0$ but nowhere else.

2.3. Is the function $f(z) = \mathrm{Re}(z)$ entire?

2.4. Let u and v be real valued functions on an open subset Ω of $\mathbb{C}$. We assume that their first partial derivatives with respect to x and y exist and are continuous.

Show that $f := u + iv$ is holomorphic in Ω if the Cauchy-Riemann equations hold.

2.5. Show that

$$f(z) := \begin{cases} \exp(-1/z^4), & z \in \mathbb{C} \setminus \{0\}, \\ 0, & z = 0, \end{cases}$$

is a discontinuous function, that nevertheless satisfies the Cauchy-Riemann equations everywhere.

2.6. Show that the pair of functions

$$u(x, y) = \frac{x}{x^2 + y^2} \quad \text{and} \quad v(x, y) = -\frac{y}{x^2 + y^2}$$

satisfies the Cauchy-Riemann equations on $\mathbb{C} \setminus \{0\}$.
Hint: You need not differentiate.

2.7. Let γ be a closed path in $\mathbb{C}$. Show that $\int_\gamma \bar{z} \, dz$ is purely imaginary.

2.8. Assume that the power series $f(z) = \sum_{n=0}^{\infty} a_n z^n$ converges in $B(0, \rho)$. Show that f is analytic on $B(0, \rho)$.
Hint:

$$z^n = \frac{1}{2\pi i} \int_{|w|=R} \frac{w^n}{w - z} \, dw, \qquad \text{when } |z| < R < \rho.$$

2.9. Let f be holomorphic on an open subset Ω of the complex plane. Let $\gamma: \,]{-1}, 1[\,\rightarrow \Omega$ be differentiable at $t = 0$.

Show that $f \circ \gamma$ is differentiable at $t = 0$ and that

$$\frac{d(f \circ \gamma)}{dt}(0) = f'(\gamma(0))\gamma'(0).$$

Chapter 3

The Exponential Function, the Logarithm and the Winding Number

3.1 The exponential function

We assume that the reader is familiar with the exponential function, the logarithm and the cos and sin functions on the real line.

We define the *exponential function* $\exp \colon \mathbb{C} \to \mathbb{C}$ by

$$\exp(z) := \sum_{n=0}^{\infty} \frac{z^n}{n!} \qquad \text{for } z \in \mathbb{C}.$$

By Theorem 1.3(c) the power series converges absolutely and uniformly on compact subsets of $\mathbb{C}$, and by Proposition 2.5 its sum $f(z) = \exp(z)$ is entire and satisfies $f' = f$, $f(0) = 1$. On the real line exp reduces to the well known real exponential function.

Let us study the exponential function on the imaginary axis in $\mathbb{C}$: For any $x \in \mathbb{R}$ we have

$$
\begin{aligned}
\exp(ix) &= \sum_{n=0}^{\infty} \frac{(ix)^n}{n!} = \sum_{m=0}^{\infty} \frac{(ix)^{2m}}{(2m)!} + \sum_{m=0}^{\infty} \frac{(ix)^{2m+1}}{(2m+1)!} \\
&= \sum_{m=0}^{\infty} (-1)^m \frac{x^{2m}}{(2m)!} + i \sum_{m=0}^{\infty} (-1)^m \frac{x^{2m+1}}{(2m+1)!} \\
&= \cos x + i \sin x.
\end{aligned}
$$

From this we see in particular that $\phi(x) := \exp(ix)$ maps the real line onto the unit circle $S^1 := \{z \in \mathbb{C} \mid |z| = 1\}$.

The most important property of the exponential function is that it couples the additive and multiplicative structures of $\mathbb{C}$ as follows:

Theorem 3.1 (The Addition Theorem).

$$\exp(a + b) = \exp(a) \exp(b) \qquad \text{for all } a, b \in \mathbb{C}.$$

Proof. When we differentiate $g(z) := \exp(z) \exp(a + b - z)$ we get $g'(z) = 0$, so g is constant. The contents of the theorem are $g(0) = g(a)$.   □

Note that exp never vanishes (by the Addition Theorem).

The power series for exp has real coefficients. Therefore $\exp(\overline{a}) = \overline{\exp(a)}$, which by the Addition Theorem implies that

$$|\exp(a)|^2 = \exp(a) \exp(\overline{a}) = \exp(a + \overline{a}) = \exp(2 \operatorname{Re} a) = \{\exp(\operatorname{Re} a)\}^2$$

so $|\exp(a)| = \exp(\operatorname{Re} a)$. From this we deduce that

$$|\exp(ia)| = 1 \quad \Leftrightarrow \quad a \in \mathbb{R}.$$

We continue with a study of the exponential function on $\mathbb{C}$: Writing $w \in \mathbb{C} \setminus \{0\}$ in the form $w = |w| \frac{w}{|w|}$ we can by the above results express w as

$$w = \exp(x) \exp(i\theta) = \exp(x + i\theta) \qquad \text{for some } x, \theta \in \mathbb{C}$$

so $\exp(\mathbb{C}) = \mathbb{C} \setminus \{0\}$. Finally,

$$\exp(a) = \exp(b) \quad \Leftrightarrow \quad a - b \in 2\pi i \mathbb{Z}.$$

Proof. "$\Leftarrow$": An immediate consequence of the Addition Theorem.

"$\Rightarrow$": Let $\exp a = \exp b$. Then $\exp(a - b) = 1$, and so $a - b \in i\mathbb{R}$, say $a - b = i\theta$. Then $\cos\theta + i\sin\theta = \exp i\theta = \exp(a - b) = 1$, so $\theta \in 2\pi\mathbb{Z}$.   □

Lemma 3.2. *Let* $\phi(\theta) := \exp(i\theta)$ *for* $\theta \in \mathbb{R}$. *Then* ϕ *is 1–1 on any half open interval of the form* $[\alpha, \alpha + 2\pi[$, *and it is a homeomorphism of the open interval* $]\alpha, \alpha + 2\pi[$ *onto the arc*

$$\{z \in \mathbb{C} \mid |z| = 1, \ z \neq \exp(i\alpha)\}.$$

A homeomorphism is a bijection which is continuous both ways.

The lemma expresses the geometrically obvious fact that the angle of a point depends continuously on the point.

Proof. The only non-trivial statement is the one about the homeomorphism property. ϕ is, however, clearly continuous, so left is just the continuity of the inverse of ϕ, i.e., the following claim:

If $\exp(i\theta_n) \to \exp(i\theta)$ where $\theta_n, \theta \in]\alpha, \alpha + 2\pi[$ then $\theta_n \to \theta$.

To see this claim, let θ_0 be any limit point of the sequence $\{\theta_n\}$. θ_0 must belong to the closed interval $[\alpha, \alpha + 2\pi]$, and $\exp(i\theta) = \exp(i\theta_0)$. Now, θ_0 cannot equal any of the end points because exp there equals $\exp(i\alpha) \neq \exp(i\theta)$, and because exp is 1–1 on $]\alpha, \alpha + 2\pi[$, we must have $\theta = \theta_0$. $\quad\square$

3.2 Logarithm, argument and power

By Section 3.1 the restriction of the exponential function to $\mathbb{R}$ is the familiar real exponential function. Its inverse mapping is (by definition) the logarithm, which here will be denoted by Log. If $z \in \mathbb{C} \setminus \{0\}$ we can write

$$z = |z|\frac{z}{|z|} = \exp(\text{Log}|z|)\exp(i\theta) = \exp(\text{Log}|z| + i\theta) \tag{3.1}$$

with θ real. From Section 3.1 we see that θ is unique if we restrict it to lie in any given half-open interval of length 2π.

Definition 3.3. *Let* $z \in \mathbb{C} \setminus \{0\}$. *The sets*

$$\arg z := \left\{\theta \in \mathbb{R} \,\middle|\, \exp(i\theta) = \frac{z}{|z|}\right\} \tag{3.2}$$

and

$$\log z := \{a \in \mathbb{C} \mid \exp(a) = z\} \tag{3.3}$$

*are called **the argument of** z and **the logarithm of** z, respectively. We will, however, use this terminology in an ambiguous way: Any real number in the set (3.2) will be called **an argument of** z and denoted by $\arg z$; any complex number in (3.3) will be called **a logarithm of** z and denoted by $\log z$. It will be clear from the context which interpretation we have in mind.*

The geometrical meaning of $\arg z$ is the following: Let l be the ray in $\mathbb{C}$ starting at the origin and passing through z. Then $\arg z$ is the set of values of the angles from the positive real semi-axis to the ray l.

We see from the definitions (3.2) and (3.3) that every non-zero complex number has an infinity of logarithms and arguments. Any two arguments differ by an integer multiple of 2π, and any two logarithms by an integer multiple of $2\pi i$. We can write

$$\log z = \text{Log}|z| + i \arg z \qquad \text{for } z \in \mathbb{C} \setminus \{0\}. \tag{3.4}$$

Definition 3.4. *If f is a never vanishing complex valued function, defined on some topological space, then by a* **continuous logarithm of** f *(or a* **branch of** $\log f$*) we mean any continuous complex valued function F such that* $\exp F = f$, *and by a* **continuous argument of** f *(or a* **branch of** $\arg f$*) we mean any continuous real valued function θ such that $f = |f|\exp(i\theta)$.*

Of course, if $f: X \to]0, \infty[$ is continuous, then $\operatorname{Log} f$ is an example of a branch of $\log f$.

From the discussion above we get a uniqueness result: If f is a never-vanishing continuous complex valued function on a connected topological space, then any two continuous logarithms of f differ by a constant integer multiple of $2\pi i$; and any two continuous arguments of f differ by a constant integer multiple of 2π.

We now turn to the question of regularity of logarithms and arguments. The following theorem tells in particular that any branch of $\log z$ automatically is holomorphic. But more is true: In Chapter 4 we shall present a generalization in which we replace $\exp$ by any non-constant holomorphic function (Theorem 4.18).

Theorem 3.5. *Let f be holomorphic on an open subset Ω of the complex plane, and let F be a continuous logarithm of f in Ω.*

Then F is also holomorphic in Ω, and $F' = f'/f$.

Proof. We shall show that F has a complex derivative at each point $z \in \Omega$. From the defining relation $\exp F(a) = f(a)$ for any $a \in \Omega$ we get

$$\frac{\exp(F(z+h)) - \exp F(z)}{h} \to f'(z) \qquad \text{as } h \to 0$$

and division through by $\exp F(z) = f(z)$ yields

$$\frac{\exp\big(F(z+h) - F(z)\big) - 1}{h} \to \frac{f'(z)}{f(z)} \qquad \text{as } h \to 0. \qquad (3.5)$$

Expansion of the exponential function in formula (3.5) shows that

$$\frac{F(z+h) - F(z)}{h} + \frac{1}{h}\sum_{n=2}^{\infty} \frac{\{F(z+h) - F(z)\}^n}{n!} \to \frac{f'(z)}{f(z)} \qquad \text{as } h \to 0$$

so it suffices to prove that

$$\frac{1}{|h|}\sum_{n=2}^{\infty} \frac{|F(z+h) - F(z)|^n}{n!} \to 0 \qquad \text{for } h \to 0. \qquad (3.6)$$

Now $|e^a - 1| \geq |a|/2$ when $|a|$ is small. Indeed,

$$\frac{e^a - 1}{a} \to 1 \qquad \text{as } a \to 0.$$

Therefore we find, putting $a = F(z+h) - F(z)$ in (3.5), that $|F(z+h) - F(z)|/|h|$ is bounded for small h, i.e., there exist $C > 0$ and $\delta \in \,]0, 1[$ such that

$$|h| < \delta \Rightarrow |F(z+h) - F(z)| \leq C|h|.$$

By help of this we may for $|h| < \delta$ estimate the left hand side of formula (3.6) as follows:

$$\frac{1}{|h|} \sum_{n=2}^{\infty} \frac{|F(z+h) - F(z)|^n}{n!} \leq \frac{1}{|h|} \sum_{n=2}^{\infty} \frac{C^n |h|^n}{n!}$$

$$\leq |h| \sum_{n=2}^{\infty} \frac{C^n}{n!} \leq |h| e^C \to 0 \qquad \text{as } h \to 0.$$

$\square$

Example 3.6. Consider, for given $\theta_0 \in \mathbb{R}$, the open set

$$\Omega := \left\{ z \in \mathbb{C} \setminus \{0\} \,\Big|\, \frac{z}{|z|} \neq \exp(i\theta_0) \right\}.$$

Let $\theta(z)$ for $z \in \Omega$ denote the value of $\arg z$ in the open interval $]\theta_0, \theta_0 + 2\pi[$. Then

(i) $z \to \theta(z)$ is a branch of $\arg z$ in Ω.

(ii) $z \to \mathrm{Log}|z| + i\theta(z)$ is a branch of $\log z$ in Ω. In particular it is a holomorphic function.

Proof. Item (i) is a consequence of the last statement of Lemma 3.2, while (ii) follows from (i). $\square$

Example 3.7. Let us take $\theta_0 = -\pi$ in Example 3.6, so that $\Omega = \mathbb{C} \setminus \{z \in \mathbb{R} \mid z \leq 0\}$. Let $\mathrm{Arg}\, z$ for $z \in \Omega$ denote the value of $\arg z$ in the interval $]-\pi, \pi[$. Then $z \to \mathrm{Arg}\, z$ is a branch of $\arg z$ in Ω. It is called the *principal branch of* $\arg z$. The corresponding continuous logarithm of z in Ω is called the *principal branch of* $\log z$ and denoted $\mathrm{Log}\, z$. So

$$\mathrm{Log}\, z = \mathrm{Log}|z| + i\,\mathrm{Arg}\, z \qquad \text{for } z \in \Omega.$$

$\mathrm{Log}\, z$ is holomorphic according to Theorem 3.5.

Note that the principal branch of $\log z$ extends the function Log on $\mathbb{R}$ from earlier so that the notation is unambiguous. If $x \in \mathbb{R}^+$ and nothing is specified to the contrary, we will very often be sloppy and write $\log x$ instead of the more precise term Log x. Note also that $(\text{Log } z)' = z^{-1}$.

Now we can define complex powers: Let $a \neq 0$. For any *fixed* choice of $\log a$ – and we have a countable infinity of choices –

$$f(z) := \exp(z \log a)$$

is called *the z'th power of a*. f is clearly holomorphic in the entire complex plane, i.e., f is entire. The Addition Theorem tells us that f satisfies the relations

$$f(z + w) = f(z)f(w), \quad f(0) = 1, \quad f(1) = a$$

In particular

$$f(n) = a^n \qquad \text{for } n = 0, 1, 2, \ldots$$

so it is natural to use the notation a^z for $f(z)$. Note also that a^n is independent of the choice of $\log a$. We see that $\exp z$ is one of the zth powers of e, viz the one with the choice $\log e = 1$, and we shall from now on often write e^z instead of $\exp z$. More generally, if $a > 0$ we shall always put

$$a^z = \exp(z \text{ Log } a) \qquad \text{for } z \in \mathbb{C}.$$

3.3 Existence of continuous logarithms

In this section we shall discuss whether branches of $\log f$ exist for given function f. Before studying general functions f we will examine the important special case $f(z) = z$. To indicate the kind of difficulties we may encounter, we present an example to the effect that $\log z$ cannot be defined in a continuous way in all of $\mathbb{C} \setminus \{0\}$. However, to make the reader happy again, we note that there do exist branches of $\log z$ in the complex plane minus any half-ray starting at the origin (Example 3.6 above).

Example 3.8. We claim that it is not possible to find a continuous $\log z$ on

$$S^1 := \{z \in \mathbb{C} \mid |z| = 1\}$$

let alone on the punctured plane $\mathbb{C} \setminus \{0\}$. It suffices to prove that there does not exist any branch of $\arg z$ on S^1 – after all $\arg z$ is the imaginary part of $\log z$. It is intuitively obvious that no branch of $\arg z$ exists on S^1: Let us keep track of the argument as we go around the unit circle counterclockwise starting at 1. From the geometrical interpretation of the argument we see that $\arg z$ increases continuously, and when we have gone all the way round it has increased by 2π. And so it does not take its former value at 1 as it should by its continuity at 1. Thus $\arg z$ cannot be defined in a continuous way all around the unit circle.

Proof. Here is a short rigorous proof of the non-existence of $\arg z$: As most proofs of non-existence it goes by contradiction, so we assume that there exists a continuous argument $\theta\colon S^1 \to \mathbb{R}$, so that

$$z = \exp(i\theta(z)) \qquad \text{for all } z \in S^1.$$

The relation shows in particular that θ is 1–1. Thus the mapping $f\colon S^1 \to \{1, -1\}$, given by

$$f(z) := \frac{\theta(z) - \theta(-z)}{|\theta(z) - \theta(-z)|} \qquad \text{for } z \in S^1$$

is well-defined and continuous. But since $f(z) = -f(-z)$ it maps the connected space S^1 onto the disconnected space $\{1, -1\}$. Contradiction! $\square$

We shall now treat the question of whether a given function f has a continuous logarithm $\log f$, i.e., the existence question. We hide most of the technicalities away in the proof of the following proposition.

Proposition 3.9 (The Homotopy Lifting Property). *Let X be a topological space and let $F\colon X \times [0,1] \to \mathbb{C} \setminus \{0\}$ be a continuous map. Assume that the restriction of F to $X \times \{0\}$ has a continuous logarithm, say $\phi\colon X \times \{0\} \to \mathbb{C}$.*
   Then F has a continuous logarithm that extends ϕ.

Proof. We use the abbreviation $\phi(x) = \phi(x, 0)$ for $x \in X$.

Step 1: We will prove that it suffices to verify the following localized version of the proposition:
   To every $x \in X$ there exists an open neighborhood $N(x)$ of x in X, and a continuous map

$$\theta_x \colon N(x) \times [0,1] \to \mathbb{C},$$

such that

$$\theta_x(y, 0) = \phi(y)d \qquad \text{for all } y \in N(x),$$

and

$$\exp(\theta_x) = F \qquad \text{on } N(x) \times [0, 1].$$

Indeed, assume that the localized version is true. Then for any $x, y \in X$ and any $x_0 \in N(x) \cap N(y)$ we have

$$\exp(\theta_x(x_0, t)) = F(x_0, t) = \exp(\theta_y(x_0, t))$$

so

$$\theta_x(x_0, t) - \theta_y(x_0, t) = 2\pi i n(t), \qquad \text{where } n(t) \in \mathbb{Z}.$$

The left side is a continuous function of t, so $t \to n(t)$ is a continuous, integer valued function on $[0, 1]$, hence a constant. Taking the condition

$$\theta_x(x_0, 0) = \phi(x_0) = \theta_y(x_0, 0)$$

into account we see that the constant is 0.

We have thus shown that θ_x and θ_y agree on their common domain of definition, and so the collection θ_x, $x \in X$ patches together to a continuous function $\theta \colon X \times [0, 1] \to \mathbb{C}$, which is the desired logarithm of F.

Step 2: We shall verify the localized version of the proposition is true, so let $x_0 \in X$. By the compactness of $[0, 1]$ we determine points $0 = t_0 < t_1 < \cdots < t_n = 1$ and a neighborhood $N = N(x_0)$ of $x_0 \in X$ such that each

$$F(N \times [t_k, t_{k+1}]), \qquad k = 0, 1, 2, \ldots, n - 1$$

is contained in a subset of $\mathbb{C} \setminus \{0\}$ on which there exists a branch of $\log z$.

We now define inductively continuous maps

$$\theta_k \colon N \times [t_k, t_{k+1}] \to \mathbb{C} \qquad \text{for } k = 0, 1, 2, \ldots, n - 1$$

in the following way:

Since there exists a branch of $\log z$ on $F(N \times [0, t_1])$ we can form $\log F$ there, i.e., we can find a continuous function $\Lambda_0 \colon N \times [0, t_1] \to \mathbb{C}$ such that

$$\exp(\Lambda_0) = F \qquad \text{on } N \times [0, t_1].$$

Noting that

$$\exp(\Lambda_0(x, 0)) = F(x, 0) = \exp(\phi[x])$$

so that

$$\exp[\phi(x) - \Lambda_0(x, 0)] = 1$$

we define
$$\theta_0(x,t) := \Lambda_0(x,t) - \Lambda_0(x,0) + \phi(x).$$

Then $\theta_0 \colon N \times [0, t_1] \to \mathbb{C}$ is a continuous function that satisfies

$$\exp(\theta_0) = F, \quad \text{and} \quad \theta_0(x,0) = \phi(x) \qquad \text{for } x \in N.$$

Arguing in exactly the same way we find successively for $k = 1, 2, \dots, n-1$ continuous functions $\theta_k \colon N \times [t_k, t_{k+1}] \to \mathbb{C}$ such that

$$\exp(\theta_k) = F, \quad \text{and} \quad \theta_k(x, t_k) = \theta_{k-1}(x, t_k) \qquad \text{for } x \in N.$$

Obviously the θ_k glue together to form a continuous function $\theta \colon N \times [0, 1] \to \mathbb{C}$ with the desired properties. $\qquad\square$

Remark. Proposition 3.9 is a special case of a result from point set topology, viz that a covering projection is a fibration. See e.g., Spanier (1966, Theorem 2.2.3 p. 67).

Proposition 3.10. *If A is a starshaped subset of $\mathbb{R}^n$ then any continuous mapping $f \colon A \to \mathbb{C} \setminus \{0\}$ has a continuous logarithm.*

Proof. With notation from Definition 2.12 we consider the continuous map $F \colon A \times [0, 1] \to \mathbb{C} \setminus \{0\}$, defined by

$$F(x, t) := f(tx + (1-t)a) \qquad \text{for } (x, t) \in A \times [0, 1].$$

Since $x \to F(x, 0) = f(a)$ is a constant function it has a continuous logarithm, hence by Proposition 3.9 so does F. A fortiori so does $x \to F(x, 1) = f(x)$. $\qquad\square$

As we shall see later (Theorem 5.9), Proposition 3.10 extends from starshaped to simply connected spaces.

Here we will as an application of Proposition 3.10 present an interesting digression: The 2-dimensional version of the famous Brouwer's fixed point theorem (Theorem 3.11(c) below).

Theorem 3.11.

(a) *There is no continuous mapping $r\colon B[0,1] \to S^1$ with the property that $r(z) = z$ for all $z \in S^1$.*

(b) *Any continuous mapping $f\colon B[0,1] \to B[0,1]$ has a fixed point.*

(c) *Let K be a non-empty, convex, compact subset of $\mathbb{C}$. Then any continuous map $f\colon K \to K$ has a fixed point.*

Proof. Ad (a): Let us assume that such a map r exists. $B[0,1]$ is starshaped, so we can by Proposition 3.10 find a continuous function $R\colon B[0,1] \to \mathbb{C}$ such that $\exp(R) = r$. Restriction to S^1 yields that $\exp(R(z)) = z$ for all $z \in S^1$, so R is a branch of $\log z$ on S^1. But such one does not exist (by Example 3.8), so we have arrived at a contradiction.

Ad (b): If f has no fixed point then we can construct a map r as in (a) as follows: Let $r(x)$ be the intersection point of S^1 with the ray from $f(x)$ through x.

Ad (c): We may without loss of generality assume that $K \subseteq B[0,1]$. Let $\phi\colon B[0,1] \to K$ be the mapping that to $z \in B[0,1]$ associates the nearest point to z in K. By (b) the mapping $f \circ \phi$ has a fixed point. And that point necessarily belongs to K, because the image of f is contained in K. $\square$

3.4   The winding number

In this paragraph we introduce the notion of winding number ($=$ index) which will turn out to be of central importance.

Definition 3.12. *Let $\gamma\colon [a,b] \to \mathbb{C}$ be a closed curve. If $z \in \mathbb{C} \setminus \gamma^*$ then the **winding number** or **index of γ relative to** z is the integer*

$$\mathrm{Ind}_\gamma(z) := \frac{\log(\gamma(b) - z) - \log(\gamma(a) - z)}{2\pi i}$$

where $t \to \log(\gamma(t) - z)$ is any continuous logarithm of $t \to \gamma(t) - z$.

Remarks 3.13.

(A) By Proposition 3.10 there exists a branch of $\log(\gamma(t) - z)$.

(B) Since any two continuous logarithms differ by a constant, we see that $\mathrm{Ind}_\gamma(z)$ does not depend on our choice of branch of logarithm.

(C) The reason why $\text{Ind}_\gamma(z)$ is called the winding number is the following: Let us write

$$\log(\gamma(t) - z) = \text{Log}|\gamma(t) - z| + i\theta(t)$$

where $t \to \theta(t) \in \arg(\gamma(t) - z)$ is a continuous argument of $t \to \gamma(t) - z$. Then

$$\text{Ind}_\gamma(z) = \frac{\theta(b) - \theta(a)}{2\pi}.$$

Now, $\theta(t)$ is geometrically the angle between the positive half of the x-axis and the ray from z through $\gamma(t)$, so $\text{Ind}_\gamma(z)$ measures the net increase in that angle as the parameter t ranges from a to b. Thus $\text{Ind}_\gamma(z)$ is the number of times (counted with sign) the ray from z through $\gamma(t)$ performs a full rotation (in the counterclockwise direction) as t increases from a to b, i.e., how often the curve γ winds around the point z.

(D) The concept of winding number of a closed curve has been generalized to higher dimensions, where it is replaced by the concept of degree of a map.

Example 3.14. Let $\triangle$ be the triangle with vertices 1, $\exp(2\pi i/3)$ and $\exp(4\pi i/3)$ (see Fig. 3.1). We will compute the winding number of its boundary $\partial\triangle$ (defined in Example 2.8(iii)) relative to the origin 0.

Note that $t \to \text{Arg}\,\partial\triangle\,(t)$ is a continuous arg z-function along the two sides $\partial\triangle\,|_{[0,1]}$ and $\partial\triangle\,|_{[2,3]}$, and $t \to \text{Arg}_1\,\partial\triangle\,(t)$ is a continuous arg z-

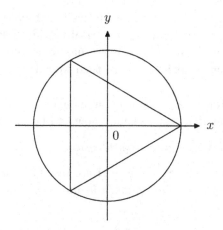

Fig. 3.1

function along the remaining side, where we let Arg_1 denote that branch of $\arg z$ that is defined in $\mathbb{C} \setminus [0, \infty[$ and takes values in $]0, 2\pi[$. Now

$$\theta(t) := \begin{cases} \text{Arg}\, \partial \triangle\, (t) & \text{for } 0 \leq t \leq 1, \\ \text{Arg}_1\, \partial \triangle\, (t) & \text{for } 1 \leq t \leq 2, \\ \text{Arg}\, \partial \triangle\, (t) + 2\pi & \text{for } 2 \leq t \leq 3, \end{cases}$$

is a continuous $\arg z$-function along $\partial\triangle$, so $\text{Ind}_{\partial\triangle}(0) = 1$.

Example 3.15. Consider the positively oriented circle around $z_0 \in \mathbb{C}$ with radius R, i.e., the closed curve $\gamma \colon [0, 2\pi] \to \mathbb{C}$, defined by

$$\gamma(t) = z_0 + Re^{it}.$$

It is easy to compute the winding number of γ relative to z_0: Indeed, $t \to \text{Log}\, R + it$ is a continuous logarithm along $\gamma(t) - z_0$, so by definition

$$\text{Ind}_\gamma(z_0) = \frac{(\text{Log}\, R + i2\pi) - (\text{Log}\, R + i0)}{2\pi i} = 1.$$

Example 3.16. The three curves $\gamma_1, \gamma_{-1}, \gamma_2 \colon [0, 2\pi] \to \mathbb{C} \setminus \{0\}$, given by

$$\gamma_j(t) := \exp(ijt) \qquad \text{for } j = 1, -1, 2$$

have the same image, viz the unit circle, but nevertheless different winding numbers relative to 0, viz 1, -1 and 2. The example shows in particular that the index of a curve depends not just on the image γ^* of the curve γ, but also on how the moving point $\gamma(t)$ traces the image γ^*.

A trivial example: A constant curve has index 0.

We have in the above examples pedantically computed the index analytically, although the results are geometrically evident. From now on we will not bother to do so. If a closed curve has so simple a look that it is geometrically obvious what its index is, then we will not elaborate on the matter.

Before we list properties of the winding number we call the attention to a topological concept that will be handy now and later:

Definition 3.17. *Let Ω be a topological space.*

(a) *Two closed curves $\gamma_0, \gamma_1 \colon [a, b] \to \Omega$ are said to be **homotopic in Ω**, if there exists a continuous map (a **homotopy**) $\Gamma \colon [a, b] \times [0, 1] \to \Omega$ such that*

$$\Gamma(t, 0) = \gamma_0(t) \quad and \quad \Gamma(t, 1) = \gamma_1(t) \qquad for \ all \ t \in [a, b]$$

and

$$\Gamma(a, s) = \Gamma(b, s) \qquad \text{for all } s \in [0, 1].$$

(b) *A closed curve is said to be **null-homotopic**, if it is homotopic to a constant curve.*

The geometrical contents of (a) are that the curve γ_0 continuously is deformed into γ_1 as the parameter s increases from 0 to 1. The curves $t \to \Gamma(t, s)$ are closed curves for each fixed $s \in [0, 1]$ and they form intermediate steps in the deformation of γ_0 into γ_1.

Theorem 3.18 (Properties of the Winding Number).

(a) *The winding number is independent of the parametrization of the basic interval as long as the orientation is unchanged: More precisely, if $\phi\colon [c, d] \to [a, b]$ is continuous and $\phi(c) = a$ and $\phi(d) = b$, then*

$$\text{Ind}_{\gamma \circ \phi}(z) = \text{Ind}_\gamma(z) \qquad \text{for all } z \in \mathbb{C} \setminus \gamma^*.$$

(b) *The winding number is a homotopy invariant, i.e., if two closed curves γ and τ are homotopic in $\mathbb{C} \setminus \{z_0\}$ then $\text{Ind}_\gamma(z_0) = \text{Ind}_\tau(z_0)$.*

(c) *Let γ be a closed curve in $\mathbb{C}$. Then the mapping $z \to \text{Ind}_\gamma(z)$ of $\mathbb{C} \setminus \gamma^*$ into $\mathbb{Z}$ is constant on each connected component of $\mathbb{C} \setminus \gamma^*$, and it is 0 on the unbounded component.*

(d) *Let γ be a closed path that does not pass through $z \in \mathbb{C}$. Then we have the following formula for the winding number:*

$$\text{Ind}_\gamma(z) = \frac{1}{2\pi i} \int_\gamma \frac{dw}{w - z}.$$

Proof. Ad (a): This is left to the reader.

Ad (b): Let $\Gamma\colon [a, b] \times [0, 1] \to \mathbb{C} \setminus \{z_0\}$ be a homotopy between γ and τ. The function $(t, s) \to \Gamma(t, s) - z_0$ has a continuous logarithm F since its domain of definition is starshaped. Now the function

$$s \to \frac{F(b, s) - F(a, s)}{2\pi i} = \text{Ind}_{\Gamma(\cdot, s)}(z_0)$$

is a continuous, integer-valued function on $[0, 1]$, hence a constant.

Ad (c): Let $\gamma\colon [a, b] \to \mathbb{C}$. Let $z_0 \in \mathbb{C} \setminus \gamma^*$ and let $r := \text{dist}(z_0, \gamma^*) > 0$. The continuous mapping $f\colon B(z_0, r) \times [a, b] \to \mathbb{C} \setminus \{0\}$ given by $f(z, t) := \gamma(t) - z$

has a continuous logarithm F since its domain of definition is starshaped. The formula

$$\text{Ind}_\gamma(z) = \frac{F(z,b) - F(z,a)}{2\pi i} \qquad \text{for } z \in B(z_0, r)$$

reveals that $\text{Ind}_\gamma(z)$ is a continuous function of z. Being also integer-valued it is constant on the connected components of $\mathbb{C} \setminus \gamma^*$.

It is left to prove that $\text{Ind}_\gamma(z_0) = 0$ for just one point z_0 in the unbounded component of $\mathbb{C} \setminus \gamma^*$, say the point $z_0 = 1 + \sup\{|\gamma(t)| \mid t \in [a,b]\}$. The homotopy

$$\Gamma(t,s) := (1-s)\gamma(t) \qquad \text{for } (t,s) \in [a,b] \times [0,1]$$

contracts γ to the constant curve $\tau = 0$, so by homotopy invariance

$$\text{Ind}_\gamma(z_0) = \text{Ind}_\tau(z_0) = 0.$$

Ad (d): If h is a continuous logarithm along $\gamma(t) - z$, i.e., if $\exp(h(t)) = \gamma(t) - z$ then we find formally by differentiation that

$$h'(t) = \frac{\gamma'(t)}{\gamma(t) - z}$$

which tempts us to define the function h by

$$h(t) := \int_a^t \frac{\gamma'(s)}{\gamma(s) - z} \, ds + c$$

where the constant $c \in \mathbb{C}$ for later purposes is chosen so that $\exp(c) = \gamma(a) - z$. Note that h is continuous in $[a,b]$ and that h is differentiable wherever γ is. An easy calculation reveals that the continuous function

$$\phi(t) := (\gamma(t) - z)e^{-h(t)}$$

has a vanishing derivative at each point where $\gamma'(t)$ exists, so ϕ is a constant. Since $\phi(a) = 1$ by our choice of c we have thus

$$e^{h(t)} = \gamma(t) - z \qquad \text{for all } t \in [a,b]$$

so h is a continuous logarithm along $\gamma(t) - z$. Finally

$$\text{Ind}_\gamma(z) = \frac{h(b) - h(a)}{2\pi i} = \frac{1}{2\pi i} \int_\gamma \frac{dw}{w - z}.$$

$\square$

3.5 Square roots

An important concept, related to the concept of logarithm, is that of n'th root. We concentrate on the special case of square roots and leave the generalization to n'th roots to the reader.

Definition 3.19. *Let $f\colon X \to \mathbb{C}$ be a complex-valued function on a topological space X. By a **continuous square root of** f (or a **branch of** $\sqrt{f}$) we mean any complex-valued continuous function $r\colon X \to \mathbb{C}$ satisfying $r^2 = f$.*

An example is the standard square root function $\sqrt{\cdot}\colon [0, \infty[\to \mathbb{R}$ with the usual convention that $\sqrt{x}$ is the unique *positive* number $r(x)$ satisfying $r(x)^2 = x$. Of course $x \to -\sqrt{x}$ is another example of a continuous square root.

The uniqueness of square roots is the topic for the next proposition. Then comes regularity and finally existence.

Proposition 3.20. *Let r_1 and r_2 be two continuous square roots of a never vanishing function, all defined on a connected topological space.*

Then either $r_1 = r_2$ or $r_1 = -r_2$. In particular, if r_1 and r_2 agree at one point they agree everywhere.

Proof.

$$\left(\frac{r_1}{r_2}\right)^2 = 1 \quad \text{so} \quad \frac{r_1(x)}{r_2(x)} \in \{-1, 1\}.$$

Thus r_1/r_2 is a continuous integer-valued function on a connected space, hence a constant, i.e., either $+1$ or -1. □

Proposition 3.21. *Let $f\colon \Omega \to \mathbb{C} \setminus \{0\}$ be a holomorphic function defined on an open subset Ω of the complex plane. Then any continuous square root of f is holomorphic on Ω.*

Remark. It is actually superfluous to assume that f never vanishes. See the remarks prior to Theorem 4.18.

Proof. Let r be any continuous square root of f in Ω. Let B be any open disc in Ω. Now f has a holomorphic logarithm F in B (Proposition 3.10 and Theorem 3.5), and so $\exp(F/2)$ is a holomorphic square root of f in B. But $r|_B$ is by Proposition 3.20 either $\exp(F/2)$ or $\exp(-F/2)$. □

Proposition 3.22. *Any holomorphic, never vanishing function on a star-shaped open subset of $\mathbb{C}$ has a holomorphic square root.*

Proof. Such a function has according to Proposition 3.10 a continuous logarithm F. A continuous square root is provided by $\exp(F/2)$, and that square root is holomorphic by Proposition 3.21. □

As we shall see later (Theorem 5.9), Proposition 3.22 extends from starshaped to simply connected sets.

Examples 3.23.

(i) The function z has on $\mathbb{C} \setminus \,]-\infty, 0]$ the continuous square root

$$\sqrt{z} = \exp\left(\frac{\text{Log}|z| + i\,\text{Arg}\,z}{2}\right) = \sqrt{|z|}\,e^{i\frac{\text{Arg}\,z}{2}}$$

which on the positive axis reduces to the ordinary square root. We call this square root for the *principal square root of z*. It is holomorphic according to Proposition 3.21.

(ii) We claim that there is no continuous square root of z on all of $\mathbb{C} \setminus \{0\}$, not even on S^1. Indeed, let $\sqrt{z}$ be one, and consider the curve $\gamma(t) := \sqrt{\exp(2\pi it)}$. It has the property that $\gamma^2 = \tau$, where

$$\tau(t) = e^{2\pi it} \qquad \text{for } t \in [0, 1].$$

Now $2\,\text{Ind}_\gamma(0) = \text{Ind}_\tau(0) = 1$, which contradicts the fact that the index is integer valued.

(iii) As a third example we will study the function $z^2 - 1$ and its holomorphic square roots which enter naturally in many considerations.

Consider for example $\arcsin z$. This is a multiple valued function defined by $w = \arcsin z \Leftrightarrow \sin w = z$. On any open subset of the complex plane where a holomorphic function w, satisfying $\sin w(z) = z$, can be defined, we must have

$$w'(z) = \frac{1}{\cos w(z)} = \frac{1}{\sqrt{1 - z^2}}$$

So a look at the holomorphic square root of $z^2 - 1$ will not be a waste of time. Our result is that the function $z^2 - 1$ on $\mathbb{C} \setminus [-1, 1]$ has exactly one holomorphic square root which is positive for $z \in \,]1, \infty[$.

To derive the result it suffices to exhibit a continuous $\sqrt{z^2 - 1}$ on $\mathbb{C} \setminus [-1, 1]$, because the statements about uniqueness and holomorphy are consequences of Propositions 3.20 and 3.21. Let $\sqrt{z}$ be the principal square root of z on $\mathbb{C}\backslash]-\infty, 0]$ (described in (i) above). If $z \in \mathbb{C}\backslash[-1, 1]$, then by a small computation

$$1 - z^{-2} \in \mathbb{C} \backslash]-\infty, 0]$$

so we can form the composite map $\sqrt{1 - z^{-2}}$. The function

$$r(z) := z\sqrt{1 - z^{-2}} \qquad \text{for } z \in \mathbb{C} \backslash [-1, 1]$$

is the desired continuous square root of $z^2 - 1$ on $\mathbb{C} \backslash [-1, 1]$.

Exercises

3.1. Show that exp maps $\{z \in \mathbb{C} \mid -\pi < \operatorname{Im} z < \pi\}$ onto $\mathbb{C} \backslash]-\infty, 0]$ bijectively, and that the inverse map is Log.

Determine the images by exp of lines (resp. line segments), parallel to the real (resp. imaginary axis).

3.2. Find the image by Log of $\{z \in \mathbb{C} \backslash \{0\} \mid |z - \frac{1}{2}| = \frac{1}{2}\}$.

3.3. Let log be the branch of $\log z$ in the snake shaped domain in Fig. 3.2 which has $\log 1 = 0$. What is $\log e$?

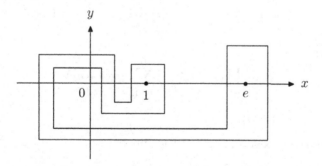

Fig. 3.2

3.4. A topological space X is said to be *contractible* if there is a continuous map $c: X \times [0, 1] \to X$ and a point $x_0 \in X$ such that

$$c(x, 0) = x \quad \text{and} \quad c(x, 1) = x_0 \qquad \text{for all } x \in X.$$

Show that the circle S^1 is *not* contractible.

3.5. Show that

$$\text{Log}(1+z) = \sum_{n=1}^{\infty} \frac{(-1)^{n-1}}{n} z^n \qquad \text{for } |z| < 1.$$

Hint: Theorem 3.5.

3.6. Show that

$$\left(1 + \frac{z}{n}\right)^n \to \exp(z) \qquad \text{as } n \to \infty$$

uniformly for z in any compact subset of $\mathbb{C}$.

3.7.

(α) (J. Bernoulli's paradox)

Is anything wrong in the following pretty argument: Taking Logs in the identity $(-z)^2 = z^2$ we get $2 \text{Log} z = 2 \text{Log}(-z)$, so $\text{Log} z = \text{Log}(-z)$.

(β) Compute $\text{Log} i$ and $\text{Log}(-i)$.

(γ) Is the formula $\text{Log}(z_1 z_2) = \text{Log}(z_1) + \text{Log}(z_2)$ correct ?

3.8. Let $a_1, a_2, \ldots, a_n$ be complex numbers such that

$$|a_j| < 1 \quad \text{and} \quad \left| \prod_{k=1}^{j} (1 + a_k) - 1 \right| < 1 \qquad \text{for } j = 1, 2, \ldots, n.$$

Show that

$$\text{Log}\left\{ \prod_{k=1}^{n} (1 + a_k) \right\} = \sum_{k=1}^{n} \text{Log}(1 + a_k).$$

3.9. Find the values of i^i.

3.10. Does there exist a branch of $\log z$ in

(α) $\{z \in \mathbb{C} \mid 1 < |z| < 2\}$?

(β) $\{z \in \mathbb{C} \mid \text{Re} \, z > \text{Im} \, z\}$?

3.11. Let $f, g \colon \mathbb{C} \to \mathbb{C}$ be continuous functions satisfying that $f^2 + g^2 = 1$. Show that there exists a continuous function $\phi \colon \mathbb{C} \to \mathbb{C}$ such that $f = \cos \phi$ and $g = \sin \phi$. Show that ϕ is unique modulo integer multiples of 2π, and that ϕ is holomorphic if f and g are holomorphic.

Hint: $f^2 + g^2 = (f + ig)(f - ig)$.

3.12. Let $\gamma, \rho: [0,1] \to \mathbb{C} \setminus \{0\}$ be closed curves. Show that

(α) $\mathrm{Ind}_{\gamma\rho}(0) = \mathrm{Ind}_\gamma(0) + \mathrm{Ind}_\rho(0)$.

(β) $\mathrm{Ind}_{-\gamma}(0) = -\mathrm{Ind}_\gamma(0)$, where $(-\gamma)(t) = \gamma(1-t)$.

(γ) $\mathrm{Ind}_{\gamma+\rho}(0) = \mathrm{Ind}_\gamma(0) + \mathrm{Ind}_\rho(0)$, where

$$(\gamma + \rho)(t) = \begin{cases} \gamma(t) & \text{if } t \in [0,1], \\ \rho(t-1) & \text{if } t \in [1,2]. \end{cases}$$

3.13. Let $\gamma: [0, 2\pi] \to \mathbb{C} \setminus \{0\}$ be the closed curve $\gamma(t) := 3\cos t + i\sin t$. Show that $\mathrm{Ind}_\gamma(0) = 1$.

3.14. Let A be a starshaped subset of $\mathbb{C}$. Let γ be a closed curve in A and let $z \in \mathbb{C} \setminus A$. Show that $\mathrm{Ind}_\gamma(z) = 0$.

3.15. Let $A: \mathbb{C} \to \mathbb{C}$ be an $\mathbb{R}$-linear isomorphism and let γ be a closed curve in $\mathbb{C} \setminus \{0\}$. Show that $\mathrm{Ind}_{A\circ\gamma}(0) = \mathrm{sign}(\det A)\,\mathrm{Ind}_\gamma(0)$.

3.16. A merry man walks with his dog on lead 10 times around a lamp post. He never allows the dog so much leash that it can get around to the other side of the lamp post of the man. Assuming that they both return to their initial positions after the man's 10 rotations, show that also the dog has been 10 times around the lamp post.

Hint: The problem is a special case of the following mathematical problem: Let $\gamma, \sigma: [a,b] \to \mathbb{C} \setminus \{0\}$ be two closed curves satisfying

$$|\gamma(t) - \sigma(t)| \leq |\gamma(t)| \qquad \text{for all } t \in [a,b].$$

Show that $\mathrm{Ind}_\gamma(0) = \mathrm{Ind}_\sigma(0)$ (note for example that γ and σ are homotopic).

3.17 (Extension of Exercise 3.16). Let $\gamma, \sigma: [a,b] \to \mathbb{C}$ be closed curves satisfying that

$$|\gamma(t) - \sigma(t)| < |\gamma(t)| + |\sigma(t)| \qquad \text{for all } t \in [a,b].$$

Show first that neither γ nor σ passes through 0, and then that $\mathrm{Ind}_\gamma(0) = \mathrm{Ind}_\sigma(0)$.

Hint: $\gamma(t)/\sigma(t)$ never takes values in $]-\infty, 0]$, so we may form $\mathrm{Log}(\gamma/\sigma)$.

3.18. Prove the:

Fundamental Theorem of Algebra. *Any complex polynomial* $p(z) = z^n + a_{n-1}z^{n-1} + \cdots + a_0$ *has at least one zero in* $\mathbb{C}$, *if* $n \geq 1$.

Hint: Assume not. Choose R so large that

$$|p(z) - z^n| < |z^n| \qquad \text{for all } |z| = R$$

consider the closed curve

$$\gamma(t) = R\exp(2\pi it) \qquad \text{for } t \in [0,1]$$

and combine Theorem 3.18(b) with Exercise 3.17.

3.19. Show that the complex valued function

$$f(z) = z^{17} - 16i\sin(93|z|^{22})z^{12} + 2$$

has a zero in the disc $B(0,2)$. (For heaven's sake, don't try to compute it!)

3.20. Let A be an invertible 3×3 matrix whose entries all are ≥ 0. Show that A has a positive eigenvalue and a corresponding eigenvector (v_1, v_2, v_3) such that $v_1 \geq 0, v_2 \geq 0, v_3 \geq 0$.

This result is known as the *Perron-Frobenius Theorem*, and it is true for $n \times n$ matrices.

Hint: Consider the compact convex set

$$\{(x_1, x_2, x_3) \in \mathbb{R}^3 \mid x_1 + x_2 + x_3 = 1, x_1 \geq 0, x_2 \geq 0, x_3 \geq 0\}$$

and the mapping

$$x \to \frac{Ax}{(Ax)_1 + (Ax)_2 + (Ax)_3}.$$

3.21. Let $\sqrt{\cdot} : \mathbb{C} \setminus]-\infty, 0] \to \mathbb{C}$ be the principal square root of z. For which values of z does the equation $\sqrt{z^2} = z$ hold?

3.22. Consider the square root $\sqrt{z^2 - 1}$ from Example 3.23(iii). Find $\sqrt{z^2 - 1}$ for z on the imaginary axis.

Show that $\sqrt{z^2 - 1}$ has a limit as $z \to 0$ through the open upper half plane, and find it. Same question when $z \to 0$ through the open lower half plane.

3.23. There are many ingenious ways of showing the non-existence of a continuous $\sqrt{z}$ on the unit circle S^1. For the benefit of the reader we will below sketch 4 different methods. They all proceed by contradiction. So we assume that there exists a branch r of $\sqrt{z}$ on S^1, i.e., a continuous function r satisfying $r(z)^2 = z$ for all $z \in S^1$. We may without loss of generality assume that $r(1) = 1$.

Method 1: The one from Example 3.23(ii).

Method 2: Show that the function $\phi(z) := r(z)/r(-z)$ is a constant (indeed $\phi(z)^2 = -1$). Derive a contradiction.

Method 3: Consider the smooth function $f(t) := r(\exp(it))$ for $t \in \mathbb{R}$. Differentiate the identity $f(t)^2 = \exp(it)$.

Method 4:

(α) Show first that r is a continuous injective mapping.

(β) Show that $r(S^1) = S^1$.

   Hint: Either use a point set topology argument ($r(S^1)$ is homeomorphic to S^1 and hence not starshaped, cf. Exercise 3.4), or by showing that $r(S^1)$ is a subgroup of the multiplicative group S^1.

(γ) Choose a point $z_0 \in S^1$ such that $r(z_0) = -1$. Obtain a contradiction using the injectivity of r^2.

3.24. Does there exist a branch of the n'th root of z on S^1 when $n > 0$? Exhibit a specimen if your answer is yes.

3.25. Let $h \colon S^1 \to \mathbb{C} \setminus \{0\}$ be continuous, and consider the closed curve $\gamma(t) := h(\exp(it))$ for $t \in [0, 2\pi]$.

(α) Show that h can be written in the form $h(z) = z^n \exp(\phi(z))$, where n is an integer and $\phi \in C(S^1)$.

   Hint: Let F be a continuous logarithm of $t \to h(e^{it})$ and $n = \mathrm{Ind}_\gamma(0)$, so that

   $$h(e^{it}) = e^{F(t)} = e^{int} e^{F(t) - int}$$

   Note that the exponent $\psi(t) := F(t) - int$ satisfies that $\psi(2\pi) = \psi(0)$, and so defines a function ϕ on S^1.

(β) Show that $\mathrm{Ind}_\gamma(0)$ is even if h is even.

(γ) Show that $\mathrm{Ind}_\gamma(0)$ is odd if h is odd. In particular, if h is odd then $\mathrm{Ind}_\gamma(0) \neq 0$.

3.26. We will in this exercise present a proof of the 2-dimensional version of the so-called

   Brouwer's Theorem on Invariance of Domain. *Let Ω be an open subset of the complex plane, and let $f \colon \Omega \to \mathbb{C}$ be continuous and 1–1. Then $f(\Omega)$ is open in $\mathbb{C}$.*

(α) Verify that the theorem is a consequence of the following

> **Lemma.** *If the function $f\colon B[0,1] \to \mathbb{C}$ is continuous and 1–1, then $f(0)$ is an interior point of $f(B[0,1])$.*

We proceed with a proof of the lemma:

(β) Show that we may assume that $f(0) = 0$, and that then $r := \mathrm{dist}(0, f(S^1)) > 0$.

(γ) Show that it suffices to lead the following assumption to a contradiction: There exists a point $z_0 \in B(0,r) \setminus f(B[0,1])$.

(δ) Consider now the closed curve $\gamma(t) := f(e^{it})$ for $t \in [0, 2\pi]$. Show that $0 = \mathrm{Ind}_\gamma(z_0) = \mathrm{Ind}_\gamma(0)$.

(ε) Show that $F\colon S^1 \times [0,1] \to \mathbb{C} \setminus \{0\}$, given by

$$F(z,s) := f\left(\frac{z}{1+s}\right) - f\left(\frac{-sz}{1+s}\right)$$

is a homotopy in $\mathbb{C} \setminus \{0\}$ between $f|_{S^1}$ and the odd function

$$h(z) := F(z,1) = f\left(\frac{z}{2}\right) - f\left(\frac{-z}{2}\right).$$

(ζ) Use Exercise 3.25(γ) to arrive at the desired conclusion.

3.27.

(α) Show that there at any given time are opposite ends of the earth which have the same weather, i.e., the same temperature and the same barometer pressure. This is a meteorological interpretation of the 2-dimensional version of

> **Borsuk-Ulam's Theorem.** *Let $S^2 := \{x \in \mathbb{R}^3 \mid |x| = 1\}$ be the unit sphere. If $f\colon S^2 \to \mathbb{C}$ is continuous then there exists an $x_0 \in S^2$ such that $f(x_0) = f(-x_0)$.*

(β) Prove the Borsuk-Ulam theorem.
Hint: Consider the map $h\colon B[0,1] \to \mathbb{C}$ given by

$$h(x+iy) := f(x,y,\sqrt{1-x^2-y^2}) - f(-x,-y,-\sqrt{1-x^2-y^2})$$

and the closed curve $\gamma(t) := h(\exp(it))$ in $\mathbb{C}$.

(γ) Prove the following: If a balloon is deflated and laid on the floor, then there exist two antipodal points which end up over the same point of the floor.

3.28 (Cf. Proposition 3.21). Let $f \in \mathrm{Hol}(\Omega)$ and assume that the zeros of f all are isolated. Let r be a continuous square root of f.

Show that $r \in \mathrm{Hol}(\Omega)$.

Chapter 4

Basic Theory of Holomorphic Functions

While Chapter 3 essentially treats continuous functions and a bit of topology
of the complex plane, this chapter specializes to holomorphic functions and
their surprising properties, derived from the Cauchy Integral Theorem. The
contents of this chapter are dealt with in all introductory text books on
complex function theory.

4.1 The Cauchy-Goursat integral theorem

The main results of this paragraph are the Cauchy-Goursat Integral Theorem
and some of its surprising consequences: The integral of a holomorphic
function f around a closed curve, the interior of which is contained in the
domain of definition of f, is zero. A holomorphic function is analytic, i.e., it
can be expanded in a power series.

Theorem 4.1 (Cauchy-Goursat). *Let f be holomorphic in an open subset
Ω of the complex plane. Let $a, b, c \in \Omega$ and assume that the triangle $\triangle$ with
vertices a, b and c is contained in Ω. Then*

$$\int_{\partial\triangle} f(z)\, dz = 0.$$

Proof. Divide the triangle $\triangle$ as in Fig. 4.1, using the vertices and the
midpoints of the segments $[a, b]$, $[b, c]$ and $[c, a]$. We get 4 smaller triangles
$\triangle^{(1)}$, $\triangle^{(2)}$, $\triangle^{(3)}$ and $\triangle^{(4)}$ as shown in the figure. Orienting all the triangles
in the counter clockwise direction it is a matter of direct verification that

$$\int_{\partial\triangle} f(z)\, dz = \sum_{i=1}^{4} \int_{\partial\triangle^{(i)}} f(z)\, dz.$$

49

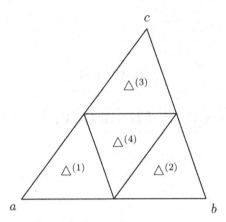

Fig. 4.1   Triangle subdivision.

Denoting the absolute value of the left hand side for α, we have for at least one small triangle $\triangle^{(i)}$, call it $\triangle_1$, that

$$\left| \int_{\partial \triangle_1} f \, dz \right| \geq \frac{\alpha}{4}$$

Note that $l(\partial \triangle_1) = \frac{1}{2} l(\partial \triangle)$ where l denotes length of the path in question.

We can repeat the process to successively obtain triangles $\triangle_n \subseteq \triangle_{n-1}$ such that $l(\partial \triangle_n) = 2^{-n} l(\partial \triangle)$ and

$$\left| \int_{\partial \triangle_n} f \, dz \right| \geq 4^{-n} \alpha$$

There is a point, say z_0, in the intersection $\bigcap_{n=1}^{\infty} \triangle_n$. f is differentiable at z_0, so given $\beta > 0$ we have from a certain n on, in the entire triangle $\triangle_n$, that

$$|f(z) - f(z_0) - f'(z_0)(z - z_0)| \leq \beta |z - z_0|. \tag{4.1}$$

The constant function $f(z_0)$ and the function $z - z_0$ have primitives, so (by Proposition 2.11(e)) the line integrals of these along any closed path vanish. Thus for any $\triangle_n$ where (4.1) holds:

$$4^{-n} \alpha \leq \left| \int_{\triangle_n} f \, dz \right| \leq \beta (l(\partial \triangle_n))^2 = \beta 4^{-n} (l(\partial \triangle))^2$$

where we have used that $z_0 \in \triangle_n$ and that $|z - z_0| \leq l(\partial \triangle_n)$ for any $z \in \partial \triangle_n$. Now, $\beta (l(\partial \triangle))^2 \geq \alpha$. Since this holds for any $\beta > 0$ we conclude that $\alpha = 0$. $\square$

By a bit of ingenuity we can get a stronger looking version of Theorem 4.1 out by allowing the function f to have singularities:

Corollary 4.2. *Let Ω be an open subset of $\mathbb{C}$, and let f be holomorphic in $\Omega \setminus \{z_0\}$. Assume furthermore that $(z - z_0)f(z) \to 0$ as $z \to z_0$. Then*

$$\int_{\partial\triangle} f(z)\, dz = 0$$

for any triangle $\triangle$ in Ω such that $z_0 \in \Omega \setminus \partial\triangle$.

Proof. Let δ be a small triangle inside of $\triangle$, containing z_0 in its interior (see Fig. 4.2). Subdividing $\triangle \setminus \delta$ into smaller triangles we get by the Cauchy-Goursat Theorem that

$$\int_{\partial\triangle} f(z)\, dz = \int_{\partial\delta} f(z)\, dz.$$

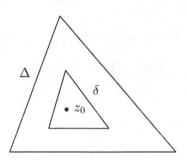

Fig. 4.2

Let us now consider a sequence $\delta_1, \delta_2, \ldots$ of equilateral triangles, each with z_0 as its center, and with diameters shrinking to 0. By elementary geometry, when we choose δ_n so that $\operatorname{dist}(z_0, \partial\delta_n) = 1/n$ then $l(\partial\delta_n) = 6\sqrt{3}/n$. By assumption

$$|f(z)| \leq \frac{c(z)}{|z - z_0|} \qquad \text{where } c(z) \to 0 \text{ as } z \to z_0.$$

Let $\varepsilon > 0$ be given. Choosing n so large that $|c(z)| \leq \varepsilon$ for all $z \in \partial\delta_n$, we get

$$\left| \int_{\partial\triangle} f \right| = \left| \int_{\partial\delta_n} f \right| \leq \sup_{\partial\delta_n}\{|f|\} l(\partial\delta_n) \leq \frac{\varepsilon}{(1/n)} \frac{6\sqrt{3}}{n} = 6\sqrt{3}\varepsilon.$$

The statement of the corollary follows since this is true for any $\varepsilon > 0$. $\qquad\square$

From Corollary 4.2 we deduce a remarkable and important regularity property: A holomorphic function cannot possess a minor singularity at a point. If it has a point singularity it is a serious one. This result is called the *Removable Singularity Theorem* or *Riemann's extension theorem*:

Theorem 4.3 (The Removable Singularity Theorem). *Let z_0 be a point in an open subset Ω of the complex plane, and let $f : \Omega \setminus \{z_0\} \to \mathbb{C}$ be holomorphic in $\Omega \setminus \{z_0\}$.*

If $(z - z_0)f(z) \to 0$ as $z \to z_0$ (in particular if f is bounded near z_0), then f can be extended to a function which is holomorphic in all of Ω.

Proof. Without loss of generality we take $z_0 = 0$ in the proof. Let $\triangle$ be a triangle, contained in Ω, with $z_0 = 0$ in its interior, and its boundary $\partial\triangle$ positively oriented (cf. Example 3.14) so that (Theorem 3.18(d))

$$1 = \frac{1}{2\pi i} \int_{\partial\triangle} \frac{dw}{w - z} \qquad \text{for any } z \in \text{int}(\triangle).$$

For any fixed $z \in \text{int}(\triangle) \setminus \{0\}$ each of the two singularities 0 and z of the function

$$w \mapsto \frac{f(w) - f(z)}{w - z} \qquad \text{on } \Omega \setminus \{0, z\}$$

satisfies the condition of Corollary 4.2. Subdividing $\triangle$ into two triangles, each with only one of the singularities, we get

$$\int_{\partial\triangle} \frac{f(w) - f(z)}{w - z} \, dw = 0$$

which implies that

$$f(z) = \frac{1}{2\pi i} \int_{\partial\triangle} \frac{f(w)}{w - z} \, dw \qquad \text{for any } z \in \text{int}(\triangle) \setminus \{0\}.$$

The right side is analytic and hence holomorphic off $\partial\triangle$ (Lemma 2.14), so the desired extension F may unambiguously be defined by

$$F(z) := \begin{cases} f(z) & \text{for } z \in \Omega \setminus \{0\}, \\[2mm] \dfrac{1}{2\pi i} \displaystyle\int_{\partial\triangle} \frac{f(w)}{w - z} \, dw & \text{for } z \in \text{int}(\triangle). \end{cases}$$

$\square$

We see that Corollary 4.2 actually does not strengthen Theorem 4.1, because the exceptional point is not a singularity after all! The corresponding result is not true in one real variable: The function $t \mapsto |t|$ is not differentiable at 0.

During the proof of Theorem 4.3 we noticed the remarkable fact that a holomorphic function is analytic. We shall in Theorem 4.8 below prove that its power series converges in the largest possible disc.

As a by-product of the *Removable Singularity Theorem* we note the following example which will be used several times, both in the near and more distant future.

Example 4.4. If f is holomorphic in the open set Ω and $a \in \Omega$ then the function

$$g(z) := \begin{cases} \frac{f(z)-f(a)}{z-a} & \text{for } z \in \Omega \setminus \{a\}, \\ f'(a) & \text{for } z = a, \end{cases}$$

is holomorphic in all of Ω.

Theorem 4.5 (The Cauchy Integral Theorem). *Let f be holomorphic in an open starshaped subset Ω of the complex plane. Then $\int_\gamma f(z)\,dz = 0$ for each closed path γ in Ω.*

Proof. When we combine the Cauchy-Goursat Theorem with Lemma 2.13 we see that f has a primitive in Ω, so we may refer to Proposition 2.11(e). $\square$

Theorem 4.6 (The Cauchy Integral Formula). *Let f be holomorphic in an open starshaped set Ω of the complex plane. Then we have for any closed path γ in Ω the formula*

$$\text{Ind}_\gamma(z)f(z) = \frac{1}{2\pi i} \int_\gamma \frac{f(w)}{w-z}\,dw \qquad \text{for all } z \in \Omega \setminus \gamma^*.$$

Proof. Fix $z \in \Omega \setminus \gamma^*$. Then the function

$$w \mapsto \begin{cases} \frac{f(w)-f(z)}{w-z} & \text{for } w \in \Omega \setminus \{z\}, \\ f'(z) & \text{for } w = z, \end{cases}$$

is according to Example 4.4 holomorphic in Ω, so by the Cauchy Integral Theorem its integral along γ is 0, i.e.,

$$\int_\gamma \frac{f(w)-f(z)}{w-z}\,dw = 0.$$

Recalling the formula (Theorem 3.18(d))

$$\text{Ind}_\gamma(z) = \frac{1}{2\pi i} \int_\gamma \frac{dw}{w - z}$$

we get the theorem.   □

As a special case we note the *Cauchy Integral Formula for a Disc:*

Theorem 4.7. *If f is holomorphic in a neighborhood of the closed disc $B[z_0, r]$, then we have for each $z \in B(z_0, r)$ the formula*

$$f(z) = \frac{1}{2\pi i} \int_{|w - z_0| = r} \frac{f(w)}{w - z} \, dw.$$

The purpose of the next chapter is to extend the Cauchy Integral Theorem and the Cauchy Integral Formula to more general domains than starshaped. The following remark (which will not be used later) points in another direction.

Remark. There is an extension of the Cauchy Integral Formula to general functions. It is often called the *Cauchy-Green Formula* since Green's theorem is used to prove it (for a proof, see e.g., Hörmander, 1973, Theorem 1.2.1):

Let ω be a bounded open domain in $\mathbb{C}$ with a smooth positively oriented boundary γ. If u is C^1 in a neighborhood of $\overline{\omega}$ then

$$u(z) = \frac{1}{2\pi i} \int_\gamma \frac{u(w)}{w - z} \, dw - \frac{1}{\pi} \int_\omega \frac{\frac{\partial u}{\partial \overline{z}}(x)}{x - z} \, dm(x) \qquad \text{for } z \in \omega,$$

where dm denotes Lebesgue measure on $\mathbb{C} = \mathbb{R}^2$.

Note that the last term on the right hand side drops out if u is holomorphic, so that the formula in that case reduces to the ordinary Cauchy Integral Formula.

A version of the Cauchy Integral Theorem can be derived from the Cauchy-Green Theorem (replace u by $(w - z)u(w)$ etc.):

$$\frac{1}{2\pi i} \int_\gamma u(z) \, dz = \frac{1}{\pi} \int_\omega \frac{\partial u}{\partial \overline{z}}(x) \, dm(x).$$

There is a striking contrast between $\mathbb{R}$ – and $\mathbb{C}$-differentiability: There are examples of functions on $\mathbb{R}$ with the property that f is continuously differentiable and f' is not differentiable anywhere. But as we noticed after Theorem 4.3 any $\mathbb{C}$-differentiable function, i.e., a holomorphic function, is in fact even analytic and hence in particular infinitely often differentiable. In the future we will use the words analytic and holomorphic interchangeably. What is new in Theorem 4.8 below is that the power series converges in the largest possible disc.

Theorem 4.8. *Let f be holomorphic in an open subset Ω of the complex plane. Then f is analytic in Ω; furthermore, for any $z_0 \in \Omega$ the power series expansion of f around z_0 converges in the largest open disc in Ω around z_0.*

In particular, f is infinitely often $\mathbb{C}$-differentiable and all derivatives $f', f'', \dots$ are holomorphic.

Also $f \in C^\infty(\Omega)$, i.e., partial derivatives of all orders with respect to the real variables x and y exist and are continuous.

Proof. Combine Lemma 2.14 and Theorem 4.7 with the uniqueness of power series expansions. For the last statement we note that

$$\frac{\partial f}{\partial x} = f' \quad \text{and} \quad \frac{\partial f}{\partial y} = i f'$$

from which we by induction get

$$\frac{\partial^{n+m} f}{\partial x^m \partial y^n} = i^n f^{(n+m)}.$$

$\square$

It may be remarked that Theorem 4.8 does not hold for an analytic function of a real variable. To take an example, the function

$$f(x) := \frac{1}{1+x^2} \quad \text{for } x \in \mathbb{R}$$

is analytic on all of the real line, but its power series expansion

$$f(x) = 1 - x^2 + x^4 - x^6 + \cdots$$

converges only for $-1 < x < 1$. The explanation is that if we view f as a function of a complex variable

$$f(z) := \frac{1}{1+z^2} \quad \text{for } z \in \mathbb{C}$$

then f is not holomorphic $=$ analytic on all of the complex plane, but only on the domain $\mathbb{C} \setminus \{i, -i\}$, and the biggest disc around 0 in that domain is $B(0,1)$.

We next present a converse to the Cauchy-Goursat Theorem, viz. Morera's theorem, which is often expedient in determining whether a given function is holomorphic. It can be used in situations where a direct resort to the definition – estimation of difference quotients etc. – is hopeless or at least very complicated. For a more thorough discussion of Morera's theorem we refer to the beautiful article Zalcman (1974). For the latest news see Globevnik (1990).

Theorem 4.9 (Morera's Theorem). *Let* $f\colon \Omega \to \mathbb{C}$ *be a continuous function on an open subset* Ω *of the complex plane. Assume that each point* $z \in \Omega$ *has a neighborhood* $U_z \subseteq \Omega$ *with the property that* $\int_\Delta f(z)\,dz = 0$ *for all triangles* Δ *contained in* U_z.
Then f *is holomorphic in* Ω.

Proof. Let $z \in \Omega$ and choose $r > 0$ so small that $B(z, r) \subseteq U_z$. It suffices to prove that f is holomorphic in $B(z, r)$ for each $z \in \Omega$. Now f has by Lemma 2.13 a primitive F in $B(z, r)$. Being $\mathbb{C}$-differentiable means that F is holomorphic. By Theorem 4.8 so is its derivative f. $\qquad\square$

4.2 Selected consequences of the Cauchy integral formula

It is a sad fact from the theory of functions of a real variable that a uniform limit of a sequence of differentiable functions need not be differentiable, although it is continuous. Indeed, there exists a continuous nowhere differentiable function on $[0, 1]$, and (by the classical Weierstrass Approximation Theorem) it is a uniform limit of a sequence of polynomials.

The situation is quite the contrary for holomorphic functions, as demonstrated by the following surprising result, in which no assumptions are made on the derivatives:

Theorem 4.10 (Weierstrass' Theorem). *Let* $f_1, f_2, \dots$ *be a sequence of functions which are holomorphic in an open subset* Ω *of the complex plane. Assume that the sequence converges, uniformly on each compact subset of* Ω, *to a function* f.
Then f *is also holomorphic in* Ω. *Furthermore, the sequence* $\{f_n'\}$ *of derivatives converges, again uniformly on each compact subset of* Ω, *to* f'.

We shall later (Exercise 8.10) meet a stronger version of Weierstrass' theorem, viz. Vitali-Porter's theorem. The assumption about uniform convergence is important; pointwise convergence does not suffice. For a discussion see Davidson (1983) and Zalcman (1974, Section 11).

Proof. Let $z_0 \in \Omega$ and choose $r > 0$ be so small that $B[z_0, r] \subseteq \Omega$. According to the Cauchy Integral Formula (Theorem 4.7) we have

$$f_n(z) = \frac{1}{2\pi i} \int_{|w-z_0|=r} \frac{f_n(w)}{w-z} \, dw \qquad \text{for all } z \in B(z_0, r) \qquad (4.2)$$

and going to the limit $n \to \infty$ we get

$$f(z) = \frac{1}{2\pi i} \int_{|w-z_0|=r} \frac{f(w)}{w-z} \, dw \qquad \text{for all } z \in B(z_0, r) \qquad (4.3)$$

which (by Lemma 2.14) shows that f is analytic and hence holomorphic in $B(z_0, r)$. Now $f \in \text{Hol}(\Omega)$, the disc being arbitrary.

Concerning the last statement we may now assume that $f = 0$ [if necessary we replace f_n by $f_n - f$].

By a compactness argument it suffices to show that $\{f_n'\}$ converges uniformly to 0 as $n \to \infty$ on any ball $B(z_0, \frac{r}{2})$ such that $B[z_0, r] \subseteq \Omega$. Differentiating formula (4.2) we get (directly or using Lemma 2.14) that

$$f_n'(z) = \frac{1}{2\pi i} \int_{|w-z_0|=r} \frac{f_n(w)}{(w-z)^2} \, dw \qquad \text{for all } z \in B(z_0, \frac{r}{2})$$

which we estimate as follows for $z \in B(z_0, \frac{r}{2})$ (cf. Proposition 2.11(a)):

$$|f_n'(z)| \leq \frac{1}{2\pi} \frac{2\pi r}{(r/2)^2} \sup_{|w-z_0|=r} \{|f_n(w)|\} = \frac{4}{r} \sup_{|w-z_0|=r} \{|f_n(w)|\}$$

so

$$\sup_{B(z_0, r/2)} \{|f_n'|\} \leq \frac{4}{r} \sup_{|w-z_0|=r} \{|f_n(w)|\}.$$

The circle $\{w \in \mathbb{C} \mid |w - z_0| = r\}$ is a compact subset of Ω, so the right hand side converges to 0 as $n \to \infty$. $\qquad \square$

An easy corollary is the fact (Proposition 2.5) that a power series in its disc of convergence defines a holomorphic function and that its derivative can be found by term by term differentiation (see Exercise 4.1). More generally one can apply Weierstrass' theorem to infinite series, the terms of which are holomorphic functions. Example: The Riemann ζ-function (Exercise 4.26 below and Chapter 11, Section 11.1).

We next generalize Proposition 1.6.

Theorem 4.11 (The Unique Continuation Theorem). *Let Ω be an open connected subset of the complex plane, and let $f, g \in \text{Hol}(\Omega)$.*

(a) *If the set of points $z \in \Omega$ where $f(z) = g(z)$ has a limit point in Ω, then $f = g$.*

(b) *In particular, the zeros of f have no limit points in Ω, unless f is identically 0.*

We remind the reader, that a point p is a *limit point* of a set A, if every neighborhood of p contains at least one point of A distinct from p.

We emphasize that the limit point from (a) and (b) must belong to Ω; limit points outside of Ω do not suffice.

Proof. Item (a) is a consequence of (b), so it suffices to prove (b). Consider the set

$$N := \{a \in \Omega \mid f^{(n)}(a) = 0 \quad \text{for all } n = 0, 1, 2, \dots \}.$$

Its complement in Ω, i.e.,

$$\Omega \setminus N = \{z \in \Omega \mid \text{There exists an } n \text{ such that } f^{(n)}(z) \neq 0\}$$

is open because each $f^{(n)}$ is continuous. On the other hand, if $a \in N$, then we see from the power series expansion

$$f(z) = \sum_{n=0}^{\infty} \frac{f^{(n)}(a)}{n!}(z - a)^n$$

that f (and hence each of its derivatives, too) is identically 0 in a neighborhood of a, so N is also open. By connectedness of Ω either $N = \Omega$ or $N = \emptyset$. It suffices to prove that the limit point $z_0 \in \Omega$ belongs to N, so that the first possibility takes place. We shall in other words show that in the power series expansion

$$f(z) = \sum_{k=0}^{\infty} a_k(z - z_0)^k$$

of f around z_0 all the coefficients $a_0, a_1, a_2, \dots$ are 0.

But that is part of Proposition 1.6.   □

The Unique Continuation Theorem is amazing: It tells us that a holomorphic function is determined by its values on e.g., any curve segment, be it ever so small; and that we cannot change the values of the function in one part of the domain without it being felt everywhere.

As an illustrative example we mention that Log thus can be characterized uniquely as the holomorphic function on $\mathbb{C}\backslash]-\infty, 0]$ that extends the ordinary real logarithm on $]0, \infty[$.

The Unique Continuation Theorem implies that holomorphic functions obey the so-called "principle of permanence of functional relationships", i.e., if holomorphic functions in a part of a region satisfy a certain relationship, then they do so everywhere. As an example we infer from the well known relation $\sin^2 x + \cos^2 x = 1$, which is true for all real x, that the same relation holds everywhere, i.e., $\sin^2 z + \cos^2 z = 1$ for all complex z (Exercise 4.16).

Definition 4.12. *Let f be holomorphic in an open subset Ω of the complex plane and let $z_0 \in \Omega$. Then f can in exactly one way be expanded in a power series (in the largest open ball B in Ω) around z_0:*

$$f(z) = \sum_{k=0}^{\infty} a_k(z - z_0)^k \qquad \text{for } z \in B.$$

*Clearly f has a zero at z_0 if and only if $a_0 = 0$. We define the **order of the zero** as*

$$\sup\{n + 1 \mid a_0 = a_1 = \cdots = a_n = 0\} \in \mathbb{N} \cup \{\infty\}$$

An equivalent characterization that does not involve the power series expansion, is that the order is the biggest n such that

$$f(z_0) = f'(z_0) = \cdots = f^{(n-1)}(z_0) = 0.$$

We leave it to the reader to prove, that if f has a zero of order $n < \infty$, then f can be factorized as

$$f(z) = (z - z_0)^n g(z) \qquad \text{where } g \in \text{Hol}(\Omega) \text{ and } g(z_0) \neq 0.$$

If f has a zero of infinite order then f vanishes identically near the zero (and hence on all of the connected component of Ω that contains z_0).

4.3 The open mapping theorem

The following continuous version of Rouché's theorem will be used in the proofs of the Open Mapping Theorem and of Bloch-Landau's theorem (Theorem 7.6). Rouché's theorem will be discussed in more details in Chapter 9, where we shall meet other versions of it.

Theorem 4.13 (Rouché's Theorem, version 1). *Let $F \in C(B[0,r])$ and let H be holomorphic on a neighborhood of $B = B[0,r]$. Assume that*

$$|F(\zeta) - H(\zeta)| \leq |H(\zeta)| \qquad \text{for all } \zeta \in \partial B \qquad (4.4)$$

and that H has a zero somewhere in $B[0,r]$.

Then F, too, has a zero in $B[0,r]$. If the inequality (4.4) is strict then F has a zero in the open ball $B(0,r)$.

Proof (adapted from Tsarpalias, 1989). We shall derive a contradiction from the assumption that F never vanishes in $B[0,r]$. Let

$$\gamma(t) := re^{it} \qquad \text{for } 0 \leq t \leq 2\pi.$$

Since F never vanishes we may divide (4.4) by F to get that

$$\left| 1 - \frac{H(\zeta)}{F(\zeta)} \right| \leq \left| \frac{H(\zeta)}{F(\zeta)} - 0 \right| \qquad \text{for } \zeta \in \partial B$$

which shows that the fraction $H(\zeta)/F(\zeta)$ is closer to 1 than to 0. Hence $\mathrm{Re}\{H(\zeta)/F(\zeta)\} \geq \frac{1}{2}$, so we may form $\mathrm{Log}\{H(\zeta)/F(\zeta)\}$ for $|\zeta| = r$. Letting ϕ be a continuous logarithm of $F \circ \gamma$ we get that $\phi + \mathrm{Log}\{H \circ \gamma / F \circ \gamma\}$ is a continuous logarithm of $H \circ \gamma$, and so $\mathrm{Ind}_{H \circ \gamma}(0) = \mathrm{Ind}_{F \circ \gamma}(0)$.

The map Γ, defined by

$$\Gamma(t,s) := F(sre^{it}) \qquad \text{for } (t,s) \in [0,2\pi] \times [0,1]$$

is a homotopy in $\mathbb{C} \setminus \{0\}$ between the constant map $F(0)$ and $F \circ \gamma$, so

$$\mathrm{Ind}_{H \circ \gamma}(0) = \mathrm{Ind}_{F \circ \gamma}(0) = \mathrm{Ind}_{F(0)}(0) = 0.$$

We arrive at the desired contradiction by proving that $\mathrm{Ind}_{H \circ \gamma}(0) \neq 0$:

We may write H in the form

$$H(z) = (z - z_1)(z - z_2) \cdots (z - z_n)g(z)$$

where $z_1, z_2, \ldots, z_n$ are the zeros of H in $B(0,r)$ and where g is holomorphic in a neighborhood of $B[0,r]$ and never vanishes in $B[0,r]$. In particular

$$\frac{H'(z)}{H(z)} = \sum_{j=1}^{n} \frac{1}{z - z_j} + \frac{g'(z)}{g(z)}$$

Now,

$$\text{Ind}_{H \circ \gamma}(0) = \frac{1}{2\pi i} \int_{H \circ \gamma} \frac{dz}{z} = \frac{1}{2\pi i} \int_\gamma \frac{H'(z)}{H(z)} \, dz$$

$$= \sum_{j=1}^n \frac{1}{2\pi i} \int_\gamma \frac{dz}{z - z_j} + \frac{1}{2\pi i} \int_\gamma \frac{g'(z)}{g(z)} \, dz = n + 0 = n > 0.$$

$\square$

Theorem 4.14 (The Open Mapping Theorem). *Let f be a non-constant holomorphic function on an open connected subset Ω of the complex plane. Then $f(\Omega)$ is open.*

Proof. It suffices to prove that there for each $z_0 \in \Omega$ exists a disc, contained in $f(\Omega)$, around $f(z_0)$.

Consider the function $H(z) := f(z) - f(z_0)$ for $z \in \Omega$. By the Unique Continuation Theorem 4.11(b) there is a ball $B[z_0, r]$ in which the only zero of H is $z = z_0$. In particular

$$R := \inf_{|\zeta - z_0| = r} \{|H(\zeta)|\} > 0.$$

Let now w be any point such that $|w - f(z_0)| < R$, and let us apply Rouché to the two functions $F(z) = f(z) - w$ and H: For $|\zeta - z_0| = r$ we get

$$|F(\zeta) - H(\zeta)| = |f(z_0) - w| < R \leq |H(\zeta)|$$

so by Rouché F has a zero in $B[z_0, r]$, i.e., w belongs to the range of f. Since w was arbitrary in $B(f(z_0), R)$, we have proved that any element of $B(f(z_0), R)$ belongs to the range of f, or in other words that $B(f(z_0), R) \subseteq f(\Omega)$. $\square$

A topological proof of Theorem 4.14 can be found in Crowe and Samperi (1989).

Corollary 4.15 (The Maximum Modulus Principle). *Let f be holomorphic in an open connected subset Ω of $\mathbb{C}$ and assume that $|f|$ has a local maximum in Ω. Then f is constant in Ω.*

Proof. Make a sketch! $\square$

In applications the following consequence is useful.

Corollary 4.16 (The Weak Maximum Principle). *Let Ω be a bounded, open subset of $\mathbb{C}$ and let f be holomorphic on Ω and continuous on the closure of Ω. Then $|f|$ takes its supremum value at a boundary point of Ω.*

Proof. Since $f \in C(\overline{\Omega})$ and $\overline{\Omega}$ is compact, $|f|$ takes its supremum value M in $\overline{\Omega} = \Omega \cup \partial\Omega$. Let us assume that it happens at an interior point $z_0 \in \Omega$ and then deduce that $|f|$ takes the same value at some boundary point, too.

By the Maximum Modulus Principle f is constant in the biggest open ball $B(z_0, R)$ in Ω with center z_0. Since the ball is the biggest, its boundary contains at least one point z' of $\partial\Omega$. By continuity $|f(z')| = M$. $\square$

We shall in the Chapters 12 and 13 extend these maximum principles to harmonic and subharmonic functions.

As an example of how to use the maximum principles we shall prove Schwarz' lemma.

Theorem 4.17 (Schwarz' Lemma). *Let $f \colon B(0,1) \to B[0,1]$ be holomorphic with $f(0) = 0$. Then*

(a) *$|f(z)| \leq |z|$ for all $z \in B(0,1)$, and $|f'(0)| \leq 1$.*

(b) *If there exists a $z_0 \in B(0,1) \setminus \{0\}$ such that $|f(z_0)| = |z_0|$, or if $|f'(0)| = 1$, then f has the form $f(z) = cz$ for all $z \in B(0,1)$, where c is a constant of absolute value 1.*

Proof. The function

$$
g(z) := \begin{cases} \frac{f(z)}{z} & \text{for } z \in B(0,1) \setminus \{0\}, \\ f'(0) & \text{for } z = 0, \end{cases}
$$

is holomorphic in $B(0,1)$ (cf. Example 4.4 above). For $|z| < r < 1$ we infer from the Weak Maximum Principle that

$$
|g(z)| \leq \max_{|w|=r}\{|g(w)|\} = \max_{|w|=r}\left\{ \frac{|f(w)|}{r} \right\} \leq \frac{1}{r}
$$

and letting $r \to 1_-$ we get $|g(z)| \leq 1$, i.e., (a).

In each of the cases of (b) the function $|g|$ assumes its supremum in $B(0,1)$, so by the Maximum Modulus Theorem g is a constant function, $g(z) = c$. And $|c| = |g(z_0)| = 1$. This proves (b). $\square$

The conclusion $|f(z)| \leq |z|$ of Schwarz' lemma expresses that f does not increase modulus. It implies in particular that if z ranges over the subdisc $B(0, r)$ then the values $f(z)$ also lie in that disc. Such considerations form the basis for the so-called *Principle of Subordination*. For more information on that topic we refer the reader to the paper MacGregor (1972).

Another important application of Schwarz' lemma is to the uniqueness question of conformal mappings, so we will return to Schwarz' lemma during our treatment of the Riemann Mapping Theorem.

We have already earlier (Theorem 3.5) observed that any continuous function g satisfying $(\exp \circ g)(z) = z$, i.e., any continuous logarithm, automatically is holomorphic. And we have claimed that an assumption in Proposition 3.21 (that f never vanishes) is superfluous. Both these results are special cases of our next theorem.

Theorem 4.18. *Let $g \colon \Omega \to \mathbb{C}$ be a continuous function on an open subset Ω of the complex plane, and let f be a non-constant holomorphic function, defined on an open connected subset of the complex plane containing $g(\Omega)$.*

If the composite map $f \circ g$ is holomorphic on Ω, then so is g.

Proof. Since holomorphy is a local property we may in the proof assume that Ω is connected. Let $a \in \Omega$ be arbitrary. We shall show that g is complex differentiable at a.

Since f is holomorphic around $g(a)$ we can write

$$f(w) - f(g(a)) = (w - g(a))^k H(w)$$

where H is holomorphic and $H(g(a)) \neq 0$. So there is an open disc D around $g(a)$ in which H never vanishes. By Proposition 3.22 H has a holomorphic k'th root f_0 on D. So we write

$$f(w) - f(g(a)) = [(w - g(a))f_0(w)]^k \qquad \text{for all } w \text{ near } g(a).$$

Here k is an integer ≥ 1, $f_0 \in \mathrm{Hol}(B(g(a), \delta))$ for some $\delta > 0$ and $f_0(g(a)) \neq 0$.

Replacing w by $g(z)$ and using the abbreviation h for the holomorphic function $h = f \circ g - f(g(a))$, we get for z in a neighborhood of a, say for z in $B(a, \varepsilon) \subseteq \Omega$, that

$$h(z) = [(g(z) - g(a))f_0(g(z))]^k$$

so we see that the bracket on the right hand side is a continuous k'th root r of the holomorphic function h on the left hand side. There are now two cases:

(1) h is identically 0 in $B(a,\varepsilon)$. In this case $g(z) = g(a)$ near $z = a$, so g is clearly complex differentiable at a.

(2) h is not identically 0 in $B(a,\varepsilon)$. In this case we know by the Unique Continuation Theorem (Theorem 4.11) that a is an isolated zero for h. According to Proposition 3.21 the root r is holomorphic in $B(a,\varepsilon)\setminus h^{-1}(0)$, so a is an isolated singularity of r. But r is continuous in $B(a,\varepsilon)$, so by the Removable Singularity Theorem (Theorem 4.3) r is actually holomorphic in a neighborhood of a.

Dividing the identity $r(z) = (g(z) - g(a))f_0(g(z))$ by $z - a$ and noting that $r(a) = 0$ we get

$$\frac{r(z) - r(a)}{z - a} = \frac{g(z) - g(a)}{z - a} f_0(g(z)).$$

The left hand side has a limit as $z \to a$, and, since $f_0(g(a)) \neq 0$, so does the difference quotient of g, i.e., g is differentiable at a.

This concludes the proof.   □

Terminology 4.19. *A function is said to be* **univalent**, *if it is holomorphic and injective.*

Theorem 4.20. *If f is univalent in an open set Ω, then $f(\Omega)$ is open and f^{-1} is holomorphic on $f(\Omega)$.*

Proof. Being open, Ω is a union of open balls. f is injective so it is not constant on any such ball. Hence $f(\Omega)$ is open by the Open Mapping Theorem (Theorem 4.14). Since f is an open map, f^{-1} is continuous, so we can apply Theorem 4.18 with $g = f^{-1}$.   □

The reader might note that a result corresponding to Theorem 4.20 is false for functions of one real variable: The inverse of a differentiable function need not be differentiable (example $f(x) = x^3$).

Theorem 4.21. *Let f be holomorphic in an open set Ω of the complex plane and let $z_0 \in \Omega$. Then there exists a neighborhood of z_0 on which f is univalent if and only if $f'(z_0) \neq 0$.*

Proof. Let us first assume that f is univalent on the open neighborhood Ω of z_0. Then $g = f^{-1}$ is holomorphic on $f(\Omega)$ by Theorem 4.20, so we may differentiate the identity $g(f(z)) = z$. We find via the chain rule that $g'(f(z_0))f'(z_0) = 1$, so $f'(z_0) \neq 0$.

Let us conversely assume that $f'(z_0) \neq 0$. Then the Jacobian of the mapping

$$f = u + iv \colon \mathbb{R}^2 \to \mathbb{R}^2$$

is $\neq 0$ (use the Cauchy-Riemann equations), so by the Inverse Function Theorem f is injective on a neighborhood of z_0. $\qquad \square$

Remark 4.22. Theorem 4.21 extends to the case of holomorphic functions of several complex variables. An induction proof on the number of variables can be found in the note Rosay (1982).

4.4 Hadamard's gap theorem

As application of Theorem 4.8 we make an unnecessary, but nice, digression back to power series.

Theorem 4.23 (Hadamard's Gap Theorem). *Let $R > 0$ be the radius of convergence of the power series*

$$f(z) = \sum_{k=0}^{\infty} a_k z^{n_k}$$

where $\{n_k \mid k = 1, 2, \dots\}$ is a sequence of natural numbers satisfying $n_{k+1} \geq (1 + \theta)n_k$ for $k = 1, 2, \dots$ for some $\theta > 0$.

Then $|z| = R$ is a natural boundary for f in the following sense: If there is a holomorphic extension of f to a connected open set $\Omega \supseteq B(0, R)$, then $\Omega = B(0, R)$.

Proof. We assume for convenience that $R = 1$, and denote the extension of f by F. It suffices to derive a contradiction from the assumption $\Omega \neq B(0, 1)$.

Under that assumption $\Omega \cap S^1 \neq \emptyset$, say $1 \in \Omega$. Choose an integer $p \geq 1/\theta$ and consider for $w \in \mathbb{C}$ the convex combination

$$z = \frac{w^p + w^{p+1}}{2}$$

We have

$$z \in B(0, 1) \subseteq \Omega \qquad \text{if } |w| \leq 1 \text{ and } z \neq 1,$$

and

$$z = 1 \qquad \text{if } w = 1$$

so $z \in \Omega$ for all w in a neighborhood of $B[0, 1]$.

The composite function

$$\phi(w) := F\left(\frac{w^p + w^{p+1}}{2}\right)$$

is therefore defined and holomorphic in that neighborhood of $B[0,1]$, and so the radius of convergence of ϕ's power series expansion around 0 is strictly bigger than 1. Let us find the power series expansion of ϕ: For $w \in B(0,1)$ we have

$$\sum_{k=0}^{\infty} a_k \left(\frac{w^p + w^{p+1}}{2}\right)^{n_k}$$

$$= \sum_{k=0}^{\infty} \frac{a_k}{2^{n_k}} \left\{ w^{pn_k} + \binom{n}{1} w^{pn_k+1} + \cdots + w^{pn_k+n_k} \right\}. \tag{4.5}$$

The exponent of the last term in the k'th bracket, viz. $pn_k + n_k$, is strictly less than the exponent of the first term in the next bracket: Indeed, since $p \geq 1/\theta$, we see that

$$pn_{k+1} - (pn_k + n_k) \geq p(1 + \theta)n_k - (pn_k + n_k) \geq (p\theta - 1)n_k > 0$$

which means that no power of w occurs more than once.

We infer that the power series expansion of ϕ is gotten by simply omitting the brackets in (4.5): Indeed that power series converges, at least for $|w| < 1$, because

$$\sum_{k=0}^{\infty} \frac{|a_k|}{2^{n_k}} \left\{ |w|^{pn_k} + \binom{n}{1}|w|^{pn_k+1} + \cdots + |w|^{pn_k+n_k} \right\}$$

$$= \sum_{k=0}^{\infty} |a_k| \left(\frac{|w|^p + |w|^{p+1}}{2}\right)^{n_k} < \infty$$

since

$$\frac{|w|^p + |w|^{p+1}}{2} < 1.$$

And by (4.5) the power series converges to ϕ.

As we mentioned above, the radius of convergence of ϕ is strictly bigger than 1. So its power series converges for some real number $r > 1$, implying that

$$\phi(r) = \sum_{k=0}^{\infty} a_k \left(\frac{r^k + r^{k+1}}{2}\right)^{n_k}$$

converges. But then the power series of f converges at

$$z = \frac{r^k + r^{k+1}}{2} > 1$$

and so the radius of convergence R of f is strictly bigger than 1.
That contradicts $R = 1$. □

As an example we mention that the function

$$f(z) := \sum_{k=1}^{\infty} \frac{z^{n!}}{n!} \qquad \text{for } z \in B(0,1)$$

extends to a continuous function on all of the complex plane, but that it cannot be prolonged to an analytic function on any domain bigger than $B(0,1)$.

Exercises

4.1. In the proof of Weierstrass' theorem (Theorem 4.10) we used the fact that f analytic implies f holomorphic, i.e., Proposition 2.5, so the remarks about power series following Weierstrass' theorem are really part of a circular argument. Repair the argument by showing directly from (4.3) in the proof of Weierstrass' theorem that f is holomorphic.

4.2. For which $n \in \mathbb{Z}$ will the function $z \mapsto z^n$ possess a primitive (1) on all of $\mathbb{C} \setminus \{0\}$ (2) locally on $\mathbb{C} \setminus \{0\}$?

4.3. Let Ω be an open convex set, and let f be holomorphic on Ω with $\operatorname{Re} f' > 0$ on Ω. Show that f is univalent on Ω.
Hint: $f(b) - f(a) = (b - a) \int_0^1 f'(a + t(b - a)) \, dt$.

4.4. Let $\gamma \colon [a, b] \to \mathbb{C} \setminus \{0\}$ be a path, and let f be continuous on γ^*. Prove the formula

$$\int_\gamma f(z) \, dz = -\int_{\gamma^{-1}} \frac{f(z^{-1})}{z^2} \, dz$$

in which the path $\gamma^{-1} \colon [a, b] \to \mathbb{C}$ is defined by $\gamma^{-1}(t) = 1/\gamma(t)$.
Show in particular that

$$\int_{|z|=r} f(z) \, dz = \int_{|z|=\frac{1}{r}} \frac{f(z^{-1})}{z^2} \, dz.$$

4.5. Let $a \in \mathbb{C}$, $r > 0$ and $|a| \neq r$. Compute

$$\frac{1}{2\pi i} \int_{|z|=r} (z - a)^n \, dz \qquad \text{for } n \in \mathbb{Z}.$$

4.6. Let $\alpha \in \mathbb{C}$. Show that the principal branch of $(1 + z)^\alpha$, i.e., the branch with $(1 + 0)^\alpha = 1$ has the following power series expansion

$$(1 + z)^\alpha = \sum_{n=0}^\infty \binom{\alpha}{n} z^n \qquad \text{for } |z| < 1.$$

4.7. Let $\alpha \in \mathbb{C}$. Is the following formula correct ?

$$\frac{d}{dz}(z^\alpha) = \alpha z^{\alpha - 1}.$$

4.8. Let f and g be continuous in the closed unit disc and analytic in the open disc. Write them as

$$f(z) = \sum_{n=0}^\infty a_n z^n \quad \text{and} \quad g(z) = \sum_{n=0}^\infty b_n z^n \qquad \text{for } |z| < 1.$$

(α) Show for any $|z| < 1$ that

$$\sum_{n=0}^\infty a_n b_n z^{2n} = \frac{1}{2\pi i} \int_{|w|=1} f(zw) g\left(\frac{z}{w}\right) \frac{dw}{w}.$$

(β) Show Parseval's identity

$$\sum_{n=0}^\infty a_n \overline{b_n} = \frac{1}{2\pi} \int_0^{2\pi} f(e^{i\theta}) \overline{g(e^{i\theta})} \, d\theta.$$

Hint: Show first for any $r \in \,]0, 1[$ that

$$\sum_{n=0}^\infty |a_n|^2 r^{2n} = \frac{1}{2\pi} \int_0^{2\pi} |f(e^{i\theta})|^2 \, d\theta.$$

4.9 (L. Fejér and F. Riesz). Let f be continuous on the closed unit disc and analytic in the interior. Show that

$$2 \int_{-1}^1 |f(x)|^2 \, dx \leq \int_0^{2\pi} |f(e^{i\theta})|^2 \, d\theta.$$

Hint: Apply the Cauchy Integral Theorem for the semi-circular disc to the function $z \mapsto f(z) \overline{f(\bar{z})}$.

4.10. Let $t \in [-1, 1]$.

(α) Show that $1 - 2tz + z^2$ does not vanish in $|z| < 1$.

(β) Show that there exists a continuous square root of $z \mapsto (1 - 2tz + z^2)^{-1}$ in $|z| < 1$ with the value 1 at $z = 0$. Call it $(1 - 2tz + z^2)^{-1/2}$.

(γ) Show that $z \mapsto (1 - 2tz + z^2)^{-1/2}$ can be expanded in a power series

$$(1 - 2tz + z^2)^{-1/2} = \sum_{n=0}^{\infty} P_n(t) z^n \qquad \text{converging for } |z| < 1.$$

(δ) Show that the P_n are polynomials in t of degree n (they are the so-called *Legendre polynomials*).

(ε) Show that P_n is a solution of the equation

$$(1 - t^2) y'' - 2ty' + n(n+1)y = 0.$$

Hint: Term by term integration yields the following explicit formula for P_n:

$$P_n(t) = \int_{|z|=r} (1 - 2tz + z^2)^{-1/2} \frac{dz}{z^{n+1}}.$$

4.11.

(α) Show that $z \mapsto \exp(-z^2 - 2zw)$ for any $w \in \mathbb{C}$ can be expanded in a power series

$$\exp(-z^2 - 2wz) = \sum_{n=0}^{\infty} H_n(w) \frac{z^n}{n!}$$

whose radius of convergence is ∞.

(β) Show that H_n is a polynomial in w of degree n (the so-called *Hermite polynomial of degree n*).

(γ) Show that the Hermite polynomials have the following properties:

$$H_n(w) = (-1)^n H_n(-w)$$
$$H'_{n+1} = -2(n+1)H_n$$
$$H''_n - 2wH'_n + 2nH_n = 0$$
$$H_n(w) = \exp(w^2) \left(\frac{d}{dw} \right)^n \{\exp(-w^2)\}$$
$$\int_{-\infty}^{\infty} H_n(x) H_m(x) \exp(-x^2) \, dx = 0 \qquad \text{if } n \neq m.$$

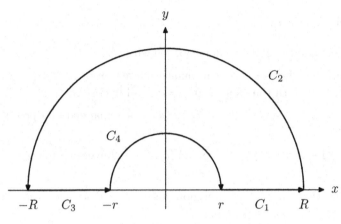

Fig. 4.3

4.12. Consider the closed path γ composed of the four paths sketched in Fig. 4.3.

(α) Show that

$$\int_\gamma e^{iz} \frac{dz}{z} = 0.$$

(β) Show that

$$\int_{C_2} e^{iz} \frac{dz}{z} \to 0 \qquad \text{as } R \to \infty,$$

and that

$$\int_{C_4} e^{iz} \frac{dz}{z} \to -\pi i \qquad \text{as } r \to 0.$$

(γ) Show finally that

$$\int_0^R \frac{\sin x}{x} dx \to \frac{\pi}{2} \qquad \text{as } R \to \infty.$$

4.13. Let $0 < v < 1$. Let $A = 1 + e^{i(-\pi+v)}$ and $B = 1 + e^{i(\pi-v)}$. Let C and c denote the circle segments indicated on Fig. 4.4.

(α) Show that

$$\int_c \frac{\text{Log } z}{z-1} dz \to 0 \qquad \text{as } v \to 0_+.$$

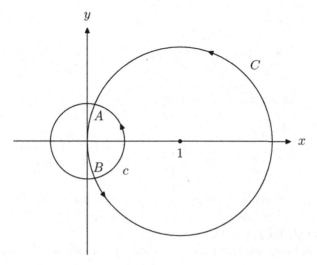

Fig. 4.4

(β) Show that

$$\int_{-\pi+v}^{\pi-v} \mathrm{Log}|1 + e^{it}|\, dt = \mathrm{Im}\left\{\int_C \frac{\mathrm{Log}\, z}{z - 1}\, dz\right\} \to 0 \qquad \text{as } v \to 0_+.$$

(γ) Show that

$$\int_\varepsilon^{\frac{\pi}{2}} \mathrm{Log}(\sin \theta)\, d\theta \to -\frac{\pi}{2}\, \mathrm{Log}\, 2 \qquad \text{as } \varepsilon \to 0_+.$$

4.14. Does there exist a holomorphic function on $B(0,1)$ that in the points $\frac{1}{2}, \frac{1}{3}, \frac{1}{4}, \frac{1}{5}, \ldots$ takes the following values:

(α) $1, 0, 1, 0, \ldots$?

(β) $\frac{1}{2}, 0, \frac{1}{4}, 0, \frac{1}{6}, 0, \ldots$?

(γ) $\frac{1}{2}, \frac{1}{4}, \frac{1}{4}, \frac{1}{6}, \frac{1}{6}, \frac{1}{8}, \frac{1}{8}, \ldots$?

(δ) $\frac{2}{3}, \frac{3}{4}, \frac{4}{5}, \frac{5}{6}, \ldots$?

4.15. Let f be holomorphic and non-constant in an open connected subset Ω of the complex plane. Show that f is non-constant in any open subset of Ω.

4.16. We recall that

$$\sin z = \sum_{n=0}^{\infty} \frac{(-1)^n}{(2n+1)!} z^{2n+1} \quad \text{and} \quad \cos z = \sum_{n=0}^{\infty} \frac{(-1)^n}{(2n)!} z^{2n}.$$

Show that $\sin^2 z + \cos^2 z = 1$ for all $z \in \mathbb{C}$.

4.17. Show that

$$\int_{-1}^{1} \frac{dt}{t-z} = \mathrm{Log}\left\{ \frac{z-1}{z+1} \right\} \qquad \text{for } z \in \mathbb{C} \setminus [-1,1].$$

4.18. Let Ω_1 and Ω_2 be two open subsets of the complex plane such that $\Omega_1 \cap \Omega_2$ is connected. Assume that every holomorphic function on Ω_j has a primitive on Ω_j for $j = 1, 2$.

   Show that every function which is holomorphic on $\Omega_1 \cup \Omega_2$, has a primitive on $\Omega_1 \cup \Omega_2$.

4.19. Let f be holomorphic on an open subset Ω of the complex plane. Let g be a continuous k'th root of f on Ω.

   Show that g, too, is holomorphic on Ω.

4.20. Let $g \colon B(a, r) \to B(g(a), R)$ be holomorphic. Prove the following inequality which generalizes Schwarz' lemma:

$$|g(z+a) - g(a)| \leq \frac{R}{r} |z| \qquad \text{for all } |z| < r.$$

Derive Liouville's theorem from the inequality.

4.21. Show that the maximum of $|\sin z|$ on the square $[0, 2\pi] \times [0, 2\pi]$ equals $\cosh(2\pi)$ and is attained at $z = \frac{\pi}{2} + 2\pi i$.

4.22. Let f be an entire, non-constant function and consider the corresponding "analytic landscape"

$$\{(x, y, |f(x+iy)|^2) \in \mathbb{R}^3 \mid x, y \in \mathbb{R}\} \qquad \text{in } \mathbb{R}^3.$$

Show that there are no highland lakes in such a landscape: When it rains the water collects in puddles around the zeros of f, not higher up, or runs off to infinity.

4.23. Prove the

Strong Maximum Principle. *Let Ω be a bounded, connected, open subset of the complex plane, and let f be a holomorphic, non-constant function on Ω. Let*

$$M := \sup_{z \in \partial\Omega} \{\limsup_{\substack{\zeta \to z \\ \zeta \in \Omega}} |f(\zeta)|\}.$$

Then $|f(z)| < M$ for all $z \in \Omega$.

4.24. Let Ω be a bounded open subset of the complex plane. Let $\{f_n\}$ be a sequence of functions which are continuous on $\overline{\Omega}$ and holomorphic on Ω. Assume that the sequence converges uniformly on the boundary of Ω.

Show that $\{f_n\}$ converges uniformly on $\overline{\Omega}$ towards a function which is continuous on $\overline{\Omega}$ and holomorphic on Ω.

4.25. Show that there exists a holomorphic function $A: B(0,1) \to \mathbb{C}$ with $A(0) = 0$ satisfying the identity $\sin(A(z)) = z$ for all $|z| < 1$.

Hint: Use Theorem 4.21 to get hold of A near 0. Differentiate the identity to get

$$A'(z) = \frac{1}{\sqrt{1 - z^2}}.$$

4.26. Show that the Riemann ζ-function

$$\zeta(z) = \sum_{n=1}^{\infty} \frac{1}{n^z}$$

is holomorphic on the open half-plane $\{z \in \mathbb{C} \mid \operatorname{Re} z > 1\}$.

4.27. Show

Schwarz' Reflection Principle. *Let Ω be a connected open set which is symmetric with respect to the real axis. Put*

$$\Omega_+ = \{z \in \Omega \mid \operatorname{Im} z > 0\},$$
$$\Omega_- = \{z \in \Omega \mid \operatorname{Im} z < 0\},$$
$$\Omega_0 = \{z \in \Omega \mid \operatorname{Im} z = 0\}.$$

Let $f: \Omega_+ \cup \Omega_0 \to \mathbb{C}$ be continuous on $\Omega_+ \cup \Omega_0$, holomorphic on Ω_+ and such that $f(\Omega_0) \subseteq \mathbb{R}$.

Then there exists exactly one function $F: \Omega \to \mathbb{C}$ which is holomorphic on Ω and equal to f on Ω_+. It satisfies

$$F(z) = \overline{F(\overline{z})} \qquad \text{for all } z \in \Omega.$$

4.28. Show

Schwarz' Reflection in the Circle. *Let Ω be a connected open set which is symmetric with respect to the unit circle, i.e.,*

$$z \in \Omega \quad \Leftrightarrow \quad 1/\bar{z} \in \Omega.$$

Define

$$\Omega_e = \{z \in \Omega \mid |z| > 1\},$$
$$\Omega_i = \{z \in \Omega \mid |z| < 1\},$$
$$\Omega_0 = \{z \in \Omega \mid |z| = 1\}.$$

Let $f \colon \Omega_e \cup \Omega_0 \to \mathbb{C} \setminus \{0\}$ be continuous on $\Omega_e \cup \Omega_0$, holomorphic on Ω_e and such that $f(\Omega_0) \subseteq \partial B(0,1)$.

Then there exists exactly one function $F \colon \Omega \to \mathbb{C}$ which is holomorphic on Ω and equal to f on Ω_e. It satisfies

$$F(z) = \overline{1/F(1/\bar{z})} \qquad \text{for all } z \in \Omega.$$

4.29. Assume that both $f \in C(\mathbb{R})$ and the function

$$x \mapsto \int_{-\infty}^{\infty} e^{-ixt} f(t)\, dt, \qquad x \in \mathbb{R}$$

are compactly supported. Show that

$$\int_{-\infty}^{\infty} e^{-ixt} f(t)\, dt = 0 \qquad \text{for all } x \in \mathbb{R}.$$

4.30. What is the maximum of the product of the four distances between a variable point in a square and the vertices of the square?

4.31. Let $f \colon B(0,1) \to B(0,1)$ be an analytic map which is bijective and maps 0 into 0.

Show that f has the form $f(z) = cz$ for some constant $c \in S^1$.

4.32. Show that $|\sqrt{z}| = \sqrt{|z|}$. Why doesn't the function $z \mapsto 1/\sqrt{z}$ constitute a counter example to the Removable Singularity Theorem?

4.33. Let f be a non-constant entire function. Show that the range of f is dense in the complex plane.

Hint: Assume not. Use Liouville.

4.34. We will say that an analytic function f, defined in $\Omega \subseteq \mathbb{C}$, is of *exponential type* if there is a $T \in \mathbb{R}$ and a constant $C > 0$ such that

$$|f(z)| \leq Ce^{T|z|} \qquad \text{for all } z \in \Omega.$$

Prove

> **Phragmén-Lindelöf's Theorem.** *If f is analytic and of exponential type in a sector Ω of angular opening strictly less than π, if f is also continuous in the closed sector $\overline{\Omega}$, and if $|f| \leq M$ for some constant M on the boundary of Ω, then $|f| \leq M$ throughout Ω.*

4.35. Let f be holomorphic on an open subset Ω of the complex plane. Let $P, Q \in \Omega$ and let $\gamma_0, \gamma_1 \colon [a, b] \to \Omega$ be paths such that $\gamma_0(a) = \gamma_1(a) = P$ and $\gamma_0(b) = \gamma_1(b) = Q$.

Assume furthermore that γ_0 and γ_1 are fixed end point homotopic in Ω, i.e., there exists a continuous map $H \colon [a, b] \times [0, 1] \to \Omega$ such that

$$H(\,\cdot\,, 0) = \gamma_0, \quad H(\,\cdot\,, 1) = \gamma_1$$
$$H(a, s) = P, \quad H(b, s) = Q \qquad \text{for all } s \in [0, 1].$$

Show that

$$\int_{\gamma_0} f(z)\, dz = \int_{\gamma_1} f(z)\, dz.$$

4.36. Let Ω be an open connected subset of $\mathbb{C}$. Let $\{f_n\}$ be a sequence of continuous functions on Ω, converging locally uniformly to $f \in \text{Hol}(\Omega)$. Assume that none of the f_n vanishes at any point.

Show that f vanishes everywhere or nowhere.

4.37. Derive Brouwer's fixed point theorem (Theorem 3.11) from Rouché's theorem.

4.38. In this exercise we will derive a formula which as special cases includes the *Poisson integral*

$$\int_0^\infty e^{-x^2}\, dx = \frac{\sqrt{\pi}}{2}$$

and the *Fresnel integrals*

$$\lim_{R \to \infty} \int_{-R}^{R} \cos(x^2)\, dx = \sqrt{\frac{\pi}{2}}$$

and

$$\lim_{R\to\infty} \int_{-R}^{R} \sin(x^2)\, dx = \sqrt{\frac{\pi}{2}}.$$

To that purpose we consider the meromorphic function

$$f(z) := \frac{e^{iz^2}}{e^{-\sqrt{2\pi}z} - 1}, \qquad z \in \mathbb{C}.$$

(α) Show, to ease your computations below, that

$$f(z) + f(-z) = -e^{iz^2}$$

and

$$f\left(i\sqrt{\tfrac{\pi}{2}} + z\right) + f\left(i\sqrt{\tfrac{\pi}{2}} - z\right) = ie^{iz^2}.$$

Consider for each $R > 0$ and fixed $\alpha \in [0, \frac{\pi}{2}[$ the following parallelogram A, B, C, D (see Fig. 4.5), in which

$$A = -B = -Re^{i\alpha}, \quad C = B + i\sqrt{\tfrac{\pi}{2}}, \quad D = A + i\sqrt{\tfrac{\pi}{2}}$$
$$E = i\sqrt{\tfrac{\pi}{2}}, \quad F = -G = re^{i\alpha}.$$

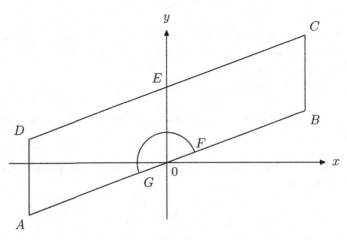

Fig. 4.5

Note that f is holomorphic on the interior of the parallelogram and on its boundary only has a singularity at $z = 0$.

(β) Show that

$$\int_{[B,C]} f(z)\, dz \to 0 \quad \text{and} \quad \int_{[A,D]} f(z)\, dz \to 0 \qquad \text{as } R \to \infty.$$

(γ) Show that

$$\int_{[C,D]} f(z)\, dz = -ie^{i\alpha} \int_0^R \exp\{it^2 e^{2i\alpha}\}\, dt$$

and that

$$\int_{[A,G]} f(z)\, dz + \int_{[F,B]} f(z)\, dz = -e^{i\alpha} \int_r^R \exp\{it^2 e^{2i\alpha}\}\, dt.$$

Hint: Use (α).

(δ) Show that the integral of f along the small half circle on the figure from G to F as $r \to 0_+$ converges to $i\sqrt{\frac{\pi}{2}}$.

(ε) Show that

$$\lim_{R\to\infty} \int_0^R \exp\{it^2 e^{2i\alpha}\}\, dt = \frac{1+i}{2}\sqrt{\frac{\pi}{2}}\, e^{-i\alpha}.$$

(ζ) Derive the Poisson and Fresnel integrals.

(η) Show the following more compact form of the result of (ε):

$$\lim_{R\to\infty} \int_0^R e^{i\pi a x^2/2}\, dx = \frac{1+i}{2} \frac{1}{\sqrt{a}} \qquad \text{for } a \neq 0,\ \text{Im}\, a \geq 0$$

where the principal value of $\sqrt{a}$ is taken on the right.

The exercise is adapted from the paper Rao (1972).

Chapter 5

Global Theory

5.1 The global Cauchy integral theorem

In this section we will state and prove a general version of the Cauchy Integral Formula and the Cauchy Integral Theorem.

Theorem 4.6 states the Cauchy Integral Formula for a closed path in a starshaped domain. We would of course like to remove the condition of starshapedness, but we must be careful, for if we do so then some kind of restriction on the paths in question is necessary:

Take as domain the annulus $A := \{z \in \mathbb{C} \mid \frac{1}{2} < |z| < 1\}$ and consider functions which are holomorphic in an open set containing the closure of A. It is simply not true that

$$f(z) = \frac{1}{2\pi i} \int_{|\zeta|=1} \frac{f(\zeta)}{\zeta - z} \, d\zeta \qquad \text{for } z \in A$$

as the function $f(z) = z^{-1}$ demonstrates.

On the positive side there exists a version of the Cauchy Integral Formula in A, viz.

$$f(z) = \frac{1}{2\pi i} \int_{|\zeta|=1} \frac{f(\zeta)}{\zeta - z} \, d\zeta - \frac{1}{2\pi i} \int_{|\zeta|=\frac{1}{2}} \frac{f(\zeta)}{\zeta - z} \, d\zeta \qquad \text{for } z \in A.$$

(Exercise: Derive it from the Cauchy Integral Formula for starshaped domains by inserting suitable auxiliary segments).

In the case of a starshaped domain Ω the Cauchy Theorem states that

$$\int_\gamma f(z) \, dz = 0 \tag{5.1}$$

for any closed path γ in Ω and any function f which is holomorphic in Ω. Let now Ω be any open subset of the complex plane, and let γ be a closed path in Ω. If (5.1) holds for all $f \in \mathrm{Hol}(\Omega)$, then in particular

$$\mathrm{Ind}_\gamma(z) = \frac{1}{2\pi i} \int_\gamma \frac{d\zeta}{\zeta - z} = 0 \qquad \text{for all } z \in \mathbb{C} \setminus \Omega.$$

Our general version of the Cauchy Integral Theorem is that the converse holds. The crucial necessary condition on the path γ, i.e., that $\mathrm{Ind}_\gamma(z) = 0$ for all $z \in \mathbb{C} \setminus \Omega$, means intuitively that γ does not circle any "hole" in Ω.

Theorem 5.1 (The Global Cauchy Theorem I). *Let γ be a closed path in an open subset Ω of the complex plane such that $\mathrm{Ind}_\gamma(z) = 0$ for all $z \in \mathbb{C} \setminus \Omega$.*

Then we have for any $f \in \mathrm{Hol}(\Omega)$ that

(a) $\displaystyle \int_\gamma f(z)\, dz = 0,$

and

(b) $\displaystyle f(z)\, \mathrm{Ind}_\gamma(z) = \frac{1}{2\pi i} \int_\gamma \frac{f(\zeta)}{\zeta - z}\, d\zeta$ *for all $z \in \Omega \setminus \gamma^*$.*

Example 5.2. Figure 5.1 show a drawing of a path that satisfies the condition of Theorem 5.1, but which nevertheless is not null-homotopic $[\Omega = \mathbb{C} \setminus \{0, 1\}]$.

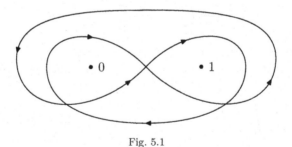

$\bullet\ 0$ $\bullet\ 1$

Fig. 5.1

Proof of Theorem 5.1 (due to Dixon, 1971). Item (a) follows from (b) when we replace f by $\zeta \mapsto (\zeta - z)f(\zeta)$, so from now on we may concentrate on (b).

By the formula for the index (Theorem 3.18(d)) (b) is equivalent to

$$\frac{1}{2\pi i} \int_\gamma \frac{f(z) - f(w)}{z - w}\, dw = 0 \qquad \text{for all } z \in \Omega \setminus \gamma^*.$$

Introducing the function $h: \Omega \times \Omega \to \mathbb{C}$ given by

$$h(z, w) := \begin{cases} \frac{f(z)-f(w)}{z-w} & \text{for } z \neq w, \\ f'(w) & \text{for } z = w, \end{cases}$$

it thus suffices to prove that

$$g(z) := \frac{1}{2\pi i} \int_\gamma h(z, w)\, dw = 0 \qquad \text{for all } z \in \Omega.$$

We start by noting some of the properties of h and g. h is continuous in all of $\Omega \times \Omega$: This is obvious off the diagonal in $\Omega \times \Omega$. To prove that h is continuous on the diagonal as well we compute

$$f(z) - f(w) = \int_0^1 \frac{d}{dt}\left\{ f(w + t(z - w)) \right\} dt$$
$$= (z - w) \int_0^1 f'(w + t(z - w))\, dt$$

which implies that h near the diagonal can be expressed as

$$h(z, w) = \int_0^1 f'(w + t(z - w))\, dt$$

a formula that also holds on the diagonal. This formula proves the continuity of h on the diagonal.

So g makes sense and is continuous in Ω. By help of Morera's theorem (Theorem 4.9) we will prove that g is even holomorphic in Ω:

If Δ is a triangle in Ω then

$$\int_{\partial\Delta} g = \frac{1}{2\pi i} \int_{\partial\Delta} \int_\gamma h(z, w)\, dw\, dz = \frac{1}{2\pi i} \int_\gamma \int_{\partial\Delta} h(z, w)\, dz\, dw.$$

Now, the function $z \to h(z, w)$ is holomorphic in Ω for fixed w (Example 4.4), so according to the Cauchy-Goursat Theorem

$$\int_{\partial\Delta} h(z, w)\, dz = 0, \quad \text{and hence} \quad \int_{\partial\Delta} g = 0$$

so according to Morera g is holomorphic in Ω.

By our assumption on the path γ the set

$$\Omega_0 := \{ z \in \mathbb{C} \setminus \gamma^* \mid \text{Ind}_\gamma(z) = 0 \}$$

is an open subset of $\mathbb{C}$ satisfying $\Omega \cup \Omega_0 = \mathbb{C}$.

For $z \in \Omega \cap \Omega_0$ we get:

$$g(z) = \frac{1}{2\pi i} \int_\gamma h(z,w)\, dw = \frac{1}{2\pi i} \int_\gamma \frac{f(z)}{z-w}\, dw - \frac{1}{2\pi i} \int_\gamma \frac{f(w)}{z-w}\, dw$$

$$= -f(z)\operatorname{Ind}_\gamma(z) + \frac{1}{2\pi i} \int_\gamma \frac{f(w)}{w-z}\, dw = \frac{1}{2\pi i} \int_\gamma \frac{f(w)}{w-z}\, dw.$$

The function

$$g_0(z) := \frac{1}{2\pi i} \int_\gamma \frac{f(w)}{w-z}\, dw \qquad \text{for } z \in \Omega_0$$

is holomorphic on Ω_0 (Lemma 2.14); the computation just made shows that g and g_0 patch together over $\Omega \cap \Omega_0$ to an unambiguously defined function G on $\Omega \cup \Omega_0 = \mathbb{C}$, i.e., G given by

$$G(z) := \begin{cases} g(z) & \text{for } z \in \Omega, \\ g_0(z) & \text{for } z \in \Omega_0, \end{cases}$$

is entire.

The unbounded component of γ^* is by Theorem 3.18(c) contained in Ω_0, so for z out there we can estimate G as follows:

$$|G(z)| = |g_0(z)| = \frac{1}{2\pi} \left| \int_\gamma \frac{f(w)}{w-z}\, dw \right| \le \frac{1}{2\pi} \frac{\sup\{|f(w)| \mid \omega \in \gamma^*\}}{\operatorname{dist}(z,\gamma^*)} L(\gamma)$$

which shows that $G(z) \to 0$ as $z \to \infty$. But then G vanishes identically by Liouville's theorem (Theorem 1.11). In particular g is identically 0 on Ω, and that fact is exactly the contents of (b).   □

An inspection of the proof reveals that it applies to the case of several paths, not just one. This comes in handy for, say, an annulus as mentioned in the introduction of this section. The precise statement is:

Theorem 5.3 (The Global Cauchy Theorem II). *Let* $\gamma_1, \gamma_2, \ldots, \gamma_n$ *be closed paths in an open subset* Ω *of the complex plane such that*

$$\sum_{j=1}^n \operatorname{Ind}_{\gamma_j}(z) = 0 \qquad \text{for all } z \in \mathbb{C} \setminus \Omega.$$

Then we have for any $f \in \operatorname{Hol}(\Omega)$ *that*

(a) $\displaystyle \sum_{j=1}^n \int_{\gamma_j} f(z)\, dz = 0,$

and

(b) $f(z) \sum_{j=1}^{n} \text{Ind}_{\gamma_j}(z) = \sum_{j=1}^{n} \frac{1}{2\pi i} \int_{\gamma_j} \frac{f(w)}{w - z} \, dw$ *for all* $z \in \Omega \setminus \{\gamma_1^* \cup \cdots \cup \gamma_n^*\}.$

Proof. Replace in the proof of Theorem 5.1 above

$$g \quad \text{by} \quad \sum_{j=1}^{n} \frac{1}{2\pi i} \int_{\gamma_j} h(z, w) \, dw,$$

$$\Omega_0 \quad \text{by} \quad \left\{ z \in \mathbb{C} \setminus \{\gamma_1^* \cup \cdots \cup \gamma_n^*\} \, \middle| \, \sum_{j=1}^{n} \text{Ind}_{\gamma_j}(z) = 0 \right\},$$

$$g_0 \quad \text{by} \quad \frac{1}{2\pi i} \sum_{j=1}^{n} \int_{\gamma_j} \frac{f(w)}{w - z} \, dw.$$

The details of the easy modification of the proof of Theorem 5.1 are left to the reader. $\qquad \Box$

Corollary 5.4. *Let f be holomorphic in an open subset Ω of the complex plane. Let $\gamma_1, \gamma_2, \ldots, \gamma_N$ and $\sigma_1, \sigma_2, \ldots, \sigma_M$ be closed paths in Ω and let $n_1, n_2, \ldots, n_N$ and $m_1, m_2, \ldots, m_M$ be integers such that*

$$\sum_{j=1}^{N} n_j \, \text{Ind}_{\gamma_j}(z) = \sum_{l=1}^{M} m_l \, \text{Ind}_{\sigma_l}(z) \qquad \text{for all } z \in \mathbb{C} \setminus \Omega.$$

Then

$$\sum_{j=1}^{N} n_j \int_{\gamma_j} f = \sum_{l=1}^{M} m_l \int_{\sigma_l} f.$$

Corollary 5.5. *Let f be holomorphic in an open subset Ω of the complex plane. Let γ and ρ be two closed paths in Ω, which are homotopic in Ω. Then*

$$\int_{\gamma} f = \int_{\rho} f.$$

In particular, if γ is null-homotopic in Ω, then $\int_{\gamma} f = 0$.

5.2 Simply connected sets

The topological concept of simple connectivity is intimately related to homotopy (Definition 3.17).

Definition 5.6. *A topological space is said to be **simply connected**, if each closed curve in it is null-homotopic.*

It is obvious that any starshaped space is simply connected. $\mathbb{R}^3 \setminus \{0\}$ is an example of a topological space which is simply connected, but not starshaped. It is intuitively obvious that $\mathbb{R}^2 \setminus \{0\}$ is connected, but not simply connected (For a rigorous proof see the remark following Theorem 5.9.) $B(0,1) \cup B(3,1)$ is simply connected, but not connected.

Theorem 5.7. *Let Ω be a simply connected open subset of $\mathbb{C}$, let γ be a closed path in Ω and let $f \in \mathrm{Hol}(\Omega)$. Then $\int_\gamma f = 0$ and*

$$f(z)\,\mathrm{Ind}_\gamma(z) = \frac{1}{2\pi i} \int_\gamma \frac{f(w)}{w-z}\, dw \qquad \text{for all } z \in \Omega \setminus \gamma^*.$$

Proof. Obvious from Corollary 5.5. □

Theorem 5.8. *Let Ω be a simply connected open subset of $\mathbb{C}$. Then any $f \in \mathrm{Hol}(\Omega)$ has a (holomorphic) primitive on Ω.*

Proof. It suffices to construct a primitive on each of the connected components of Ω, so we may as well from the start assume that Ω is connected. Choose $z_0 \in \Omega$ and let γ_z be a path in Ω from z_0 to $z \in \Omega$. Then

$$F(z) := \int_{\gamma_z} f(w)\, dw$$

is according to Theorem 5.7 independent of the choice of path from z_0 to z, so F is well defined on Ω. Proceeding as in the proof of Lemma 2.13 we see that $F' = f$. □

We can now as promised extend Proposition 3.22 and Proposition 3.10 to simply connected sets.

Theorem 5.9. *Let Ω be a simply connected open subset of $\mathbb{C}$ and let f be a never vanishing holomorphic function on Ω. Then f has a holomorphic logarithm and a holomorphic square root on Ω.*

Proof. It suffices to construct the logarithm and the square root in each of the connected components of Ω, so we may as well assume that Ω is connected. Let F be a primitive of f'/f in Ω; such one exists by Theorem 5.8. The function fe^{-F} is constant because

$$(fe^{-F})' = f'e^{-F} - fF'e^{-F} = \left(f' - f\frac{f'}{f} \right) e^{-F} = 0$$

so $f = ce^F$ for some constant c. Writing $c = e^w$ we see that $F - w$ is a holomorphic logarithm of f.

A holomorphic square root is $\exp(F - w/2)$. $\qquad\square$

In passing we note that this gives us a rigorous proof that $\mathbb{R}^2 \setminus \{0\}$ is not simply connected: Combine Example 3.23(ii) and Theorem 5.9.

For other results on simply connected sets see the discussion around the Riemann Mapping Theorem (Theorem 8.20), Chapter 8, Section 8.4, Corollary 9.12 and Exercise 9.20.

Exercises

5.1. Consider the path γ in Example 5.2. What is

$$\int_\gamma \frac{dz}{z}?$$

$$\int_\gamma \frac{dz}{z - \varepsilon} \qquad \text{when } \varepsilon > 0 \text{ is small?}$$

$$\int_\gamma \frac{e^z - e^{-z}}{z^6}\, dz?$$

5.2. Let f be holomorphic in $\mathbb{C} \setminus \{0\}$. Show that

$$\int_{|z|=R} f = \int_{|z|=1} f \qquad \text{for all } R > 0.$$

5.3. Show that the function $1 - z^2$ has a holomorphic square root in any simply connected open subset of the complex plane which does not contain $\{-1, 1\}$. Find the possible values of

$$\int_\gamma \frac{dz}{\sqrt{1 - z^2}}$$

when γ is a closed path in such a domain.

5.4. Let Ω be an open, connected and simply connected subset of the complex plane, and let $z_0, a_1, a_2, \ldots, a_N \in \Omega$ be distinct points. Choose $\varepsilon > 0$ so small that the closed balls $B[a_1, \varepsilon], \ldots, B[a_N, \varepsilon]$ are contained in Ω and so that $a_j \notin B[a_k, \varepsilon]$ when $j \neq k$. For $h \in \mathrm{Hol}(\Omega \setminus \{a_1, a_2, \ldots, a_N\})$ we introduce the constants

$$c_j = \frac{1}{2\pi i} \int_{|w-a_j|=\varepsilon} h(w)\, dw \in \mathbb{C}.$$

Let $\gamma = \gamma_z$ be any path in $\Omega \setminus \{a_1, a_2, \ldots, a_N\}$ from z_0 to z. Show that the following function of z:

$$\int_\gamma \left\{ h(w) - \frac{c_1}{w - a_1} - \cdots - \frac{c_N}{w - a_N} \right\} dw$$

is independent of the choice of path from z_0 to z.

5.5. Theorem 5.9 deals with holomorphic functions, but the result is really a topological one. The purpose of this exercise is to establish the following topological version of Theorem 5.9:

> **Theorem 5.10.** *Let Ω be a simply connected open subset of the complex plane and let $f\colon \Omega \to \mathbb{C} \setminus \{0\}$ be continuous. Then f has a continuous logarithm on Ω.*

It suffices to prove the theorem for each of the connected components of Ω, so we may as well assume that Ω is connected.

(α) Prove that $\mathrm{Ind}_{f \circ \gamma}(0) = 0$ for all closed curves γ in Ω.

(β) Let $\gamma_1, \gamma_2 \colon [0,1] \to \Omega$ be continuous curves such that $\gamma_1(0) = \gamma_2(0)$ and $\gamma_1(1) = \gamma_2(1)$. Let $\phi_1, \phi_2 \colon [0,1] \to \mathbb{C}$ be continuous logarithms of $f \circ \gamma_1$ and $f \circ \gamma_2$ respectively, starting at the same point $\phi_1(0) = \phi_2(0)$. Show that $\phi_1(1) = \phi_2(1)$.

(γ) Fix $z_0 \in \Omega$ and $w_0 \in \mathbb{C}$ such that $\exp w_0 = f(z_0)$. Find for any $z \in \Omega$ a curve γ such that $\gamma(0) = z_0$ and $\gamma(1) = z$ and take a continuous logarithm ϕ of $f \circ \gamma$ such that $\phi(0) = w_0$.

Show that $\phi(1)$ does not depend on the choice of the curve γ, so that we unambiguously may define $F(z) := \gamma(1)$.

(δ) Show that F is a logarithm of f.

(ε) Show that $F \circ \gamma$ is continuous for any curve γ from (γ).

(ζ) Show that F is a continuous logarithm of f.

Chapter 6

Isolated Singularities

We begin the present chapter with a study of functions which are holomorphic in an annulus; we show that they can be expanded in Laurent series in much the same spirit as functions which are holomorphic in a disc can be expanded in power series. The special case of a holomorphic function, defined in a punctured disc is the topic of the remainder of the chapter. The center of the disc is in that case said to be an isolated singularity of the function. We classify isolated singularities into removable singularities, poles and essential singularities. We finally prove the Residue Theorem and use it to evaluate definite integrals of various types; this is certainly one of the high points of any introductory course on complex analysis.

A deeper study of essential singularities can be found in the next chapter in which the two Picard theorems are derived.

6.1 Laurent series

Definition 6.1. *A **Laurent series** is a series of the form*

$$\sum_{n=-\infty}^{\infty} a_n z^n \tag{6.1}$$

where the coefficients a_n for $n \in \mathbb{Z}$ are complex numbers, and where z is a complex number different from 0.

If the two series

$$\sum_{n=-\infty}^{-1} a_n z^n = \sum_{m=1}^{\infty} a_{-m} \left(\frac{1}{z} \right)^m \tag{6.2}$$

and

$$\sum_{n=0}^{\infty} a_n z^n \tag{6.3}$$

both are convergent with sums S_- and S_+ respectively, then we will say that the Laurent series (6.1) converges and that its sum is $S_- + S_+$. If at least one of the series (6.2) and (6.3) diverges, then we will say that the Laurent series (6.1) diverges.

The concepts of absolute convergence and uniform convergence are defined similarly.

By definition the convergence of the Laurent series (6.1) means convergence of the two series (6.2) and (6.3) which are power series in $1/z$ and z respectively. So it is to be expected that results, analogous to those for power series, hold for Laurent series.

Theorem 6.2. *Let*

$$r := \limsup_{m \to \infty} |a_{-m}|^{1/m} \quad and \quad R := \frac{1}{\limsup_{n \to \infty} |a_n|^{1/n}}.$$

Assume that $0 \le r < R \le \infty$, and let A denote the annulus

$$A := \{z \in \mathbb{C} \mid r < |z| < R\}.$$

The Laurent series (6.1) converges absolutely at each point of A, and uniformly on any compact subset of A. When $|z| < r$ or $|z| > R$ the Laurent series (6.1) diverges.

The function f, given by

$$f(z) := \sum_{n=-\infty}^{\infty} a_n z^n \quad for \ z \in A$$

is holomorphic in A, and its derivative can be found by term by term differentiation, i.e.,

$$f'(z) = \sum_{n=-\infty}^{\infty} n a_n z^{n-1} \quad for \ z \in A.$$

We know that a function which is holomorphic in an open disc may be expanded in a power series there. We will now show the corresponding result for a function which is holomorphic in an annulus.

Theorem 6.3. *Let f be holomorphic in an annulus $A := \{z \in \mathbb{C} \mid r < |z| < R\}$ where $0 \leq r < R \leq \infty$.*

Then there exists exactly one Laurent series (6.1) such that it converges at each point of A and such that

$$f(z) = \sum_{n=-\infty}^{\infty} a_n z^n \qquad for \ z \in A.$$

The coefficients of this Laurent series can be found from f as follows

$$a_n = \frac{1}{2\pi i} \int_{|w|=\rho} \frac{f(w)}{w^{n+1}} \, dw \qquad for \ n \in \mathbb{Z}$$

where ρ is arbitrary in the open interval $]r, R[$.

Proof. Assume that f is the sum of a Laurent series (6.1) in A. By the previous theorem the series converges uniformly on the circle $|w| = \rho$, so

$$\frac{1}{2\pi i} \int_{|w|=\rho} \frac{f(w)}{w^{n+1}} \, dw = \frac{1}{2\pi i} \, sum_{m=-\infty}^{\infty} a_m \int_{|w|=\rho} \frac{dw}{w^{n-m+1}}$$

$$= \sum_{m=-\infty}^{\infty} a_m \delta_{mn} = a_n.$$

This proves the uniqueness of a possible Laurent expansion and that the coefficients will be the ones indicated in the statement of the theorem. It is left to show that f is the sum of a Laurent series in A.

Choose r' and R' such that $r < r' < R' < R$. If $r' < |z| < R'$ we have from the Global Cauchy Formula (Theorem 5.3) that

$$f(z) = \frac{1}{2\pi i} \int_{|w|=R'} \frac{f(w)}{w - z} \, dw - \frac{1}{2\pi i} \int_{|w|=r'} \frac{f(w)}{w - z} \, dw.$$

The first integral, i.e., the function

$$G(z) = \frac{1}{2\pi i} \int_{|w|=R'} \frac{f(w)}{w - z} \, dw$$

is holomorphic in $\{z \in \mathbb{C} \mid |z| \neq R'\}$, hence in particular in the disc $B(0, R')$, and so it has a power series expansion

$$G(z) = \sum_{m=0}^{\infty} b_m z^m \qquad for \ |z| < R'.$$

The second integral, i.e., the function

$$g(z) = -\frac{1}{2\pi i} \int_{|w|=r'} \frac{f(w)}{w-z} \, dw$$

may for $|z| > r'$ be treated as follows:

$$g(z) = -\frac{1}{2\pi i} \int_{|w|=r'} \frac{f(w)}{w-z} \, dw = \frac{1}{2\pi i} \frac{1}{z} \int_{|w|=r'} \frac{f(w)}{1-\frac{w}{z}} \, dw$$

$$= \frac{1}{2\pi i} \frac{1}{z} \int_{|w|=r'} f(w) \sum_{n=0}^{\infty} \frac{w^n}{z^n} \, dw$$

$$= \frac{1}{2\pi i} \frac{1}{z} \sum_{n=0}^{\infty} \frac{1}{z^n} \int_{|w|=r'} f(w) w^n \, dw$$

which exhibits a Laurent expansion of g. We conclude that $f = G + g$ has a Laurent expansion in the annulus $\{z \in \mathbb{C} \,|\, r' < |z| < R'\}$. But by uniqueness the Laurent series will be the same irregardless of the values of r' and R', so it converges in all points of A and has the sum f there. $\qquad\square$

Laurent series constitute an important tool in digital signal processing in form of the Z-transform: A signal is a two-sided sequence $\{a_n\}_{n\in\mathbb{Z}}$ of complex numbers; its Z-transform is by definition the sum of the Laurent series (6.1). This is a generalization of the corresponding Fourier series

$$\sum_{n=-\infty}^{\infty} a_n e^{-in\theta}$$

but the Laurent series may converge even if the Fourier series does not. The Z-transform has many pleasant properties. For instance that the Z-transform of a convolution is the product of the Z-transforms. For more information on the Z-transform and its applications the reader may look in any book on digital signal processing, say the classic monograph Oppenheim and Schafer (1975).

6.2   The classification of isolated singularities

Definition 6.4. *If a function f is holomorphic on an open set of the form $B(a,r) \setminus \{a\}$, we say that a is an **isolated singularity of** f.*

As examples of isolated singularities the reader can have the following three functions in mind:

$$\frac{\sin z}{z}, \quad \frac{\cos z}{z} \quad \text{and} \quad \exp\left(\frac{1}{z}\right).$$

Each of them has an isolated singularity at the point $z = 0$.

The classification of an isolated singularity of f is based on the behavior of f at the singularity a. There are three mutually excluding cases:

In the first case f remains bounded near a. This case is taken care of by the Removable Singularity Theorem (Theorem 4.3): f extends across a to an analytic function on all of $B(a,r)$. We say that a is a *removable singularity* of f. An example: $z = 0$ is a removable singularity of the function $\frac{\sin z}{z}$. We will always let f denote also the extended function, even if we do not say so explicitly. So for example we give the function $\frac{\sin z}{z}$ the value 1 at $z = 0$.

In the second case f has a *pole* at $z = a$ meaning by definition that $|f(z)| \to \infty$ as $z \to a$. Here the function $g(z) := 1/f(z)$ has a removable singularity at $z = a$ where it takes the value 0, so we can write it in the form

$$g(z) = (z - a)^n h(z)$$

where $n \in \{1, 2, \ldots\}$, and where h is analytic near a and $h(a) \neq 0$. Thus $(z - a)^n f(z)$ remains bounded near $z = a$. The smallest integer $n > 0$ such that $(z - a)^n f(z)$ remains bounded near a is called the *order of the pole*. It equals the order of the zero of $1/f$. If a pole has order 1 we call it a *simple pole*. An example of a simple pole is $\frac{\cos z}{z}$. A more general example of a simple pole is a quotient of the form $f(z) = g(z)/h(z)$ where g and h are holomorphic near $z = a$, $h(a) = 0$, $h'(a) \neq 0$ and $g(a) \neq 0$.

The third and final case is the one in which $\lim |f(z)|$ as $z \to a$ does not exist in $\mathbb{R} \cup \{\infty\}$. We say that f has an *essential singularity* at a. An example is $f(z) = \exp(z^{-1})$.

The beautiful relation between the coefficients of the Laurent expansion of f and the behavior of f at the isolated singularity is taken up in Exercise 6.1 below.

Definition 6.5. *Let a be an isolated singularity of the holomorphic function $f \colon B(a,r) \backslash \{a\} \to \mathbb{C}$. By the **residue of f** at a we mean the complex number*

$$\text{Res}(f; a) := \frac{1}{2\pi i} \int_{|z-a|=\varepsilon} f(z)\, dz$$

where ε is any number in the interval $]0, r[$. By Corollary 5.5 the residue of f at a is independent of the choice of ε.

The next sections will show that residues are important for the evaluation of definite integrals. So we must be able to compute residues. The simplest instance occurs when f has a removable singularity at a. Here $\mathrm{Res}(f; a) = 0$ by the Cauchy Integral Theorem (Theorem 4.5). Another particularly simple case – which even occurs quite often – is the case of a simple pole. This is singled out as a special instance in the following recipe for computing the residue at a pole.

Proposition 6.6.

(a) *If f has a pole of order $n < \infty$ at $z = a$ then the function $(z - a)^n f(z)$ has a removable singularity at $z = a$, and*

$$\mathrm{Res}(f; a) = \frac{1}{(n-1)!} \left(\frac{d}{dz} \right)^{n-1} \{ (z - a)^n f(z) \} \Big|_{z=a}.$$

(b) *In particular, if f has a simple pole at $z = a$ then*

$$\mathrm{Res}(f; a) = \{ (z - a) f(z) \}_{z=a}.$$

(c) *If f and g are holomorphic near $z = a$, $g(a) = 0$ and $g'(a) \neq 0$, then*

$$\mathrm{Res} \left(\frac{f}{g}; a \right) = \frac{f(a)}{g'(a)}.$$

Proof. We prove only (b) and (c) and leave the easy generalization (a) to the reader.

Ad (b): The definition of a simple pole implies that $g(z) := (z - a) f(z)$ remains bounded near $z = a$, so that g has a removable singularity at $z = a$. Now, let's write

$$f(z) = \frac{g(z)}{z - a} = \frac{g(z) - g(a)}{z - a} + \frac{g(a)}{z - a}$$

and observe that the first term on the right hand side is holomorphic across the singularity $z = a$ by the Removable Singularity Theorem. So

$$\int_{|z-a|=\varepsilon} f(z)\, dz = \int_{|z-a|=\varepsilon} \frac{g(z) - g(a)}{z - a}\, dz + g(a) \int_{|z-a|=\varepsilon} \frac{dz}{z - a}$$

$$= 0 + g(a) 2\pi i$$

proving (b).

Ad (c): Since f/g has a simple pole (if $f(a) \neq 0$) or a removable singularity (if $f(a) = 0$) at $z = a$ we get from (b) that

$$\operatorname{Res}\left(\frac{f}{g}; a\right) = \lim_{z \to a}\left\{(z - a)\frac{f(z)}{g(z)}\right\}$$

$$= \lim_{z \to a}\frac{f(z)}{(g(z) - g(a))/(z - a)} = \frac{f(a)}{g'(a)}.$$

$\square$

As a final example of how to compute residues we consider the function $\exp(z^{-1})$ which has an essential singularity at $z = 0$. Since the series

$$\exp(z^{-1}) = \sum_{n=0}^{\infty}\frac{1}{n!}z^{-n}$$

converges uniformly on compact subsets of $\mathbb{C} \setminus \{0\}$ we can integrate term by term and find

$$\operatorname{Res}\left(\exp(z^{-1}); 0\right) = \frac{1}{2\pi i}\sum_{n=0}^{\infty}\frac{1}{n!}\int_{|z|=\varepsilon} z^{-n}\, dz = \sum_{n=0}^{\infty}\frac{1}{n!}\delta_{1n} = 1.$$

Of course, by the same method we see more generally that any function f which around $z = a$ has the Laurent expansion

$$f(z) = \sum_{n=-\infty}^{\infty} a_n(z - a)^n$$

has $\operatorname{Res}(f; a) = a_{-1}$.

6.3 The residue theorem

6.3.1 *The statement.*

The Residue Theorem generalizes both the Cauchy Integral Theorem and the Cauchy Integral Formula. Our version of it runs as follows:

Theorem 6.7 (The Cauchy Residue Theorem). *Let Ω be an open subset of the complex plane and let γ be a closed path in Ω such that $\operatorname{Ind}_\gamma(z) = 0$ for all $z \in \mathbb{C} \setminus \Omega$. Let f be holomorphic in Ω except for finitely many isolated singularities $a_1, a_2, \ldots, a_N \in \Omega \setminus \gamma^*$. Then*

$$\frac{1}{2\pi i}\int_\gamma f(z)\, dz = \sum_{k=1}^{N}\operatorname{Ind}_\gamma(a_k)\operatorname{Res}(f; a_k).$$

The hypothesis on γ is clearly satisfied if γ is homotopic in Ω to a constant curve, say if Ω is simply connected. In most examples γ will be a simple closed path traversed counterclockwise, and $\mathrm{Ind}_\gamma(z)$ will be 1 or 0 according to whether z is inside or outside γ.

Proof. Choose closed paths $\gamma_1, \gamma_2, \ldots, \gamma_N$ in Ω such that each γ_j circles around the singularity a_j once and does not contain any of the other singularities of f. More precisely choose them such that

$$\mathrm{Ind}_{\gamma_j}(z) = 0 \qquad \text{for all } z \in \mathbb{C} \setminus \Omega \text{ and } j = 1, 2, \ldots, N$$

and

$$\mathrm{Ind}_{\gamma_j}(a_k) = \delta_{jk} \qquad \text{for } j, k = 1, 2, \ldots, N.$$

Then

$$\mathrm{Ind}_\gamma(z) = \sum_{j=1}^N \mathrm{Ind}_\gamma(a_j)\, \mathrm{Ind}_{\gamma_j}(z) \qquad \text{for all } z \in \mathbb{C} \setminus (\Omega \setminus \{a_1, a_2, \ldots, a_N\})$$

so by the Global Cauchy Theorem (Theorem 5.3 or better its Corollary 5.4)

$$\int_\gamma f = \sum_{j=1}^N \mathrm{Ind}_\gamma(a_j) \int_{\gamma_j} f.$$

Dividing through by $2\pi i$ we get the Residue Theorem. $\square$

The Cauchy Residue Theorem can be used to evaluate definite integrals. How successful one is depends largely on correct choice of contour and representative function. The techniques will be illustrated below by various examples that will provide us with recipes for evaluation of definite integrals of different types.

6.3.2   *Example A.*

An integral of the form

$$\int_0^{2\pi} F(\cos\theta, \sin\theta)\, d\theta$$

is by the substitution $z = \exp(i\theta)$ converted into a complex line integral as follows:

$$\int_0^{2\pi} F(\cos\theta, \sin\theta)\, d\theta = \int_{|z|=1} F\left(\frac{z + z^{-1}}{2}, \frac{z - z^{-1}}{2i}\right) \frac{dz}{iz}.$$

As a non-trivial example we evaluate

$$\int_0^{2\pi} \frac{\sin^2 \theta}{a + \cos \theta} \, d\theta \qquad \text{for } a \in \mathbb{C} \setminus [-1, 1]$$

where the restriction on a is chosen so that the denominator in the integrand does not vanish anywhere.

Using the above substitution we find

$$\int_0^{2\pi} \frac{\sin^2 \theta}{a + \cos \theta} \, d\theta = -\frac{1}{2i} \int_{|z|=1} \frac{(z^2 - 1)^2}{z^2(z^2 + 2az + 1)} \, dz$$

where the integrand, i.e., the function

$$f(z) = \frac{(z^2 - 1)^2}{z^2(z^2 + 2az + 1)}$$

has a pole of order 2 at $z = 0$ and simple poles at

$$z = -a + \sqrt{a^2 - 1} \quad \text{and} \quad z = -a - \sqrt{a^2 - 1}$$

with corresponding residues $-2a$, $2\sqrt{a^2 - 1}$ and $-2\sqrt{a^2 - 1}$, where we to avoid ambiguity agree to take $\sqrt{a^2 - 1}$ as that branch of the square root of $a^2 - 1$ which is positive for $a > 1$. (Cf. Example 3.23(iii).) To apply the Residue Theorem none of the poles

$$h(a) := -a - \sqrt{a^2 - 1} \quad \text{and} \quad g(a) := -a + \sqrt{a^2 - 1}$$

may belong to the unit circle $|z| = 1$. Since $h(a)$ and $g(a)$ are roots of the equation $z^2 + 2az + 1 = 0$, we have $h(a)g(a) = 1$ and $h(a) + g(a) = -2a$. Now if, say, $|h(a)| = 1$, then $g(a)$ is the complex conjugate of $h(a)$, so

$$\text{Re}(h(a)) = \frac{h(a) + \overline{h(a)}}{2} = \frac{h(a) + g(a)}{2} = -a.$$

But this means that a is real and $\sqrt{a^2 - 1}$ is purely imaginary, and that happens only in the excluded cases $a \in [-1, 1]$. Since $|h(a)|$ never takes the value 1 and is > 1 for $a > 1$ we see that $|h(a)| > 1$ for all a considered, and so $|g(a)| < 1$ for all $a \in \mathbb{C} \setminus [-1, 1]$. By the Residue Theorem

$$\int_0^{2\pi} \frac{\sin^2 \theta}{a + \cos \theta} \, d\theta = -\frac{1}{2i} 2\pi i \{\text{Res}(f; 0) + \text{Res}(f; g(a))\}$$

$$= -\pi \{-2a + 2\sqrt{a^2 - 1}\} = 2\pi \{a - \sqrt{a^2 - 1}\}.$$

The method of this example can be applied to any integral of the form

$$\int_0^{2\pi} \frac{P(\cos t, \sin t)}{Q(\cos t, \sin t)} \, dt$$

where P and Q are polynomials in two variables and $Q(\cos t, \sin t) \neq 0$ for all $t \in [0, 2\pi]$.

6.3.3 *Example B.*

In this example we want to evaluate integrals of the form $\int_{-\infty}^{\infty} R(x) \, dx$ where R is a rational function. We state our result as a proposition:

Proposition 6.8. *Let P and Q be polynomials in one variable with $\deg P < \deg Q$ and assume that Q has no zeros on the real line.*

(a) *If $\mu > 0$ then*

$$\lim_{R \to \infty} \int_{-R}^{R} \frac{P(x)}{Q(x)} e^{i\mu x} \, dx$$

exists, and

$$\lim_{R \to \infty} \int_{-R}^{R} \frac{P(x)}{Q(x)} e^{i\mu x} \, dx = 2\pi i \sum_w \operatorname{Res}\left(\frac{P(z)}{Q(z)} e^{i\mu z}; w \right)$$

where the sum ranges over $\{w \in \mathbb{C} \mid \operatorname{Im} w > 0, \ Q(w) = 0\}$.

(b) *If $\deg(P) < \deg(Q) - 1$, then*

$$\int_{-\infty}^{\infty} \frac{P(x)}{Q(x)} \, dx$$

converges absolutely, and

$$\int_{-\infty}^{\infty} \frac{P(x)}{Q(x)} \, dx = 2\pi i \sum_w \operatorname{Res}\left(\frac{P}{Q}; w \right)$$

where the sum ranges over $\{w \in \mathbb{C} \mid \operatorname{Im} w > 0, \ Q(w) = 0\}$.

Proof. Ad (a): Note that there exist two positive constants A and B such that

$$\left| \frac{P(z)}{Q(z)} \right| \leq \frac{A}{|z|}, \qquad \text{when } |z| \geq B.$$

Choose $R > B\sqrt{2}$ so large that the triangle in Fig. 6.1 contains all the zeros of Q in the upper half plane.

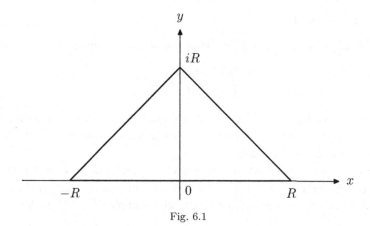

Fig. 6.1

The distance from the origin to each of the two skew sides of the triangle is $R/\sqrt{2}$, so

$$\left|\frac{P(z)}{Q(z)}\right| \le \frac{A\sqrt{2}}{R} \qquad \text{for } z \in [R, iR]^* \cup [iR, -R]^*.$$

With the usual parametrization of the segment $\gamma := [R, iR]$, i.e.,

$$\gamma(t) = [R, iR](t) = R(it + 1 - t) \qquad \text{for } 0 \le t \le 1$$

we find

$$\left|\int_{[R,iR]} \frac{P(z)}{Q(z)} e^{i\mu z}\, dz\right| = \left|\int_0^1 \frac{P(\gamma(t))}{Q(\gamma(t))} e^{i\mu R((i-1)t+1)} R(i-1)\, dt\right|$$

$$\le \frac{A\sqrt{2}}{R} \int_0^1 e^{-\mu R t} R\sqrt{2}\, dt = \frac{2A}{\mu R}(1 - e^{-\mu R})$$

$$\le \frac{2A}{\mu R} \to 0 \qquad \text{as } R \to \infty.$$

In exactly the same way we see that the integral along the other skew side $[iR, -R]$ of the triangle tends to zero as $R \to \infty$. The result (a) is then an immediate consequence of the Residue Theorem.

Ad (b): Proceed in the same way except for noting the stronger estimate

$$\left|\frac{P(z)}{Q(z)}\right| \le \frac{A}{|z|^2} \qquad \text{for } |z| \text{ large.}$$

$\square$

6.3.4   *Example C.*

As an example of an integral involving a multivalued function we consider

$$\int_0^\infty \frac{P(x)}{x^\alpha Q(x)}\, dx.$$

We assume that $0 < \alpha < 1$ and that P and Q are polynomials in one variable such that $\deg P < \deg Q$ and such that Q has no zeros on $[0, \infty[$. Of course, $x^\alpha = \exp(\alpha \operatorname{Log} x)$ in the integral.

It turns up to be a good idea to introduce that branch of the logarithm which is defined in $\Omega := \mathbb{C} \setminus [0, \infty[$ and has its imaginary part in the open interval $]0, 2\pi[$. So for the remainder of this example, if $z \in \Omega$ we let

$$\log z = \operatorname{Log}|z| + i\theta(z), \qquad \text{where } 0 < \theta(z) < 2\pi,$$

and

$$z^{-\alpha} = \exp(-\alpha \log z) = \exp(-\alpha \operatorname{Log}|z| - i\alpha\theta(z)).$$

We let f denote the function

$$f(z) := z^{-\alpha} \frac{P(z)}{Q(z)} \qquad \text{for } z \in \Omega$$

It is holomorphic in Ω except at the zeros of Q.

Consider the following contour in which $R > 0$ is chosen so large, and r and ε, where $r > \varepsilon > 0$, are chosen so small that the area inside contains all the zeros of Q. $\gamma_{\pm\varepsilon}$ denote the two horizontal paths

$$\gamma_{\pm\varepsilon}(t) = t \pm i\varepsilon \qquad \text{for } r - \varepsilon \leq t \leq R - \varepsilon$$

in Fig. 6.2.

Then

$$\int_{\gamma_\varepsilon} f\, dz = \int_{r-\varepsilon}^{R-\varepsilon} (t + i\varepsilon)^{-\alpha} \frac{P(t + i\varepsilon)}{Q(t + i\varepsilon)}\, dt$$

$$= \int_{r-\varepsilon}^{R-\varepsilon} e^{-\alpha \operatorname{Log}(t+i\varepsilon)} \frac{P(t + i\varepsilon)}{Q(t + i\varepsilon)}\, dt$$

and similarly

$$\int_{\gamma_{-\varepsilon}} f\, dz = e^{-2\pi i\alpha} \int_{r-\varepsilon}^{R-\varepsilon} e^{-\alpha \operatorname{Log}(t-i\varepsilon)} \frac{P(t - i\varepsilon)}{Q(t - i\varepsilon)}\, dt$$

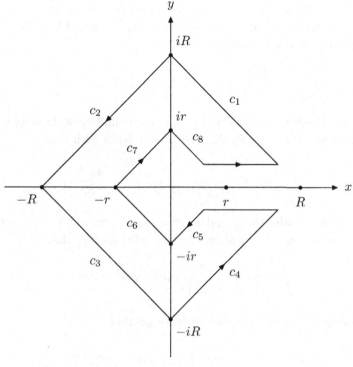

Fig. 6.2

so that

$$\lim_{\varepsilon \to 0_+} \left\{ \int_{\gamma_\varepsilon} f dz - \int_{\gamma_{-\varepsilon}} f\,dz \right\} = (1 - e^{-2\pi i\alpha}) \int_r^R x^{-\alpha} \frac{P(x)}{Q(x)}\,dx. \qquad (6.4)$$

Let $c_1, c_2, \ldots, c_8$ denote the paths indicated in the figure. Using the abbreviation

$$\Sigma := \sum_{z \in \Omega} \operatorname{Res}(f; z)$$

we get from the Residue Theorem that

$$\left| \int_{\gamma_\varepsilon} f\,dz - \int_{\gamma_{-\varepsilon}} f\,dz - \Sigma \right| \le \sum_{j=1}^{8} \left| \int_{c_j} f\,dz \right| \qquad (6.5)$$

so we shall estimate the eight integrals

$$\int_{c_j} f\,dz, \qquad j = 1, 2, \ldots, 8.$$

Let us first consider the paths c_1, c_2, c_3, c_4 on the boundary of the large square. Here we infer from the assumption $\deg P < \deg Q$ that there exist constants $A > 0$, $B > 0$ such that

$$\left| \frac{P(z)}{Q(z)} \right| \leq \frac{A}{|z|} \qquad \text{for all } |z| \geq B.$$

Noting that the distance from the origin to the boundary of the large square is $R/\sqrt{2}$ we get [choosing $R/\sqrt{2} \geq B$] for any outer path that

$$\left| \int_{c_j} f\, dz \right| \leq \frac{1}{(R/\sqrt{2})^\alpha} \frac{A}{R\sqrt{2}} R\sqrt{2} = \frac{A2^{1+\frac{\alpha}{2}}}{R^\alpha}.$$

We shall next consider the boundary of the small square. Here we estimate P/Q by a constant $C > 0$ and find for any inner path c_j that

$$\left| \int_{c_j} f\, dz \right| \leq \frac{C}{(r/\sqrt{2})^\alpha} r\sqrt{2} \leq C2^{\frac{1+\alpha}{2}} r^{1-\alpha}.$$

Substituting these estimates into (6.5) we get that

$$\left| \int_{\gamma_\varepsilon} f dz - \int_{\gamma_{-\varepsilon}} f\, dz \right| \leq 4[A2^{1+\frac{\alpha}{2}} R^{-A} + C2^{\frac{1+\alpha}{2}} r^{1-\alpha}]$$

$$= C_1 R^{-\alpha} + C_2 r^{1-\alpha}$$

for some constants C_1 and C_2. Letting $\varepsilon \to 0_+$ we read from (6.4) that

$$\left| (1 - e^{-2\pi i \alpha}) \int_r^R x^{-\alpha} \frac{P(x)}{Q(x)}\, dx - \Sigma \right| \leq C_1 R^{-\alpha} + C_2 r^{1-\alpha}.$$

Finally, letting $r \to 0$ and $R \to \infty$ we see that the right hand side converges to zero, so

$$(1 - e^{-2\pi i \alpha}) \int_0^\infty x^{-\alpha} \frac{P(x)}{Q(x)}\, dx - \Sigma = 0.$$

The *final result* is now

$$\int_0^\infty x^{-\alpha} \frac{P(x)}{Q(x)}\, dx = \frac{\pi e^{\pi i \alpha}}{\sin(\pi \alpha)} \sum_{z \in \Omega} \operatorname{Res}(f; z).$$

Exercises

6.1. Let $z = 0$ be an isolated singularity of a function f, and

$$f(z) = \sum_{n=-\infty}^{\infty} a_n z^n$$

its Laurent expansion. Show that

(α) 0 is a removable singularity of f iff $a_n = 0$ for all $n < 0$.

(β) 0 is a pole of order m of f iff $a_{-m} \neq 0$ and $a_n = 0$ for all $n < -m$.

(γ) 0 is an essential singularity of f iff $a_n \neq 0$ for infinitely many negative n.

6.2. Evaluate

$$\int_0^{2\pi} e^{(e^{-i\theta})} \, d\theta.$$

6.3. Find the value of

$$\frac{1}{2\pi i} \int_{|z|=1} \sin\left(\tfrac{1}{z}\right) dz.$$

6.4. Let N be the year you were born. Find the value of

$$\frac{i}{4} \int_{|z|=N} \tan(\pi z) \, dz.$$

6.5. Show that

$$\int_0^{2\pi} e^{\cos\theta} \cos(n\theta - \sin\theta) \, d\theta = \frac{2\pi}{n!} \qquad \text{for } n = 0, 1, \dots$$

6.6. Let P be a polynomial of degree ≥ 2. Show that

$$\int_{|z|=R} \frac{dz}{P(z)} = 0,$$

when R is chosen so large that the circle $|z| = R$ surrounds all the zeros of P.

6.7. Let P be a polynomial of degree $n \geq 2$ with distinct zeros $z_1, z_2, \dots, z_n$. Show that

$$\sum_{j=1}^{n} \frac{1}{P'(z_j)} = 0.$$

6.8. Let f be holomorphic in a neighborhood of $z = a$, and assume that $f(a) = 0$ and $f'(a) \neq 0$. Compute for $\varepsilon > 0$ small the integral

$$\int_{|z-a|=\varepsilon} z \frac{f'(z)}{f(z)} \, dz.$$

6.9. Let m and n be integers ≥ 0. Find

$$\frac{1}{2\pi i} \int_{|z|=1} \left(z - a + \frac{1}{z - a} \right)^m (z - a)^{n-1} \, dz, \qquad \text{when } |a| \neq 1.$$

Deduce *Wallis' formula*

$$\frac{1}{2\pi} \int_0^{2\pi} (2 \cos \theta)^{2m} \, d\theta = \frac{(2m)!}{(m!)^2}.$$

6.10. Show that

$$\int_0^{2\pi} \frac{d\theta}{1 - 2r \cos \theta + r^2} = \frac{2\pi}{|1 - r^2|} \qquad \text{when } r \in \mathbb{R} \setminus \{1, -1\}.$$

6.11. Show that

$$\int_0^{2\pi} \frac{d\theta}{1 + a \cos \theta} = \frac{2\pi}{\sqrt{1 - a^2}} \qquad \text{for } 0 < a < 1.$$

6.12. Show that

$$\int_0^{2\pi} \frac{d\theta}{1 + \mu^2 \cos^2 \theta} = \frac{2\pi}{\sqrt{1 + \mu^2}} \qquad \text{for } \mu \in \mathbb{R}$$

and derive the formula

$$\int_0^{2\pi} \frac{d\theta}{a^2 \cos^2 \theta + b^2 \sin^2 \theta} = \frac{2\pi}{ab} \qquad \text{for } a > 0, b > 0.$$

6.13. Show that

$$\int_0^{\pi} \frac{d\theta}{(a + \sin^2 \theta)^2} = \pi (a + \tfrac{1}{2})(a^2 + a)^{-\frac{3}{2}}, \qquad \text{when } a > 0.$$

6.14. Show that

$$\int_{-\infty}^{\infty} \frac{e^{iax}}{1 + x^2} \, dx = \pi e^{-|a|} \qquad \text{for } a \in \mathbb{R}.$$

6.15. Show that

$$\int_0^\pi \tan(\theta + ib)\, d\theta = \pi i \operatorname{sign}(b) \qquad \text{for } b \in \mathbb{R} \setminus \{0\}.$$

6.16. Pinpoint the error in the following reasoning: Let γ_R be the half circle in the upper half plane with center 0 and radius $R > 0$. Since the function (called *Sinus Cardinalis*)

$$\operatorname{Sinc}(z) := \frac{\sin z}{z}$$

is entire (i.e., holomorphic in all of $\mathbb{C}$), it follows from the Cauchy Integral Theorem with $f(z) = (\operatorname{Sinc} z)^2$ that

$$\int_{-\infty}^\infty f(x)\, dx = \lim_{R \to \infty} \int_{-R}^R f(x)\, dx = -\lim_{R \to \infty} \int_{\gamma_R} f(z)\, dz.$$

Estimating sinus by 1 we get

$$\left| \int_{\gamma_R} (\operatorname{Sinc} z)^2\, dz \right| \le \int_0^\pi R^{-2} R\, d\theta = \frac{\pi}{R} \xrightarrow[R \to \infty]{} 0$$

so that

$$\int_{-\infty}^\infty (\operatorname{Sinc} x)^2\, dx = 0.$$

Show that the correct value of the integral isn't 0, but π.

6.17. Show for $n = 1, 2, \ldots$ that

$$\int_{-\infty}^\infty (\operatorname{Sinc} x)^n\, dx = \frac{\pi}{2^{n-1}(n-1)!} \sum_{k < \frac{n}{2}} \binom{n}{k} (-1)^k (n-2k)^{n-1}.$$

6.18. Show that

$$\int_{-\infty}^\infty \frac{x^3 \sin x}{(x^2+1)^2}\, dx = \frac{\pi}{2e}.$$

6.19. Show that

$$\int_0^\infty \frac{x \sin x}{x^4 + 1}\, dx = \tfrac{\pi}{2} e^{-\frac{1}{\sqrt{2}}} \sin(\tfrac{1}{\sqrt{2}}).$$

6.20. Show that

$$\int_{-\infty}^\infty \frac{\cos x}{\cosh x}\, dx = \frac{\pi}{\cosh(\frac{\pi}{2})}.$$

Hint: Consider the contour of height π, see Fig. 6.3.

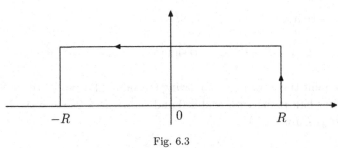

Fig. 6.3

6.21. Show for $n = 2, 3, \ldots$ that

$$\int_0^\infty \frac{dx}{1 + x^n} = \frac{\frac{\pi}{n}}{\sin\left(\frac{\pi}{n}\right)}$$

by help of the contour in Fig. 6.4 in which $P = R \exp(2\pi i/n)$.

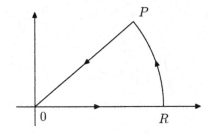

Fig. 6.4

Compute the more general expression

$$\int_0^\infty \frac{x^\alpha}{1 + x^n} dx \qquad \text{where } -1 < \alpha < n - 1.$$

Answer:

$$\frac{\pi/n}{\sin((1 + \alpha)\pi/n)}.$$

6.22. Show that

$$\int_0^\infty \frac{\text{Log } x}{(x^2 + 1)^2} dx = -\frac{\pi}{4}.$$

Hint: Consider the contour in Fig. 6.5.

An elementary deduction of the result can be found in Arora *et al.* (1988).

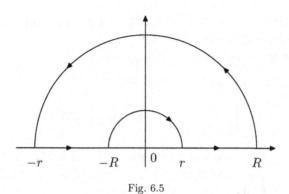

Fig. 6.5

6.23. Show that

$$\int_0^\infty \frac{\text{Log}\,x}{x^2 - 1}\,dx = \frac{\pi^2}{4}.$$

Hint: Consider the contour in Fig. 6.6.

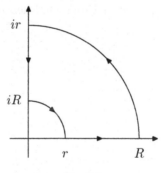

Fig. 6.6

6.24. In this problem we will deduce the following

Theorem. *If f is a rational function such that $|f(z)| = 1$ whenever $|z| = 1$, then f has away from its poles the form*

$$f(z) = cz^m \prod_{k=1}^n \frac{z - a_k}{1 - \overline{a_k}z}$$

where c is a constant of modulus 1, m is an integer, $n \in \{0, 1, 2, \ldots\}$, and where $a_1, a_2, \ldots, a_n$ are complex constants such that $|a_k| \neq 1$ for all k.

Proceed as follows for the proof:

(α) Show that we in the proof without loss of generality may assume that 0 is neither a zero nor a pole of f.

Hint: Divide by an appropriate power of z.

(β) Let $a \in \mathbb{C}$. Show that

$$\left| \frac{z - a}{1 - \overline{a}z} \right| = 1 \qquad \text{for all } z \in S^1 \setminus \{a\}.$$

(γ) Let $z_1, z_2, \ldots, z_r$ denote the zeros of f inside the unit disc, and $p_1, p_2, \ldots, p_s$ its poles there, zeros as well as poles repeated according to their multiplicity. Show that the function

$$F(z) := f(z) \prod_{k=1}^{r} \frac{1 - \overline{z_k}z}{z - z_k} \prod_{l=1}^{s} \frac{z - p_l}{1 - \overline{p_l}z}$$

is holomorphic and zero-free in a neighborhood of $B[0, 1]$ and that $|F(z)| = 1$ whenever $|z| = 1$.

(δ) Show that F is a constant in $B(0, 1)$ and hence everywhere.

(ε) Finish the proof.

Chapter 7

The Picard Theorems

In this chapter we shall prove Picard's "big" theorem on the strange behavior of holomorphic functions near essential singularities, and as a consequence derive his "little" theorem on entire functions. These two theorems have been the spur for much development in complex analysis, in particular for the so-called Nevanlinna theory which is concerned with the distribution of the values of meromorphic functions. Let us cite the two theorems:

Theorem 7.1 (Picard's Little Theorem). *The range of a nonconstant entire function omits at most one complex number.*

Theorem 7.2 (Picard's Big Theorem). *Let f be holomorphic in $\Omega \backslash \{z_0\}$, where Ω is an open subset of the complex plane and $z_0 \in \Omega$.*

If f has an essential singularity at z_0 then $f(\Omega \setminus \{z_0\})$ is either $\mathbb{C}$ or $\mathbb{C}$ with one point removed.

7.1 Liouville's and Casorati-Weierstrass' theorems

Liouville's theorem (Theorem 1.11) gives a first weak hint to the truth of Picard's little theorem, so we repeat it here:

Theorem 7.3 (Liouville's Theorem). *Let f be entire.*

(a) *If f is bounded then f is constant.*

(b) *More generally, if f for some real constants a, b and m satisfies the estimate*

$$|f(z)| \leq a|z|^m \qquad \text{for all } |z| > b$$

then f is a polynomial of degree $\leq m$.

As an important application of Liouville's theorem we derive the Fundamental Theorem of Algebra which asserts that every algebraic equation over the field of complex numbers has a root. More generally that it takes any complex value and thus shows the validity of Picard's little theorem for polynomials. The Fundamental Theorem of Algebra is important because it implies that any polynomial can be written as a product of linear factors.

Corollary 7.4 (The Fundamental Theorem of Algebra). *Let $a_0, a_1, \ldots, a_n \in \mathbb{C}$ where $n \geq 1$ and $a_n \neq 0$, and let p denote the polynomial*

$$p(z) := a_0 + a_1 z + \cdots + a_n z^n \qquad \text{for } z \in \mathbb{C}.$$

Then there exists a $z_0 \in \mathbb{C}$ such that $p(z_0) = 0$.

Although the Fundamental Theorem of Algebra looks like an algebraic result, being concerned with polynomials, it also involves the completeness of the reals: It is obviously false if $\mathbb{C} = \mathbb{R} + i\mathbb{R}$ is replaced by $\mathbb{Q} + i\mathbb{Q}$, where $\mathbb{Q}$ denotes the rational numbers. For a discussion and many references see p. 117–118 of the monography Burckel (1979).

Proof. The proof goes by contradiction, so we assume that p never vanishes. Then the function $f(z) := 1/p(z)$ is entire and converges to 0 as $|z| \to \infty$. In particular f is bounded and so a constant by Liouville's theorem. But then p is constant, contradicting that the degree of p is ≥ 1. □

The Casorati-Weierstrass' theorem is an immediate consequence of Picard's big theorem. We give a direct proof though, since it is quite simple. In the Russian literature the theorem is attributed to Y.V. Sokhotskii. For a detailed history of it see Neuenschwander (1978).

Theorem 7.5 (Casorati-Weierstrass' Theorem). *Let f have an essential singularity at $z_0 \in \mathbb{C}$. Then for any $w \in \mathbb{C}$ there exists a sequence $z_1, z_2, \ldots$ such that $z_n \to z_0$ and $f(z_n) \to w$ as $n \to \infty$.*

Proof. The proof goes by contradiction: If the assertion were false then there would exist a $w_0 \in \mathbb{C}$ and an $\varepsilon > 0$ and a neighborhood Ω of z_0 such that

$$|f(z) - w_0| \geq \varepsilon \qquad \text{for all } z \in \Omega \setminus \{z_0\}.$$

The function

$$g(z) := \frac{1}{f(z) - w_0}, z \in \Omega \setminus \{z_0\}$$

is then analytic in $\Omega \setminus \{z_0\}$ and bounded (by $1/\varepsilon$), so z_0 is a removable singularity for g. Thus

$$f(z) = w_0 + \frac{1}{g(z)}$$

is either analytic at $z = z_0$ (if $g(z_0) \neq 0$), or has a pole at $z = z_0$ (if $g(z_0) = 0$). In any case it contradicts our assumption about z_0 being an essential singularity. $\qquad \square$

7.2 Picard's two theorems

Our key to the proof of Picard's big theorem is the following remarkable result on the range of functions which are analytic in the unit disc.

Theorem 7.6 (Bloch-Landau's Theorem). *There exists a positive number A with following property: Let f be any function, holomorphic in the unit disc and such that $|f'(0)| \geq 1$. Then the range of f contains a disc of radius A.*

The surprising feature of the theorem is of course the existence of the universal constant A in spite of the vast class of functions involved. For our purposes any positive constant will suffice.

Remark. Let us for

$$f \in \Phi := \{g \in \text{Hol}(B(0,1)) \mid |g'(0)| \geq 1\}$$

put

$$L(f) := \sup\{r > 0 \mid f(B(0,1)) \text{ contains a disc of radius } r\}.$$

The theorem above then tells us that *Landau's constant*

$$L := \inf_{f \in \Phi} \{L(f)\}$$

is positive. Our proof of Theorem 7.6 reveals that $L \geq \frac{1}{13}$. The exact value of Landau's constant is not known, but it has been ascertained that $0.5 \leq L \leq 0.56$.

Theorem 7.6 has an immediate corollary of an even more surprising quantitative discovery on the range of an analytic function:

Theorem 7.7 (Bloch's Theorem). *Let f be any function which is holomorphic in $B(0,1)$ and has $|f'(0)| \geq 1$.*

Then there exists a subdisc D_f of $B(0,1)$ such that $f(D_f)$ contains a disc of radius 0.43 and f is univalent on D_f.

Remark. Let $B(f)$ be the supremum of all $r > 0$ for which there exists a region D on which f is univalent and such that $f(D)$ contains a disc of radius r. Then Bloch's theorem tells us that *Bloch's constant*

$$B := \inf\{B(f) \mid f \in \mathrm{Hol}(B(0,1)) \text{ and } |f'(0)| \geq 1\}$$

is bigger than 0.43. The exact value of Bloch's constant is not known, although it is known that $0.43 \leq B \leq 0.47$.

For interesting gossip on André Bloch's life see the essay Campbell (1985) and the follow up Cartan and Ferrand (1988).

Proof of Theorem 7.6. We may assume that $|f'(0)| = 1$. We will first treat the special case where f is holomorphic on a neighborhood of the closed disc $B[0,1]$.

The function $a\colon [0,1] \to [0,\infty[$ given by

$$a(r) := (1 - r)\sup_{|z|=r}\{|f'(z)|\} \qquad \text{for } 0 \leq r \leq 1$$

is continuous. Since $a(0) = 1$ and $a(1) = 0$, there exists a largest number $s \in [0,1[$ such that $a(s) = 1$. Choose $\zeta \in B(0,1)$ such that $|\zeta| = s$ and $|f'(\zeta)| = \sup_{|z|=s}\{|f'(z)|\}$.

Consider for $R := (1 - s)/2$ the function

$$F(z) := 2(f(Rz + \zeta) - f(\zeta)) \qquad \text{for } |z| \leq 1$$

F is holomorphic on $B(0,1)$, $F(0) = 0$, and

$$|F'(0)| = 2R|f'(\zeta)| = \frac{2Ra(s)}{1 - s} = 1.$$

Furthermore, since $a(r) < 1$ when $s < r$, we get

$$\frac{|F'(z)|}{2} = R|f'(Rz + \zeta)| \leq R\sup\{|f'(w)| \mid |w| \leq s + R\}$$

$$= R\sup\{|f'(w)| \mid |w| = s + R\}$$

$$= \frac{R}{1 - s - R}a(s + R) < \frac{R}{1 - s - R} = 1$$

so $|F'(z)| \leq 2$.

In the following lemma we shall show that the range of F contains a disc of radius $1/6$. From the definition of F we see that the range of f then contains a disc of radius $1/12$. This takes care of the special case.

In the general case we consider for any $0 < \rho < 1$ the function $z \to f(\rho z)/\rho$. It satisfies the conditions of the special case, so its range contains a disc of radius $1/12$. But that means that the range of f contains a disc of radius $\rho/12$. Taking $\rho = 12/13$ we see that the range of f contains a disc of radius $1/13$. $\qquad\square$

Lemma 7.8. *Let $F \in \mathrm{Hol}(B(0,1))$ satisfy that $F(0) = 0$, $F'(0) = 1$ and $|F'| \leq M$, where M is a constant. Then*

$$F(B(0,1)) \supseteq B\left(0, \frac{1}{2(M+1)}\right).$$

Proof. The function

$$z \to \frac{F'(z) - 1}{M + 1}$$

satisfies the conditions of Schwarz' lemma, so

$$|F'(z) - 1| \leq (M + 1)|z| \qquad \text{for all } |z| < 1.$$

Integrating $F'(z) - 1$ along the segment $[0, z]$ we get the inequality

$$|F(z) - z| \leq \frac{M+1}{2}|z|^2.$$

It says in particular that we for ζ on the circle $|\zeta| = 1/(M+1)$ have $|F(\zeta) - \zeta| \leq \frac{1}{2(M+1)}$. If furthermore $w \in B(0, \frac{1}{2(M+1)})$ then

$$|(F(\zeta) - w) - (\zeta - w)| = |F(\zeta) - \zeta| \leq \frac{1}{2(M+1)} < |\zeta - w|$$

so Rouché's theorem (Theorem 4.13) tells us that the function $z \to F(z) - w$ has a zero in the ball $B(0, 1/(M+1))$. Since w was arbitrary we have shown that

$$F\left(B\left(0, \frac{1}{M+1}\right)\right) \supseteq B\left(0, \frac{1}{2(M+1)}\right)$$

which is more than required. $\qquad\square$

We need one result more prior to Picard's big theorem, viz. Schottky's theorem.

Theorem 7.9 (Schottky's Theorem). *Let $M > 0$ and $r \in \,]0,1[$ be given. Then there exists a constant $C > 0$ such that the following implication holds:*

If F is holomorphic in $B(0,1)$, omits 0 and 1 from its range, and if $|F(0)| \leq M$, then $|F(z)| \leq C$ for all $z \in B(0,r)$.

Proof. Since F does not assume the value 0 in the starshaped set $B(0,1)$, F has a holomorphic logarithm $\log F$ [Proposition 3.10], which we choose such that $|\mathrm{Im}(\log F(0))| \leq \pi$. The function $A := (2\pi i)^{-1} \log F$ does not assume integer values, because F never takes the value 1. Let $\sqrt{A}$ and $\sqrt{A-1}$ be holomorphic square roots of A and $A - 1$ (they exist by Proposition 3.22). Then $B := \sqrt{A} - \sqrt{A-1}$ is holomorphic in $B(0,1)$, vanishes nowhere and cannot assume the values $\sqrt{n} \pm \sqrt{n-1}$ for $n = 1, 2, \ldots$ (Indeed, if

$$\sqrt{A(z)} - \sqrt{A(z) - 1} = \sqrt{n} \pm \sqrt{n-1} \qquad \text{for some } z \text{ and } n$$

then taking reciprocals we get

$$\sqrt{A(z)} + \sqrt{A(z) - 1} = \sqrt{n} - \sqrt{n-1} \qquad \text{resp. } \sqrt{n} + \sqrt{n-1}.$$

Adding and squaring shows that $A(z) = n$, which is excluded.)

Since B never vanishes there is a holomorphic branch H of $\log B$, and H cannot assume the values

$$a_{n,m} := \mathrm{Log}\{\sqrt{n} \pm \sqrt{n-1}\} + 2\pi i m \qquad \text{for } n = 1, 2, \ldots \text{ and } m \in \mathbb{Z}.$$

We leave it to the reader to verify that every disc of radius 10 contains at least one of the points $a_{n,m}$, so that the range of H cannot cover any disc of radius 10.

If $z \in B(0,1)$ and $H'(z) \neq 0$, then the values of the function

$$\zeta \to \frac{H(\zeta) - H(z)}{H'(z)} \qquad \text{for } \zeta \in B(z, 1 - |z|)$$

must fill a disc of radius $(1 - |z|)/13$ (by our proof of the Bloch-Landau Theorem), so the values of H fill a disc of radius $|H'(z)|(1 - |z|)/13$. This quantity cannot exceed 10, so

$$|H'(z)|(1 - |z|) \leq 130. \tag{7.1}$$

Although (7.1) was derived under the assumption $H'(z) \neq 0$, it is clearly also valid when $H'(z) = 0$. Since $H(z) - H(0)$ is the integral of H' along the segment $[0, z]$, the estimate (7.1) leads to

$$|H(z)| \leq |H(0)| + 130 \,\mathrm{Log}(1 - |z|)^{-1} \leq |H(0)| - 130 \log(1 - r) \tag{7.2}$$

for $|z| \leq r$. Now recall the definition of H :

$$\exp H = \sqrt{\frac{\log F}{2\pi i}} - \sqrt{\frac{\log F}{2\pi i} - 1}.$$

Taking reciprocals and adding the result we get

$$\sqrt{\frac{\log F(z)}{2\pi i}} = \frac{e^{H(z)} + e^{-H(z)}}{2} \tag{7.3}$$

so that

$$F = -\exp\{\tfrac{\pi i}{2}(e^{2H} + e^{-2H})\} \quad \text{and} \quad |F| \leq \exp\{\pi e^{2|H|}\}.$$

In view of (7.2) the theorem follows once we establish that $|H(0)| \leq C_1$ where C_1 is a constant depending only on the bound M on $F(0)$.

We will first treat those functions that satisfy the extra requirement $|F(0)| \geq \frac{1}{2}$. For such a function F we get from (7.3) that there is a constant C_2, depending only on M such that

$$C_2 \geq \left| \frac{e^{H(0)} + e^{-H(0)}}{2} \right| \geq \frac{e^{\mathrm{Re}(H(0))} - e^{-\mathrm{Re}(H(0))}}{2} = \sinh\big(\mathrm{Re}(H(0))\big)$$

which gives us an upper bound of the desired type on $\mathrm{Re}(H(0))$. Similarly for a lower bound on $\mathrm{Re}(H(0))$. The imaginary part poses no problems since we always may choose $H = \log B$ such that $|\mathrm{Im}(H(0))| \leq \pi$. We have now proved the theorem under the extra requirement that $|F(0)| \geq \frac{1}{2}$.

If $|F(0)| < \frac{1}{2}$ then we apply the just derived result to the function $1 - F$ instead of F. $\qquad \square$

We finally arrive at Picard's big theorem, which is a remarkable generalization of the Casorati-Weierstrass theorem.

Theorem 7.10 (Picard's Big Theorem). *If F has an essential singularity then the range of F omits at most one complex number.*

The example $F(z) = \exp(1/z)$ shows that it is too much to hope that the range of F is all of $\mathbb{C}$.

Proof. By translation in $\mathbb{C}$ we may assume that the singularity is situated at $z = 0$, and by dilation that F is holomorphic in $B(0, e^{2\pi}) \setminus \{0\}$. We will show that if F omits two complex numbers, say a and b, then 0 is either a pole or a removable singularity; that will prove the theorem. We may even assume that F omits the special values 0 and 1 – if not we replace F by

$$z \to \frac{F(z) - a}{b - a}.$$

Since 0 is not a pole of F there exists a sequence $\{z_n\}$ such that $z_n \to 0$ as $n \to \infty$ and $\{|F(z_n)|\}$ remains bounded, say $|F(z_n)| \leq M < \infty$ for all n.

Passing to a subsequence we may assume that

$$1 > |z_1| > \cdots > |z_n| > |z_{n+1}| > \cdots \xrightarrow[n \to \infty]{} 0.$$

Consider for fixed n the function $\zeta \to F(z_n e^{2\pi i \zeta})$ which is holomorphic in $B(0, 1)$, omits the values 0 and 1 and has $|F(z_n e^{2\pi i 0})| \leq M$. By Schottky's theorem there exists a constant C, depending only on the bound M, such that

$$|F(z_n e^{2\pi i \zeta})| \leq C \qquad \text{for all } \zeta \in B[0, \tfrac{1}{2}].$$

In particular

$$|F(z_n e^{2\pi i t})| \leq C \qquad \text{for all } t \in [-\tfrac{1}{2}, \tfrac{1}{2}]$$

so that $|F|$ is bounded by C on the circle $\{z \in \mathbb{C} \mid |z| = |z_n|\}$.

Since the constant C is independent of n we get by the Weak Maximum Principle that $|F| \leq C$ on all of $B(0, |z_1|) \setminus \{0\}$. But then 0 is a removable singularity of F. $\qquad\square$

The next result, Picard's little theorem, is an extension of the Fundamental Theorem of Algebra and of Liouville's theorem.

Theorem 7.11 (Picard's Little Theorem). *If F is a non-constant entire function then the range of F omits at most one complex number.*

The function $F(z) = \exp z$ demonstrates that it does happen that a complex number is absent from the range of F.

Proof. Consider the function

$$G(z) := F(z^{-1}) \qquad \text{for } z \in \mathbb{C} \setminus \{0\}.$$

If 0 is an essential singularity for G then we are done by Picard's big theorem. If 0 is a pole of order m or a removable singularity of G (a pole of order $m = 0$) then G can be written in the form $G(z) = z^{-m} H(z)$ where H is holomorphic in all of $\mathbb{C}$, also at $z = 0$. Now $F(z) = z^m H(z^{-1})$ so that

$$|F(z)| \leq (|H(0)| + 1)|z|^m \qquad \text{for } |z| \text{ large.}$$

By Liouville's theorem F is a polynomial, and so its range includes all of $\mathbb{C}$ by the Fundamental Theorem of Algebra. $\qquad\qquad\qquad\qquad\qquad\square$

7.3 Alternative treatment

The following alternative treatment of Schottky's and Picard's theorems is borrowed from the paper Minda and Schober (1983).

$$\Delta = \frac{\partial^2}{\partial x^2} + \frac{\partial^2}{\partial y^2}$$

denotes the *Laplace operator* = the *Laplacian* on $\mathbb{R}^2$. In polar coordinates (r, θ) it takes the form

$$\Delta = \frac{\partial^2}{\partial r^2} + \frac{1}{r}\frac{\partial}{\partial r} + \frac{1}{r^2}\frac{\partial^2}{\partial \theta^2}.$$

We shall a couple of times use that $\Delta F = 0$ when F is a holomorphic function. That is a consequence of the Cauchy-Riemann equations (see Theorem 12.2 for more information on this point).

Definition 7.12. *Let Ω be an open subset of the complex plane. A continuous function $\rho\colon \Omega \to [0, \infty[$ is said to be a **pseudo-metric** if it is of class C^2 on $\{z \in \Omega \mid \rho(z) > 0\}$. A **metric** is a function in $C^2(\Omega,]0, \infty[)$, so a metric is also a pseudo-metric.*

*The **(Gaussian) curvature** of a pseudo-metric ρ is the function κ defined by*

$$\kappa(z, \rho) := -\frac{\Delta(\operatorname{Log}\rho(z))}{\rho(z)^2}$$

in the subset of Ω where $\rho(z) > 0$.

Examples 7.13.

(i)

$$\lambda_R(z) := \frac{2R}{R^2 - |z|^2}$$

is a metric on $B(0, R)$ with constant curvature -1.

(ii)

$$\sigma(z) := \frac{2}{1 + |z|^2}$$

is a metric on $\mathbb{C}$ with constant curvature $+1$.

The formula in the following lemma expresses that the curvature is a conformal invariant, i.e., is preserved under holomorphic mappings, and so it is a natural object that ought to be studied.

Lemma 7.14. *Let $f\colon \Omega \to \Omega'$ be a holomorphic map between two open subsets Ω and Ω' of the complex plane, and let ρ be a metric on Ω'. Then $f^*(\rho) := \rho \circ f|f'|$ is a pseudo-metric on Ω and*

$$\kappa(z, f^*(\rho)) = \kappa(f(z), \rho) \qquad \text{for all } z \in \Omega \text{ for which } f'(z) \neq 0.$$

Proof. It is obvious that $f^*(\rho)$ is a pseudo-metric, so left is the formula for the curvature, i.e., that

$$-\frac{\Delta(\text{Log}(\rho \circ f|f'|))}{\rho \circ f|f'|^2}(z) = -\frac{(\Delta \text{Log}\,\rho)(f(z))}{\rho(f(z))^2}$$

or in other words that

$$\Delta\{\text{Log}(\rho \circ f) + \text{Log}|f'|\}(z) = \{\Delta \text{Log}\,\rho\}(f(z))|f'(z)|^2.$$

Here the term $\Delta(\text{Log}|f'|)$ on the left hand side vanishes, because $\text{Log}|f'|$ (locally) is the real part of a holomorphic function (branches of $\log f'$), so it is left to show the formula

$$\{\Delta(F \circ f)\}(z) = (\Delta F)(f(z))|f'(z)|^2$$

a formula that may be left to the reader. $\square$

Theorem 7.15 (Ahlfors' Lemma). *λ_1 is the biggest pseudometric on $B(0,1)$ among the ones with curvature at most -1.*

Proof. Let us for arbitrary but fixed $r \in\,]0,1[$ consider the function $v := \rho/\lambda_r$ on $B[0,r]$. It suffices to show that $v \leq 1$, i.e., that $\rho \leq \lambda_r$ because we then obtain the desired conclusion by letting $r \to 1$.

Since v is non-negative and $v(z) \to 0$ as $|z| \to r$, we see that the continuous function v attains its supremum m at some point z_0 in the open ball $B(0,r)$. We shall show that $m \leq 1$. If $m = 0$ we are done, so we may as well assume that $\rho(z_0) > 0$. Since z_0 is a maximum for v and hence for $\text{Log}\,v$, we get that $0 \geq (\Delta \text{Log}\,v)(z_0)$. Now,

$$\begin{aligned}
0 \geq (\Delta \text{Log}\,v)(z_0) &= (\Delta \text{Log}\,\rho)(z_0) - (\Delta \text{Log}\,\lambda_r)(z_0) \\
&= -\kappa(z_0,\rho)\rho(z_0)^2 + \kappa(z_0,\lambda_r)\lambda_r(z_0)^2 \\
&= -\kappa(z_0,\rho)\rho(z_0)^2 - \lambda_r(z_0)^2.
\end{aligned}$$

By the assumption on the curvature of ρ we find $0 \geq \rho(z_0)^2 - \lambda_r(z_0)^2$, so $m = \rho(z_0)/\lambda_r(z_0) \leq 1$. $\qquad\qquad\qquad\qquad\qquad\qquad\qquad\qquad\qquad$ $\square$

Proposition 7.16. *There exists a metric ρ on $\mathbb{C} \setminus \{0, 1\}$ with the following two properties:*

(a) ρ *has curvature at most -1.*

(b) $\rho \geq c\sigma$ *for some constant $c > 0$.*

Proof. Using the Laplacian in polar coordinates we find by brute force computations for any $\alpha \in \mathbb{R}$ and $z \in \mathbb{C} \setminus \{0\}$ that

$$\Delta(\mathrm{Log}(1 + |z|^\alpha)) = \alpha^2 \frac{|z|^{\alpha-2}}{(1 + |z|^\alpha)^2}.$$

Furthermore $\Delta(\mathrm{Log}|z|) = 0$ because $\mathrm{Log}|z|$ locally is the real part of a holomorphic function (branches of $\log z$). Combining these two facts we get for any $\alpha, \beta \in \mathbb{R}$ and $z \in \mathbb{C} \setminus \{0\}$ that

$$\Delta\left\{\mathrm{Log}\, \frac{1 + |z|^\alpha}{|z|^\beta}\right\} = \alpha^2 \frac{|z|^{\alpha-2}}{(1 + |z|^\alpha)^2}.$$

Since the Laplacian $\Delta = \partial^2/\partial x^2 + \partial^2/\partial y^2$ has constant coefficients it satisfies that

$$(\Delta F)(z - z_0) = \Delta(w \to F(w - z_0))(z)$$

so in particular for $z \in \mathbb{C} \setminus \{1\}$

$$\Delta\left\{\mathrm{Log}\, \frac{1 + |z-1|^\alpha}{|z-1|^\beta}\right\} = \alpha^2 \frac{|z-1|^{\alpha-2}}{(1 + |z-1|^\alpha)^2}.$$

We claim that for suitable $\alpha > 0$, $\beta > 0$ a positive multiple of the metric

$$\tau(z) := \frac{1 + |z|^\alpha}{|z|^\beta} \cdot \frac{1 + |z-1|^\alpha}{|z-1|^\beta} \qquad \text{for } z \in \mathbb{C} \setminus \{0, 1\}$$

can be used as the desired metric ρ.

By a straightforward calculation

$$\kappa(z, \tau) = -\alpha^2 \frac{|z|^{\alpha+2\beta-2}}{(1 + |z|^\alpha)^4} \cdot \frac{|z-1|^{2\beta}}{(1 + |z-1|^\alpha)^2}$$

$$-\alpha^2 \frac{|z-1|^{\alpha+2\beta-2}}{(1 + |z-1|^\alpha)^4} \cdot \frac{|z|^{2\beta}}{(1 + |z|^\alpha)^2}$$

so the curvature is everywhere negative.

Taking $\alpha = 1/6$ and $\beta = 3/4$ we find that

$$\lim_{|z| \to \infty} \frac{\tau(z)}{\sigma(z)} = +\infty \tag{7.4}$$

and

$$\lim K(z, \tau) = -\infty, \qquad \text{when } z \to 0, 1 \text{ or } \infty. \tag{7.5}$$

Combining (7.5) with the fact that the curvature is negative we find that there exists a constant $k > 0$ such that $\kappa(z, \tau) \leq -k$ on $\mathbb{C} \setminus \{0, 1\}$. Then $\rho := \sqrt{k}\tau$ has curvature at most -1, which is the first property.

From (7.4) we see that

$$\lim_{|z| \to \infty} \frac{\rho(z)}{\sigma(z)} = +\infty$$

which implies the second property. $\qquad \square$

Theorem 7.17 (Schottky's Theorem). *To each $M > 0$ and $r \in {]0, 1[}$ there exists a constant $C > 0$ such that the following implication holds:*

If $f \in \mathrm{Hol}(B(0, 1))$, $|f(0)| \leq M$ and the range of f omits 0 and 1, then $\sup_{|z| \leq r}\{|f(z)|\} \leq C$.

Proof. Let ρ and c be as in Proposition 7.16. Then $f^*(\rho)$ is a pseudo-metric on $B(0, 1)$ with curvature at most -1 (Lemma 7.14), by Ahlfors' lemma $f^*(\rho) \leq \lambda_1$, and so $f^*(\sigma) \leq c^{-1}f^*(\rho) \leq c^{-1}\lambda_1$, i.e.,

$$\frac{|f'(z)|}{1 + |f(z)|^2} \leq \frac{c^{-1}}{1 - |z|^2} \qquad \text{for all } z \in B(0, 1)$$

and so

$$\frac{|f'(z)|}{1 + |f(z)|^2} \leq c_1 \qquad \text{for all } z \in B(0, r)$$

where c_1 denotes the constant $c_1 = [c(1 - r^2)]^{-1}$.

Since f never takes the value 0, the function $t \to |f(tz)|$ is continuously differentiable for any fixed $z \in B(0, r)$ and

$$\left| \frac{d}{dt}(\arctan|f(tz)|) \right| \leq \frac{|f'(tz)||z|}{1 + |f(tz)|^2} \leq c_1$$

so

$$\big| \arctan|f(z)| - \arctan|f(0)| \big| \leq \int_0^1 \left| \frac{d}{dt}(\arctan|f(tz)|) \right| dt \leq c_1$$

from which we get

$$\arctan|f(z)| \le c_1 + \arctan|f(0)| \le c_1 + \arctan M$$

which is the theorem with $C = \tan(c_1 + \arctan M)$. $\square$

From here we proceed exactly as before to prove Picard's big theorem etc.

Exercises

7.1. Let F be an entire function with the property that

$$\frac{F(z)}{z} \to 0 \qquad \text{as } |z| \to \infty.$$

Show that F is a constant.

7.2. Let F be an entire function with the property that

$$|F(z)| \ge 1 \qquad \text{for all } z \in \mathbb{C}.$$

Show that F is a constant.

7.3. Let F be an entire function with the property that $\operatorname{Re} F$ is bounded from above. Show that F is a constant
Hint: Consider $\exp F$.

7.4 (Injectiveness of the Fourier Transform, cf. Newman, 1974).
Let f be continuous and absolutely integrable over $\mathbb{R}$ and assume that

$$\int_{-\infty}^{\infty} e^{-ix\xi} f(x)\, dx = 0 \qquad \text{for all } \xi \in \mathbb{R}.$$

(α) Show for each fixed $a \in \mathbb{R}$ that

$$\int_{-\infty}^{a} e^{-i\xi(x-a)} f(x)\, dx = -\int_{a}^{\infty} e^{-i\xi(x-a)} f(x)\, dx$$

and let $F_a(\xi)$ denote this common value.

(β) Show that the left hand side for fixed $a \in \mathbb{R}$ can be extended to a function which is continuous and bounded in the upper half plane $\operatorname{Im} \xi \ge 0$ and is holomorphic in the open half plane $\operatorname{Im} \xi > 0$.
   Show similar statements for the right hand side.

(γ) Show successively that F_a for fixed $a \in \mathbb{R}$ is holomorphic in all of $\mathbb{C}$, constant, even 0.

(δ) Show that $f = 0$.

(ε) Show by induction on n that the same result holds for functions f which are integrable over $\mathbb{R}^n$.

7.5. Let F be an entire function that never takes values in $]0, \infty[$. Show that F is a constant.

7.6. Let F be a rational function, i.e., a quotient between two polynomials. Assume that its poles occur at $a_1, a_2, \ldots, a_n$ with orders $m_1, m_2, \ldots, m_n$.

Show that there exist a polynomial p and constants $c_{jk} \in \mathbb{C}$ such that

$$F(z) = p(z) + \sum_{j=1}^{n} \sum_{k=1}^{m_j} \frac{c_{jk}}{(z - a_j)^k} \qquad \text{for } z \in \mathbb{C} \setminus \{a_1, a_2, \ldots, a_n\}.$$

7.7. Let F be holomorphic in $\mathbb{C}$ with the exception of finitely many poles. Assume that $F(1/z)$ has an inessential singularity at $z = 0$. Show that F is a rational function, i.e., a quotient between two polynomials.

7.8. Show the following version of Picard's little theorem:

Let F be a non-polynomial, entire function. Then there exists a $z_0 \in \mathbb{C}$ such that F takes every value in $\mathbb{C} \setminus \{z_0\}$ infinitely often.

Hint: Assume the conclusion fails. Show by Picard's big theorem that the function $z \mapsto F(\frac{1}{z})$ only has an inessential singularity at $z = 0$. Apply Liouville's theorem.

7.9. Let F be a non-constant, entire function. Assume that F does not take the value $a \in \mathbb{C}$. Let $b \in \mathbb{C} \setminus \{a\}$. Show that F takes the value b infinitely often.

7.10. Let F be entire and injective. Show that F is a polynomial of degree 1.

7.11. Define for fixed $\varepsilon > 0$ the function $f \colon B(0, 1) \to \mathbb{C}$ by

$$f(z) := \frac{\varepsilon}{2} \left\{ \left(\frac{1+z}{1-z} \right)^{1/\varepsilon} - 1 \right\} \qquad \text{for } z \in B(0, 1).$$

(α) Show that f has $f'(0) = 1$ and hence satisfies the conditions in Bloch-Landau's theorem.

(β) Show that $-\varepsilon/2$ does not belong to the range of f.

(γ) Taking $\varepsilon = 1/13$ we see that $f(B(0,1))$ does not contain a ball around 0 with radius $1/13$. Doesn't that contradict Bloch-Landau's theorem?

7.12. Let $\mathcal{S}_0$ denote the set of functions $f \in \text{Hol}(B(0,1))$ which satisfy $f(0) = 0$, $f'(0) = 1$, and $f(z) = 0$ only if $z = 0$.

We want to show that there exists a universal constant $\alpha > 0$ such that $f \in \mathcal{S}_0$ implies that $B(0, \alpha) \subseteq f(B(0,1))$.

We prove it by contradiction. So assume that the implication is false.

(α) Show that there exist sequences $\{\alpha_n\} \subseteq \mathbb{C} \setminus \{0\}$ such that $\alpha_n \to 0$ as $n \to \infty$, and $\{f_n\} \subseteq \mathcal{S}_0$ such that $\alpha_n \notin f_n(B(0,1))$.

(β) Show that

$$1 - \frac{f_n}{\alpha_n} : B(0,1) \to \mathbb{C} \setminus \{0\}$$

has a holomorphic logarithm L_n such that $L_n(0) = 0$.

(γ) Show that the sequence $\{L_n\}$ is uniformly bounded on $B(0, 1/2)$.

(δ) Show that $\{L_n'(0)\}$ is a bounded sequence.

(ε) Derive a contradiction!

The following exercises follow the alternative treatment (cf. Section 7.3).

7.13. Show that the metric ρ on $B(a,r)$, given by

$$\rho(z) := \frac{2r}{r^2 - |z - a|^2} \qquad \text{for } z \in B(a,r)$$

has curvature -1.

7.14. Derive Schwarz' lemma from Ahlfors' lemma.

Hint: Consider the metric $f^*(\lambda_1)$ and note that

$$\frac{d}{dx}(\arcsin \phi(x)) = \frac{\phi'(x)}{1 - \phi(x)^2}.$$

7.15. In this exercise we shall give a proof of Liouville's theorem based on Ahlfors' lemma.

Let $f \in \text{Hol}(\mathbb{C})$ be bounded, say $f(\mathbb{C}) \subseteq B(0,r)$ for some $r \in \,]0, \infty[$.

(α) Show that $f^*(\lambda_r) \leq \lambda_R$ for any $R > 0$.

(β) Show that $f^*(\lambda_r) = 0$.

(γ) Derive Liouville's theorem.

7.16. In this exercise we give a proof of Picard's little theorem, based on Ahlfors' lemma and the metric ρ from Proposition 7.16(a). So let $f\colon \mathbb{C} \to \mathbb{C} \setminus \{0,1\}$ be holomorphic.

(α) Show that $f^*(\rho) \leq \lambda_R$ on any ball $B(0,R)$.

(β) Show that $f^*(\rho) = 0$.

(γ) Derive Picard's little theorem.

Chapter 8

Geometric Aspects and the Riemann Mapping Theorem

We start this chapter by studying geometric aspects of analytic maps, in particular of the so-called Möbius transformations. We finish by proving the Riemann mapping theorem which states that simply connected regions contained in $\mathbb{C}$ but different from $\mathbb{C}$, are conformally equivalent.

8.1 The Riemann sphere

We have earlier studied analytic functions near isolated singularities. To handle such singularities it is aesthetically pleasing and often even convenient to introduce the Riemann sphere = the extended complex plane. As an example consider the map $z \to z^{-1}$ which has an isolated singularity at $z = 0$ and whose image does not contain 0. Adding the point ∞ to $\mathbb{C}$ and imposing the rules $\frac{1}{0} = \infty$ and $\frac{1}{\infty} = 0$ will make the map into a homeomorphism of the extended complex plane.

Definition 8.1. *The **extended complex plane** $\mathbb{C}_\infty$ consists of the complex plane plus an extra point ∞ (called infinity). We give $\mathbb{C}_\infty$ the topology which is defined by keeping the topology on $\mathbb{C}$ and by letting the sets $(\mathbb{C} \setminus K) \cup \{\infty\}$, where K runs through the compact subsets of $\mathbb{C}$, be the neighborhoods of ∞. The sets*

$$B(\infty; r) := \{z \in \mathbb{C} \mid |z| > r\} \cup \{\infty\}, \qquad 0 < r < \infty,$$

then constitute a neighborhood basis for the point ∞.

It may be remarked that $\mathbb{C}_\infty$ is the usual 1-point compactification of $\mathbb{C}$ as found in any book on point set topology.

We adopt the following natural rules for computing with the symbol ∞ and a complex number z:

$$\frac{z}{\infty} = 0, \quad \infty + z = z + \infty = \infty \quad \text{for any } z \in \mathbb{C}$$

$$\frac{z}{0} = \infty, \quad z \cdot \infty = \infty \cdot z = \infty \quad \text{for any } z \in \mathbb{C} \setminus \{0\}$$

$$0 \cdot \infty = \infty \cdot 0 = 0$$

We interpret $\mathbb{C}_\infty$ geometrically by exhibiting an explicit homeomorphism between the unit sphere

$$S^2 = \{(x, y, z) \in \mathbb{R}^3 \mid x^2 + y^2 + z^2 = 1\}$$

in $\mathbb{R}^3$ and the extended complex plane $\mathbb{C}_\infty$: As usual we identify the complex plane with the xy -plane in $\mathbb{R}^3$ such that the real axis is identified with the x-axis and the imaginary axis with the y-axis in $\mathbb{R}^3$. Stereographic projection from the north pole $N = (0, 0, 1)$ establishes a bijection π of $S^2 \setminus \{N\}$ onto $\mathbb{C}$. Putting $\pi(N) := \infty$ we see that π is a bijection of S^2 onto $\mathbb{C}_\infty$ which in coordinates is given by

$$\pi(x, y, z) = \begin{cases} \frac{x+iy}{1-z} & \text{for } (x, y, z) \in S^2 \setminus \{N\}, \\ \infty & \text{for } (x, y, z) = N. \end{cases}$$

Its inverse $\pi^{-1} \colon \mathbb{C}_\infty \to S^2$ has in coordinates the expression

$$\pi^{-1}(z) = \begin{cases} \frac{(2\,\mathrm{Re}\, z,\ 2\,\mathrm{Im}\, z,\ |z|^2 - 1)}{|z|^2 + 1} & \text{for } z \in \mathbb{C}, \\ (0, 0, 1) & \text{for } z = \infty. \end{cases}$$

From Fig. 8.1 it is geometrically obvious (and also easily verified from the explicit expressions above) that both π and π^{-1} are continuous so that π indeed is a homeomorphism of S^2 onto $\mathbb{C}_\infty$.

We call S^2 for the *Riemann sphere* when we have in mind the connection between S^2 and $\mathbb{C}_\infty$ which is established above by stereographic projection from the north pole.

Definition 8.2. *A* **circle** *in* $\mathbb{C}_\infty$ *is either an ordinary circle in* $\mathbb{C}$ *or a straight line in* $\mathbb{C}$ *with the point* ∞ *added.*

A pretty, geometric relation between S^2 and $\mathbb{C}_\infty$ is expressed by

Proposition 8.3. *Under the stereographic projection* π *the circles of the sphere* S^2 *are mapped onto the circles of* $\mathbb{C}_\infty$.

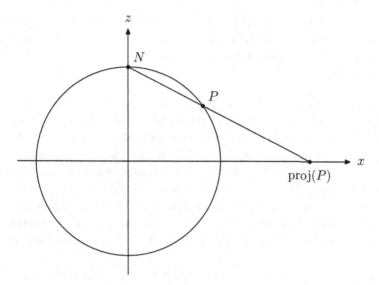

Fig. 8.1   Stereographic projection.

Proof. The verification consists really just of elementary computations that are better left to the reader. We shall here only note that a circle on S^2 is the intersection between S^2 and a plane in $\mathbb{R}^3$.   □

Let us return to the map $z \to z^{-1}$ which is a bijection of $\mathbb{C}_\infty$ onto itself. The corresponding bijection of the Riemann sphere onto itself is rotation by the angle π around the x-axis (seen by elementary computations from the formulas for π and π^{-1}); in particular it maps neighborhoods of 0 onto neighborhoods of ∞ and vice versa. In the light of these remarks we introduce the following extensions of previous definitions:

Definition 8.4.

(a) *A function $f : B(\infty; r) \to \mathbb{C}$ is said to be* **holomorphic around** ∞ *if the function $w \to f(w^{-1})$ is holomorphic around 0.*

(b) *A function $f : B(\infty; r) \setminus \{\infty\} \to \mathbb{C}$ is said to have an* **isolated singularity at** ∞ *if f is holomorphic in $B(\infty; r) \setminus \{\infty\}$.*

   Such a singularity at ∞ is said to be removable/a pole of order k/an essential singularity if $w \to f(w^{-1})$ has a removable singularity/a pole of order k/an essential singularity at 0.

(c) *Let Ω be an open subset of $\mathbb{C}_\infty$. A function f which is holomorphic on Ω except for isolated singularities, all of which are poles, is said* to be **meromorphic in** Ω. More precisely, to each $z \in \Omega$ there must exist an open disc $B(z,r) \subseteq \Omega$ such that either f is holomorphic on $B(z,r)$ or f is holomorphic on the punctured disc $B(z,r) \setminus \{z\}$ and z is a pole.

A holomorphic function is in particular meromorphic. In short, a meromorphic function is holomorphic except for poles. Essential singularities are not allowed. Meromorphic functions occur naturally as quotients between holomorphic functions: If f and g are holomorphic in an open subset Ω of $\mathbb{C}_\infty$ and if furthermore g does not vanish identically on any of the connected components of Ω, then f/g is meromorphic in Ω. We shall later see that the converse is true as well, i.e., any meromorphic function can be written as such a quotient (Exercise 10.9). A rational function, i.e., a function of the form P/Q where P and Q are polynomials in z, is meromorphic in all of $\mathbb{C}_\infty$.

If f is meromorphic in an open subset Ω of $\mathbb{C}_\infty$ then we shall from now on ascribe the value ∞ to f at any point of Ω at which f has a pole. Then f is defined on all of Ω and is a continuous mapping of Ω into $\mathbb{C}_\infty$. In particular, any rational function is a continuous mapping of $\mathbb{C}_\infty$ into $\mathbb{C}_\infty$.

8.2   The Möbius transformations

In this paragraph we introduce and study the Möbius transformations of the extended complex plane. They are important examples of rational maps with surprising and beautiful properties.

Definition 8.5. *A mapping $f \colon \mathbb{C}_\infty \to \mathbb{C}_\infty$ of the form*

$$f(z) = \begin{cases} \frac{az+b}{cz+d} & \text{for } z \in \mathbb{C}, \\ \frac{a}{c} & \text{for } z = \infty, \end{cases} \tag{8.1}$$

where the four complex numbers $a, b, c, d \in \mathbb{C}$ satisfy

$$\det \begin{Bmatrix} a & b \\ c & d \end{Bmatrix} = ad - bc \neq 0$$

*is called a **Möbius transformation** (other authors use terms like fractional linear transformation, bilinear transformation, homographic transformation or homography).*

Theorem 8.6. *The Möbius transformation* (8.1) *is a homeomorphism and a biholomorphic map of* $\mathbb{C}_\infty$ *onto* $\mathbb{C}_\infty$.

If $c = 0$ *then* f *is a biholomorphic map of* $\mathbb{C}$ *onto* $\mathbb{C}$. *If* $c \neq 0$ *then* f *is a biholomorphic map of* $\mathbb{C} \setminus \{-\frac{d}{c}\}$ *onto* $\mathbb{C} \setminus \{\frac{a}{c}\}$.

Proof. As is easy to check, the Möbius transformation

$$g(z) := \frac{ez + f}{gz + h} \qquad \text{where} \quad \begin{Bmatrix} e & f \\ g & h \end{Bmatrix} = \begin{Bmatrix} a & b \\ c & d \end{Bmatrix}^{-1}$$

is the inverse of f. Being rational both f and g are continuous from $\mathbb{C}_\infty$ into $\mathbb{C}_\infty$. □

The next two theorems are included simply because they are pretty. We won't need them later on.

Theorem 8.7. *If* $A = \begin{Bmatrix} a & b \\ c & d \end{Bmatrix}$ *we let* f_A *denote the Möbius transformation*

$$f_A(z) := \frac{az + b}{cz + d} \qquad \text{for } z \in \mathbb{C}_\infty.$$

The map $A \to f_A$ *is a group-homomorphism of* $\mathrm{GL}(2, \mathbb{C})$ *into the group of homeomorphisms of* $\mathbb{C}_\infty$. *So the Möbius transformations form a subgroup of the group of homeomorphisms of* $\mathbb{C}_\infty$.

Proof. Left to the reader. □

Theorem 8.8. *The set of Möbius transformations equals the set of bijective meromorphic mappings of* $\mathbb{C}_\infty$ *onto* $\mathbb{C}_\infty$.

Proof. The only statement which is not yet verified, is that any bijective meromorphic mapping $f : \mathbb{C}_\infty \to \mathbb{C}_\infty$ is a Möbius transformation. We verify it:

Since ∞ is either a pole or a removable singularity, f can near ∞ be written as

$$f(z) = z^N h\left(\tfrac{1}{z}\right) \qquad \text{for some } N \in \{0, 1, 2, \ldots\}$$

where h is holomorphic near 0. In particular there exist constants C and R such that

$$|f(z)| \leq C|z|^N \qquad \text{for } |z| \geq R.$$

Since f has only finitely many singularities in $B[0, R]$ and since these are poles, there exists a polynomial $q \neq 0$ such that qf is entire. Combining this with the estimate above we get by Liouville's theorem that qf is a polynomial, say p. So $f = p/q$ is rational. We may of course assume that p and q have no first order factor in common.

If the degree of the polynomial p is bigger than 1, then p takes the value 0 at two different points or has a double root; in any case f is not injective (cf. Theorem 4.21). Thus p is a polynomial of degree 1 or 0. Similar arguments reveal that q is of degree 1 or 0. So $f = p/q$ has the form

$$f(z) = \frac{az + b}{cz + d} \qquad \text{where } a, b, c, d \in \mathbb{C}.$$

If $\det\left\{ \begin{smallmatrix} a & b \\ c & d \end{smallmatrix} \right\} = 0$ then (a, b) is proportional to (c, d), so f reduces to a constant. But f was a bijection, so the determinant is not 0.   □

Theorem 8.9. *Any Möbius transformation maps circles in $\mathbb{C}_\infty$ onto circles in $\mathbb{C}_\infty$.*

Proof. Let us note that any Möbius transformation (8.1) is composed of rotations

$$z \mapsto e^{i\theta} z \qquad \text{where } \theta \in \mathbb{R}$$

dilations

$$z \mapsto az \qquad \text{where } a > 0$$

translations

$$z \mapsto z + z_0 \qquad \text{where } z_0 \in \mathbb{C}$$

and the inversion $z \mapsto z^{-1}$. This is trivial if $c = 0$, and for $c \neq 0$ it is apparent from the easily verified formula

$$f(z) = \frac{a}{c} - \frac{ad - bc}{c} \frac{1}{cz + d}.$$

Obviously rotations, dilations and translations map circles in $\mathbb{C}_\infty$ onto circles in $\mathbb{C}_\infty$, so it is just left to show that the inversion does too. But that is a consequence of Proposition 8.3 and the remark following it.

Here is an analytic proof for the reader who dislikes geometry: Lines and circles in $\mathbb{C}$ are sets of the form

$$\{z \in \mathbb{C} \mid \alpha|z|^2 + \beta z + \overline{\beta}\overline{z} + \gamma = 0\}$$

where α and γ range over $\mathbb{R}$ and β over $\mathbb{C}$ subject to the condition $|\beta|^2 > \alpha\gamma$. Replacement of z by $1/z$ transforms the equation

$$\alpha|z|^2 + \beta z + \overline{\beta}\overline{z} + \gamma = 0$$

into

$$\alpha + \beta\bar{z} + \bar{\beta}z + \gamma|z|^2 = 0$$

which is an equation of the same type. □

Another important and useful property of Möbius transformations is that they preserve angles. To formulate that properly and to clarify the concepts we introduce a definition:

Definition 8.10. *Let* $\alpha\colon I \to \mathbb{C}$ *and* $\beta\colon J \to \mathbb{C}$ *be two paths and let* $(t_0, s_0) \in I \times J$. *Assume that* $\alpha(t_0) = \beta(s_0)$ *and that* $\alpha'(t_0) \neq 0$ *and* $\beta'(s_0) \neq 0$.

*The **angle from** α **to** β at (t_0, s_0) is the discrete set*

$$\arg \beta'(s_0) - \arg \alpha'(t_0) = \arg \frac{\beta'(s_0)}{\alpha'(t_0)}$$

of real numbers.

Theorem 8.11. *Let* Ω *be an open subset of* $\mathbb{C}$ *and let* $f\colon \Omega \to \mathbb{C}$ *be complex differentiable at* $z_0 \in \Omega$ *with* $f'(z_0) \neq 0$. *Then* f *is angle preserving at* z_0 *in the following sense:*

Let $\alpha\colon I \to \mathbb{C}$ *and* $\beta\colon J \to \mathbb{C}$ *be two paths with* $z_0 = \alpha(t_0) = \beta(s_0)$ *and* $\alpha'(t_0) \neq 0$ *and* $\beta'(s_0) \neq 0$.

Then the angle from $f \circ \alpha$ *to* $f \circ \beta$ *at* (t_0, s_0) *equals the angle from* α *to* β *at* (t_0, s_0).

Proof. The computation

$$\frac{(f \circ \beta)(s_0)}{(f \circ \alpha)(t_0)} = \frac{f'(\beta(s_0))\beta'(s_0)}{f'(\alpha(t_0))\alpha'(t_0)} = \frac{f'(z_0)\beta(s_0)}{f'(z_0)\alpha'(t_0)} = \frac{\beta'(s_0)}{\alpha'(t_0)}$$

reveals that the proof is simpler than the statement. □

Corollary 8.12. *The Möbius transformation* (8.1) *is angle preserving everywhere in* $\mathbb{C} \setminus \{-\frac{d}{c}\}$.

Proof. Differentiation of (8.1) yields

$$f'(z) = \frac{ad - bc}{(cz + d)^2} \neq 0.$$

A more sophisticated argument runs as follows: f is injective (Theorem 8.6), hence $f'(z) \neq 0$ (Theorem 4.21). □

The above results enable us to examine any given Möbius transformation by a minimal amount of computations, using in particular that Möbius transformations preserve circles in $\mathbb{C}_\infty$ and angles.

Example 8.13. The Möbius transformation

$$f(z) := \frac{z-i}{z+i} \qquad \text{for } z \in \mathbb{C}_\infty$$

occurs as the so-called *Cayley transformation* in the theory of operators on a Hilbert space. Since f maps circles in $\mathbb{C}_\infty$ onto circles in $\mathbb{C}_\infty$ we get from $f(0) = -1$, $f(1) = -i$ and $f(\infty) = 1$ that $f(\mathbb{R} \cup \{\infty\})$ is the unit circle.

Combining the injectivity of f with $f(i) = 0$ and $f(-i) = \infty$ we conclude that f maps the upper half plane $\{z \in \mathbb{C} \mid \operatorname{Im} z > 0\}$ onto $B(0,1)$ and the lower half plane onto $B(\infty, 1)$. The mapping of the upper half plane onto the disc is also interesting from another point of view: It establishes a connection between two models of non-Euclidean geometry. In fact f is an isometry with respect to the non-Euclidean distances.

We proceed with another interesting example of a Möbius transformation that we shall need below in our proof of the Riemann Mapping Theorem.

Example 8.14. Let $A(a; \cdot)$ for $a \in B(0,1)$ denote the Möbius transformation

$$A(a; z) := \frac{z-a}{1-\overline{a}z} \qquad \text{for } z \in \mathbb{C}_\infty.$$

We claim that $A(a; \cdot)$ is a univalent map of $B(0,1)$ onto itself with $A(a; a) = 0$. And that any other such map is a constant (of modulus 1) multiple of $A(a; \cdot)$.

Note for later reference that $A(0; \cdot)$ is the identity map, that $A(-a; \cdot) = A(a; \cdot)^{-1}$, that $A'(a; a) = (1 - |a|^2)^{-1}$ and $A'(a; 0) = 1 - |a|^2$.

Proof of the claim. When $z \in S^1 = \partial B(0,1)$ then

$$|A(a; z)| = \left| \frac{z-a}{1-\overline{a}z} \right| = |z| \left| \frac{1-az^{-1}}{1-\overline{a}z} \right| = \left| \frac{1-a\overline{z}}{1-\overline{a}z} \right| = 1$$

so by the Weak Maximum Principle $A(a; B(0,1)) \subseteq B[0,1]$, and further by the Open Mapping Theorem $A(a; B(0,1)) \subseteq B(0,1)$.

Since the inverse mapping $A(a; \cdot)^{-1} = A(-a; \cdot)$ of $A(a; \cdot)$ is of the same form as $A(a; \cdot)$ we see that $A(a; \cdot)$ maps $B(0,1)$ univalently onto $B(0,1)$.

Let g be a univalent map of $B(0,1)$ onto itself with $g(a) = 0$. Then both $f := g \circ \{A(a; \cdot)^{-1}\}$ and its inverse satisfy the conditions of Schwarz' lemma, so $|f(z)| \leq |z|$ and $|f^{-1}(z)| \leq |z|$. But then

$$|f(z)| \leq |z| = |f^{-1}(f(z))| \leq |f(z)|$$

so by the equality part of Schwarz' lemma there exists a constant $c \in \mathbb{C}$ with $|c| = 1$, such that $f(z) = cz$ for all $z \in B(0,1)$. And then $g(z) = cA(a; z)$ for all $z \in B(0,1)$. $\square$

8.3   Montel's theorem

An important ingredient in our proof of the Riemann Mapping Theorem is the following compactness result which in the literature is often called the *Condensation Principle*. In functional analytic terms it states that the holomorphic functions on an open set constitute a Montel space.

Theorem 8.15 (Montel's Theorem, (1907)). *Let $\mathcal{F}$ be a sequence of holomorphic functions on an open subset Ω of $\mathbb{C}$, and assume that $\mathcal{F}$ is locally uniformly bounded in Ω.*

Then $\mathcal{F}$ has a subsequence which converges locally uniformly in Ω to a holomorphic function.

The corresponding theorem is not true on $\mathbb{R}$ as the example $\{\sin nx \mid n = 1, 2, \ldots\}$ shows.

Proof. Let A be a countable dense subset of Ω. By the standard diagonal sequence argument we extract a subsequence $\{f_n\}$ from $\mathcal{F}$ which converges at each point $a \in A$. The program is to show that $\{f_n\}$ converges uniformly on each compact subset K of Ω. This will prove Montel's theorem because the limit function automatically will be analytic (by Weierstrass' Theorem 4.10). Since $C(K)$ is complete it suffices to show that $\{f_n|_K\}$ is a Cauchy sequence in $C(K)$. So let $\varepsilon > 0$ and a compact subset K of Ω be given.

Since K is compact, $\mathrm{dist}(K, \mathbb{C} \setminus \Omega) > 0$, so therefore we can fix $R \in {]0, \mathrm{dist}(K, \mathbb{C} \setminus \Omega)[}$, and let M be a bound for $\mathcal{F}$ on the compact subset $\{z \in \mathbb{C} \mid \mathrm{dist}(z, K) \leq R\}$.

For $a \in \mathbb{C}, z \in K, |z - a| < R$ and $f \in \mathcal{F}$ we get by applying Schwarz' lemma to the function $w \to [f(z + wR) - f(z)]/2M$ that

$$|f(z) - f(a)| \leq 2M \frac{|z - a|}{R}.$$

Combining $a \in \mathbb{C}$, $z \in K$, $|z - a| < \min\{R, \frac{R\varepsilon}{8M}\}$ and $f \in \mathcal{F}$ we get that $|f(z) - f(a)| < \varepsilon/4$.

By the compactness of K we can cover it by finitely many discs $B(z_1, r)$, $B(z_2, r), \ldots, B(z_s, r)$ with centers $z_1, z_2, \ldots, z_s$ in K and with (the same) radius $r = \min\{R, \frac{R\varepsilon}{8M}\}/2$. Choose $a_j \in A \cap B(z_j, r)$ for $j = 1, 2, \ldots, s$. Then for any $z \in K$ there is an a_j such that

$$|z - a_j| < 2r = \min\left\{R, \frac{R\varepsilon}{8M}\right\}$$

so by the estimate above

$$
\begin{aligned}
|f_n(z) &- f_m(z)| \\
&\leq |f_n(z) - f_n(a_j)| + |f_n(a_j) - f_m(a_j)| + |f_m(a_j) - f_m(z)| \\
&\leq \frac{\varepsilon}{2} + |f_n(a_j) - f_m(a_j)| \\
&\leq \frac{\varepsilon}{2} + \sum_{j=1}^{s} |f_n(a_j) - f_m(a_j)|
\end{aligned}
$$

which shows that

$$\|f_n - f_m\|_{\infty, K} \leq \frac{\varepsilon}{2} + \sum_{j=1}^{s} |f_n(a_j) - f_m(a_j)|.$$

The sequence $\{f_n(a_j)\}$ converges for each fixed j by the very choice of $\{f_n\}$, so $\|f_n - f_m\|_{\infty, K} \leq \varepsilon$ when n and m both are sufficiently large. We have thus proved that $\{f_n|_K\}$ is a Cauchy sequence in $C(K)$ as desired. $\qquad\square$

The following useful result is a corollary of Montel's theorem. It says that analyticity is preserved when the parameter in a family of holomorphic functions is integrated away.

Proposition 8.16. *Let $f \in C(I \times \Omega)$ where I is an interval on the real line and Ω is an open subset of the complex plane. If*

$$f(t, \cdot) \in \mathrm{Hol}(\Omega) \qquad \text{for each fixed } t \in I,$$

and

$$\sup_{z \in K} \int_I |f(t, z)|\, dt < \infty \qquad \text{for each compact subset } K \text{ of } \Omega$$

then

$$z \mapsto \int_I f(t, z)\, dt$$

is holomorphic on Ω.

Proof. If I is compact then Fubini and Morera (Theorem 4.9) secure the conclusion, so left is the non-compact case. Here we take an increasing sequence $\{I_n\}$ of compact subintervals I_n of I, such that their union is I. Then as just stated

$$F_n(z) := \int_{I_n} f(t, z)\, dt$$

is holomorphic on Ω. By the assumptions $\{F_n\}$ is a locally uniformly bounded sequence. It converges pointwise to

$$F(z) := \int_I f(t, z)\, dt$$

by the Dominated Convergence Theorem. Montel's theorem implies that F is holomorphic on Ω. □

8.4 The Riemann mapping theorem

Definition 8.17. *Two open subsets V and W of the complex plane are said to be **conformally equivalent** if there exists a **conformal equivalence** $\phi\colon V \to W$, i.e., a univalent map ϕ of V onto W.*

Conformal equivalence is an equivalence relation [by Theorem 4.20].

Examples 8.18.

(i) The Cayley transformation

$$z \mapsto \frac{z - i}{z + i}$$

is a conformal equivalence of the open upper half plane onto the unit disc.

(ii) $z \to e^z$ is a conformal equivalence of the strip $0 < \operatorname{Im} z < \pi$ onto the upper half plane and takes $i\pi/2$ into i. Combining with the Cayley transformation we see that

$$z \mapsto \frac{1 + ie^z}{e^z + i}$$

is a conformal equivalence of the strip onto the unit disc.

(iii) $z \mapsto e^{-z}$ maps the strip $0 < \operatorname{Im} z < 2\pi$ q conformally onto the plane cut along the negative real axis.

Definition 8.19. *A region is an open, connected and non-empty subset of* $\mathbb{C}$.

V and W are of course homeomorphic if they are conformally equivalent. So topological properties may prohibit domains to be conformally equivalent. For example, an annulus and a disc are not homeomorphic, and so a fortiori not conformally equivalent. But topology alone is not the complete story either: $\mathbb{C}$ and the unit disc $B(0,1)$ are homeomorphic, but $\mathbb{C}$ is not conformally equivalent to any bounded domain (Liouville's theorem), in particular not to the unit disc. So much the more chocking is the fantastic statement that B. Riemann enunciated in his Göttingen dissertation of 1851:

Theorem 8.20 (The Riemann Mapping Theorem). *Every simply connected region other than* $\mathbb{C}$ *is conformally equivalent to the open unit disc.*

The Riemann Mapping Theorem does not provide us with a formula for the conformal equivalence ϕ between the given simply connected domain V and the unit disc. For an explicit solution of a problem it will be necessary to have more information about ϕ than its mere existence. Explicit formulas for ϕ when V is bounded by polygons can be found in Section 17.6 of Hille (1962). For a discussion of an possible extension of ϕ to a homeomorphism of $\overline{V}$ onto $B[0,1]$ see Novinger (1975).

An easy, but surprising consequence of the Riemann Mapping Theorem is that any two simply connected regions of $\mathbb{C}$ are homeomorphic.

If $\phi\colon V \to B(0,1)$ is a conformal equivalence then $f \mapsto f \circ \phi$ is an algebra isomorphism between $\mathrm{Hol}(B(0,1))$ and $\mathrm{Hol}(V)$. So any problem about the algebra $\mathrm{Hol}(V)$ can be carried over to a problem on the unit disc, and the (possible) solution then carried back from there to $\mathrm{Hol}(V)$.

We will prove the following version of the Riemann Mapping Theorem. The earlier one is of course a corollary.

Theorem 8.21 (The Riemann Mapping Theorem). *Any connected non-empty open set* $\Omega \subseteq \mathbb{C}$ *which is not the entire complex plane, and on which every never vanishing holomorphic function has a holomorphic square root, is conformally equivalent to the open unit disc. In fact, given* $a \in \Omega$ *there exists exactly one conformal equivalence* $H\colon \Omega \to B(0,1)$ *such that* $H(a) = 0$ *and* $H'(a) > 0$.

Proof. For the sake of clarity we divide the proof into 5 steps. We let $\mathcal{J}$ be the set of all univalent maps of Ω into $B(0,1)$.

Step 1. We show that $\mathcal{J}$ is not empty.
Choose any $z_0 \in \mathbb{C} \backslash \Omega$ (recall that Ω is not all of $\mathbb{C}$). The function $z \mapsto z - z_0$
does not vanish anywhere in Ω, so it has a holomorphic square root g:

$$[g(z)]^2 = z - z_0 \qquad \text{for all } z \in \Omega.$$

Clearly g must be univalent. It is also immediately verified that $-g(\Omega)$
and $g(\Omega)$ have empty intersection. $-g(\Omega)$ is open by the Open Mapping
Theorem. Let $z_1 \in -g(\Omega)$ and $B(z_1, r) \subseteq -g(\Omega)$. The map $\psi(z) := r/(z - z_1)$
is univalent and maps the complement of $B(z_1, r)$ (in particular $g(\Omega)$) into
the unit disc $B(0, 1)$. The composite map $\psi \circ g$ then belongs to $\mathcal{J}$. Thus $\mathcal{J}$
is not empty.

Step 2. We find H.
Every $g \in \mathcal{J}$ is univalent on Ω and therefore g' does not vanish on Ω. Thus

$$\eta := \sup_{g \in \mathcal{J}} |g'(a)| > 0.$$

Choose $f_n \in \mathcal{J}$ for $n = 0, 1, 2, \ldots$ such that $\lim |f_n'(a)| = \eta$. The f_n are
uniformly bounded on Ω so Montel's theorem applies. By choosing a subsequence if necessary we may assume that $\{f_n\}$ converges locally uniformly
on Ω to a holomorphic function H. By Weierstrass' theorem (Theorem 4.10)
$\{f_n'\}$ converges locally uniformly to H', so that $|H'(a)| = \eta$. In particular
H is not constant.

Step 3. $H \in \mathcal{J}$.
Suppose not, so that there exist $x, y \in \Omega, x \neq y$ such that $H(x) = H(y)$.
Choose a disc $B(x, r)$ whose closure is in Ω, that does not contain y and
satisfies that

$$\inf\{|H(z) - H(y)| \mid |z - x| = r\} > 0.$$

This is possible because zeros of nonconstant holomorphic functions are
isolated. If the sequence $\{f_n\}$ is as in Step 2 we get for large enough n that

$$|f_n(x) - f_n(y)| < \inf\{|f_n(z) - f_n(y)| \mid |z - x| = r\}.$$

If $h_n(w) := f_n(w) - f_n(y)$ does not vanish in $B(x, r)$ then $1/h_n$ is a coun-
terexample to the Weak Maximum Principle. So

$$h_n(w) = f_n(w) - f_n(y) = 0 \qquad \text{for some } w \in B(x, r).$$

But that contradicts that f_n is univalent.

Since $f_n(\Omega) \subseteq B(0,1)$ and $\{f_n\}$ converges to H, we see that $H(\Omega) \subseteq B[0,1]$. Since H is not a constant function, $H(\Omega)$ is open. Thus $H(\Omega) \subseteq B(0,1)$ and $H \in \mathcal{J}$.

Step 4. $H(a) = 0$ and H has range $B(0,1)$.
Put now $\alpha = H(a)$ and let g be the composite map $g := A(\alpha, H)$. Then $g \in \mathcal{J}$. By a small computation

$$|g'(a)| = \frac{|H'(a)|}{1 - |\alpha|^2} = \frac{\eta}{1 - |\alpha|^2} > \eta \qquad \text{unless } \alpha = 0.$$

Thus $H(a) = 0$.

Suppose H is not onto $B(0,1)$ and let $b \in B(0,1) \setminus H(\Omega)$. The function $A(b, H)$ does not vanish in Ω, so it has a holomorphic square root ϕ in Ω: $[\phi(z)]^2 = A(b, H(z))$. Clearly $\phi \in \mathcal{J}$. Recalling that $H(a) = 0$ and $|H'(a)| = \eta$, we compute that

$$|\phi'(a)| = \eta \frac{1 - |b|^2}{2\sqrt{|b|}}.$$

Now, $\psi := A(\phi(a), \phi)$ also belongs to $\mathcal{J}$, and we find that

$$|\psi'(a)| = \eta \frac{1 + |b|}{2\sqrt{|b|}} > \eta$$

which is a contradiction. Hence H is onto.

Step 5. The uniqueness.
The uniqueness of H is an easy consequence of the uniqueness statement in Example 8.14 above.   □

8.5 Primitives

We have seen in Theorem 5.8 that every holomorphic function on a simply connected domain has a primitive there. There are open sets for which this fails. For example the plane punctured at the origin has not got this property: The function $1/z$ defined on $\mathbb{C} \setminus \{0\}$ does not possess a primitive (because if it did, the line integral of $1/z$ along the unit circle would be zero, but it is $2\pi i$). It is a most remarkable fact that a connected open set has this property if and only if it is homeomorphic to the open unit disc.

Theorem 8.22. *For any non-empty connected open set Ω the following are equivalent:*

(a) *Every holomorphic function on Ω has a primitive.*

(b) *Ω is homeomorphic to the unit disc.*

Proof. *Item (a) $\Rightarrow$ item (b):* If $\Omega = \mathbb{C}$, the map

$$z \mapsto (1 - e^{-|z|})\frac{z}{|z|}$$

is a homeomorphism of $\mathbb{C}$ onto the open unit disc. If $\Omega \neq \mathbb{C}$ then the hypotheses of Theorem 8.21 are satisfied: Let f be a non-vanishing holomorphic function on Ω. By assumption f'/f has a primitive, say ϕ. It follows that $f = const \cdot e^{\phi}$. Adjusting ϕ by adding a suitable constant we see that f has a holomorphic logarithm, and in particular a holomorphic square root. By Theorem 8.21 Ω is conformally equivalent to the unit disc.

Item (b) $\Rightarrow$ item (a): If Ω is homeomorphic to the unit disc then it is simply connected, and so we may appeal to Theorem 5.8. $\qquad\square$

Exercises

8.1. When is a Möbius transformation T a projection, i.e., when is $T^2 = T$?

8.2. Show that the Möbius transformation which takes -1 to 0, 0 to 1 and 1 to ∞, takes the unit circle (except 1) to the imaginary axis.

8.3. Find a Möbius transformation that maps the semi-disc $\{z \in B(0,1) \mid \text{Im } z > 0\}$ onto the positive quadrant.

8.4. Find a Möbius transformation ϕ that maps the open right half plane onto $B(0,1)$ in such a way that $\phi(1) = 0$ and $\phi'(1) > 0$.

8.5. Let f and g be holomorphic mappings of $B(0,1)$ onto a subset Ω of the complex plane. Assume that f is a bijection of $B(0,1)$ onto Ω and that $f(0) = g(0)$. Show that

$$g(B(0,r)) \subseteq f(B(0,r)) \qquad \text{for any } r \in]0,1[.$$

8.6. Let $f \colon B(0,1) \to B(0,1)$ be holomorphic and fix more than one point. Show that $f(z) = z$ for all $z \in B(0,1)$.

8.7.

(α) Why can't you display an analytic function $f\colon B(0,1) \to B(0,1)$ such that $f(1/2) = 3/4$ and $f'(1/2) = 2/3$?

(β) Can you find an analytic function $f\colon B(0,1) \to B(0,1)$ such that $f(0) = 1/2$ and $f'(0) = 3/4$? How many are there?

8.8. Let π_+ be the open right half plane, and let $f\colon \pi_+ \to \pi_+$ be analytic. Assume that $f(a) = a$ for some $a \in \pi_+$. Show that $|f'(a)| \leq 1$.

8.9. Let $f, g\colon B(0,1) \to \mathbb{C}$ be holomorphic functions, g univalent. Assume $f(0) = g(0)$ and $f(B(0,1)) \subseteq g(B(0,1))$. Show that $|f'(0)| \leq |g'(0)|$ and that equality sign implies $f(B(0,1)) = g(B(0,1))$.

8.10. Prove Vitali-Porter's theorem:

> **Theorem (Vitali 1903, Porter 1904).** *Let Ω be a connected open subset of $\mathbb{C}$ and let $\{f_n\}$ be a sequence of holomorphic functions on Ω. Assume that the sequence $\{f_n\}$ is locally uniformly bounded in Ω, and that $\{f_n(z)\}$ has a limit as $n \to \infty$ for a set of $z \in \Omega$ with an accumulation point in Ω. Then $\lim_{n\to\infty} f_n(z)$ exists locally uniformly in Ω.*

Hint: Montel's theorem.

8.11. Let $\{f_n\}$ be a locally uniformly bounded sequence of holomorphic functions in a connected open subset Ω of $\mathbb{C}$, and assume that f_n is zero-free in Ω for each n.

Show that $\lim f_n = 0$ locally uniformly in Ω if there exists a $z_0 \in \Omega$ such that $\lim f_n(z_0) = 0$.

Hint: Apply the Maximum Principle to $1/f_n$ and refer to the Vitali-Porter Theorem.

8.12. In this problem we generalize a part of Schwarz' lemma.

Let Ω be a bounded, connected, open subset of $\mathbb{C}$, let $a \in \Omega$ and let $f\colon \Omega \to \Omega$ be a holomorphic function with $f(a) = a$.

We put

$$f_1 = f, f_{n+1} = f \circ f_n \qquad \text{for } n = 1, 2, \ldots$$

(α) Show that there exists a constant K, depending on Ω and a but not on f, such that $|f'(a)| \leq K$.

(β) Show that $|f'(a)| \leq 1$.

Hint: Compute $f'_n(a)$.

(γ) Assuming that $f'(a) = 1$, prove that $f(z) = z$ for all $z \in \Omega$.
Hint: Write

$$f(z) - z = c_m(z - a)^m + (z - a)^{m+1} h(z)$$

and compute the coefficient c_m.

(δ) Assuming that $|f'(a)| = 1$, prove that f is a bijection of Ω onto Ω.
Hint: Find a sequence of integers $n_k \nearrow \infty$ as $k \to \infty$ and a function g such that
$$f'(a)^{n_k} \to 1 \qquad \text{and } f_{n_k} \to g.$$

8.13. Show that $z \to e^{-z}$ is a conformal equivalence of the strip $-\pi < \operatorname{Im} z < \pi$ onto the plane cut along the negative real axis.

8.14. Find a conformal equivalence of

$$\{z \in \mathbb{C} \mid |z| < 1 \text{ and } |z - (1 + i)| < 1\}$$

onto the unit disc.

8.15. Find a conformal equivalence between the semi-disc

$$\{z \in B(0, 1) \mid \operatorname{Im} z > 0\}$$

and the unit disc.

8.16. Find an explicit formula for that conformal equivalence H between

$$\{z \in \mathbb{C} \mid -1 < \operatorname{Re} z < 1\}$$

and $B(0, 1)$ which satisfies $H(0) = 0$ and $H'(0) > 0$.

8.17. Show that $\{z \in \mathbb{C} \mid 0 < |z| < 1\}$ is not conformally equivalent to any annulus $\{z \in \mathbb{C} \mid r < |z| < R\}$ where $0 < r < R < \infty$. For more information on general annuli we refer the reader to Theorem 3 in Chapter 7, Section 1 of Narasimhan (1985).

8.18. Let Ω be an open connected subset of $\mathbb{C}$. Let $\{f_n\}$ be a sequence of univalent analytic functions on Ω, and assume that $\{f_n\}$ converges locally uniformly to a function f.

Show that f is either 1–1 or a constant (Hurwitz, 1888), and show that both possibilities can occur.

8.19. Let f be holomorphic and bounded in $\{z \in \mathbb{C} \mid -1 < \operatorname{Im} z < 1\}$ and assume that $f(x) \to 0$ as $x \to \infty$.

Prove that

$$\lim_{x \to \infty} f(x + iy) = 0 \qquad \text{for each } y \in \,]{-}1, 1[$$

and that the passage to the limit is uniform when y is confined to any interval of the form $[-\alpha, \alpha]$ where $0 < \alpha < 1$.

Hint: Consider $f_n(z) := f(z + n)$ for $z = x + iy$ in the square $\{x + iy \mid |x| < 1, |y| < 1\}$.

8.20. Show the following generalization of Schwarz' lemma:

Let f be analytic in a neighborhood of $B[0, 1]$ and let $|f(z)| \le M$ for $|z| = 1$ and let $f(0) = a$. Then

$$|f(z)| \le \frac{M|z| + |a|}{|a||z| + M} \qquad \text{for } |z| < 1.$$

8.21. Show (using suitable Möbius transformations) that the Schwarz' Reflection Principle and Schwarz' Reflection in the circle (cf. Exercise 4.27 and Example 4.28) are equivalent.

Chapter 9

Meromorphic Functions and Runge's Theorems

9.1 The argument principle

In this section we derive an important and beautiful result called the argument principle, which expresses the number of zeros of an analytic function inside a closed curve in terms of a winding number. More generally, it is a formula for the difference between the number of zeros and poles of a meromorphic function.

Observation 9.1. *Let f be meromorphic in an open set $\Omega \subseteq \mathbb{C}$, and assume that there is a compact subset of Ω that contains all the zeros (resp. poles) of f in Ω.*

Then the number of zeros (resp. poles) of f in Ω is finite.

Proof. Left to the reader. □

We remind the reader that index = argument increment divided by 2π (Remark 3.13(C)).

Theorem 9.2 (The Argument Principle). *Let Ω be a simply connected open subset of $\mathbb{C}$, γ a closed curve in Ω and f meromorphic in Ω. Let $z_1, z_2, \ldots, z_r$ and $p_1, p_2, \ldots, p_s$ be all zeros and poles of f with multiplicities $n_1, n_2, \ldots, n_r$ and $m_1, m_2, \ldots, m_s$ respectively. Suppose that none of the zeros and poles lie on γ.*

Then
$$\operatorname{Ind}_{f \circ \gamma}(0) = \sum_{j=1}^{r} n_j \operatorname{Ind}_{\gamma}(z_j) - \sum_{k=1}^{s} m_k \operatorname{Ind}_{\gamma}(p_k).$$

If γ is even a closed path then
$$\frac{1}{2\pi i} \int_{\gamma} \frac{f'(z)}{f(z)} \, dz = \sum_{j=1}^{r} n_j \operatorname{Ind}_{\gamma}(z_j) - \sum_{k=1}^{s} m_k \operatorname{Ind}_{\gamma}(p_k).$$

141

See also Exercise 9.1.

Proof. Write

$$f(z) = \prod_{j=1}^{r}(z - z_j)^{n_j} \prod_{k=1}^{s}(z - p_k)^{-m_k} g(z)$$

where g is holomorphic and vanishes nowhere in Ω. Choose any continuous logarithms for $\gamma - z_j$, $\gamma - p_k$ and g (recall that g does not vanish in Ω). Then

$$\sum_{j=1}^{r} n_j \log(\gamma - z_j) - \sum_{k=1}^{s} m_k \log(\gamma - p_k) + (\log g) \circ \gamma$$

is a continuous logarithm for $f \circ \gamma$. The proof follows from the definition of Ind (Definition 3.12) and the formula for the index of a path (Theorem 3.18(d)). □

To understand what the above means, let γ be a circle. For w inside γ the winding number $\mathrm{Ind}_\gamma(w) = 1$ and for w outside it is zero. The Argument Principle thus says that the change in $\arg f(z)$ as z traces γ equals $Z - P$ where Z is the number of zeros and P the number of poles of f inside γ, multiplicities counted.

Example 9.3. The Argument Principle can give information about the location of the zeros of a polynomial. We present here a typical example of the kind of problems that can be solved:

The polynomial $f(z) := z^8 + az^3 + bz + c$, where $a \geq 0, b \geq 0$ and $c > 0$ has exactly 2 zeros in the first quadrant.

If $t \geq 0$ then $f(t) > 0$ and $\mathrm{Re}(f(it)) > 0$, so f has no zeros on the positive semi-axes.

Let $R > 0$ and consider the quarter circle

$$z = Re^{i\theta}, \qquad 0 \leq \theta \leq \tfrac{\pi}{2}.$$

On this quarter circle f is

$$f(Re^{i\theta}) = R^8 e^{8i\theta}(1 + aR^{-5}e^{-5i\theta} + bR^{-7}e^{-7i\theta} + cR^{-8}e^{-8i\theta}).$$

For R sufficiently large the parenthesis has positive real part, so a continuous argument function for f along the quarter circle is

$$\theta \to 8\theta + \mathrm{Arg}\{R^{-8}e^{-8i\theta}f(Re^{i\theta})\}$$

where Arg denotes the principal determination of the argument.

Consider now the curve indicated on Fig. 9.1.

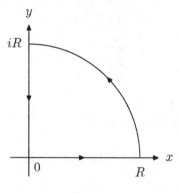

Fig. 9.1

We have observed that $\mathrm{Re}(f(it)) > 0$ when $t \geq 0$. So as z varies from iR to 0 along the imaginary axis the argument of f increases by $\mathrm{Arg}\, f(0) - \mathrm{Arg}\, f(iR) = -\mathrm{Arg}\, f(iR)$, and so the argument increment is in absolute value less than π.

The argument of f is constant (viz. equal to 0) from 0 to R. Finally from R to iR along the circle it increases by $\frac{8\pi}{2} + \Delta = 4\pi + \Delta$, where $|\Delta| < \pi$. So along the curve on the figure the argument of f increases by $4\pi + something$, where $|\,something\,| < 2\pi$. But the argument increment is an integer multiple of 2π, so here it is exactly 4π.

By the Argument Principle there are exactly 2 zeros of f (counted with multiplicity) inside the contour, and hence (since $R > 0$ can be arbitrarily large) in the first quadrant.

9.2   Rouché's theorem

The following clever lemma will be useful in the proofs of Rouché's and Runge's theorems. It is an integral representation theorem like the Cauchy Integral Formula. In contrast to the earlier results it asserts the existence of a curve with certain properties, not that every curve has those properties. The power of integral representations has already been demonstrated in connection with the Cauchy Integral Formula; it led to local power series expansions.

Lemma 9.4. *Let Ω be an open subset of $\mathbb{C}$ and let $F \subseteq \Omega$ be compact. Then there exists a finite set of piecewise linear closed paths $\gamma_1, \gamma_2, \ldots, \gamma_n$ in $\Omega \setminus F$ such that for any function $f \in \text{Hol}(\Omega)$:*

$$0 = \sum_{j=1}^{n} \int_{\gamma_j} f(z) \, dz$$

and

$$f(a) = \frac{1}{2\pi i} \sum_{j=1}^{n} \int_{\gamma_j} \frac{f(z)}{z-a} \, dz \qquad \text{for all } a \in F.$$

Proof. Consider the following grid (i.e., set of squares) on $\mathbb{R}^2$:

$$\left\{ (x,y) \in \mathbb{R}^2 \,\middle|\, \frac{i}{N} \leq x \leq \frac{i+1}{N}, \frac{j}{N} \leq y \leq \frac{j+1}{N} \right\}$$

where i and j range over the integers, and where the positive integer N is chosen so large that any square of the grid having a point in common with F is contained in Ω. We orient each of the squares in the counter clockwise direction.

Denote by $Q_1, Q_2, \ldots, Q_m$ the set of those squares in the grid that have a point in common with F. The union of these Q_i contains F – that much is clear. Note also that if for example a vertex of a square belongs to F, then all the four squares having this point as a vertex are included in the system $Q_1, Q_2, \ldots, Q_m$. A similar remark applies to a point on an edge. Let $L_1, L_2, \ldots, L_r$ denote the boundary segments of this system of squares, i.e., $L_1, L_2, \ldots, L_r$ are those edges of the system which are edges of exactly one square. Then L_j must for each j be contained in $\Omega \setminus F$. Since each square is oriented, each segment is oriented, too.

Now let f be holomorphic in Ω. Since $Q_k \subseteq \Omega$ we get by Cauchy's theorem that

$$\int_{\partial Q_k} f(z) \, dz = 0 \qquad \text{for } 1 \leq k \leq m. \tag{9.1}$$

And if a belongs to the interior of the square Q_k (it can belong to only one)

$$\frac{1}{2\pi i} \int_{\partial Q_k} \frac{f(z)}{z-a} \, dz = f(a), \qquad a \in \text{int}(Q_k) \tag{9.2}$$

and for the other squares

$$\frac{1}{2\pi i} \int_{\partial Q_l} \frac{f(z)}{z-a} \, dz = f(a), \qquad a \in \text{int}(Q_k), \ k \neq l. \tag{9.3}$$

Adding (9.1) over all k we find that

$$\sum_k \int_{\partial Q_k} f(z)\, dz = 0.$$

If an edge is common to two squares, then the orientations induced on it are opposite and so the net contribution is zero. We get

$$\sum_j \int_{L_j} f(z)\, dz = 0.$$

Using the same reasoning we get from (9.2) and (9.3) that if a belongs to the interior of one of the squares then

$$\sum_j \frac{1}{2\pi i} \int_{L_j} \frac{f(z)}{z-a}\, dz = f(a.) \tag{9.4}$$

Now every $b \in F$ is a positive distance from $L_1 \cup L_2 \cup \cdots \cup L_r$ and is a limit of a's that belong to the interior of one of the squares. (9.4) is therefore by continuity valid for all $a \in F$.

We shall finally show that the oriented line segments

$$L_j = [a_j, b_j], \qquad \text{where } j = 1, 2, \ldots, r$$

can be grouped together to a collection of closed paths $\gamma_1, \gamma_2, \ldots, \gamma_n$.

For any polynomial p we get from what we have already shown that

$$0 = \sum_{j=1}^{r} \int_{L_j} p'(z)\, dz = \sum_{j=1}^{r} \{p(b_j) - p(a_j)\}$$

so

$$\sum_{j=1}^{r} p(b_j) = \sum_{j=1}^{r} p(a_j) \qquad \text{for any polynomial } p. \tag{9.5}$$

For $z_0 \in \mathbb{C}$ let

$$n_a := \text{the number of } j \text{ for which } a_j = z_0$$
$$n_b := \text{the number of } j \text{ for which } b_j = z_0.$$

Choose a polynomial p such that $p(z_0) = 1$ and such that $p = 0$ at the other points, i.e., at

$$\{a_1, a_2, \ldots, a_r, b_1, b_2, \ldots, b_r\} \setminus \{z_0\}.$$

We conclude that the a's are the same as the b's; more precisely that there exists a permutation π of $\{1, 2, \ldots, r\}$ such that $b_j = a_{\pi(j)}$ for $j = 1, 2, \ldots, r$, and so $L_j = [a_j, a_{\pi(j)}]$ for $j = \ldots, r$.

Any permutation is a product of disjoint cycles, and each of the cycles of π gives rise to a closed path of L_j's.   □

Remark 9.5. The paths γ_j, $j = 1, \ldots, n$ in Lemma 9.4 may be chosen to be simple.

This can be seen as follows: From the proof we get that there exists a permutation π of $\{1, \ldots, r\}$ such that $b_j = a_{\pi(j)}$. We choose π_0 to be such a permutation having the largest possible number of disjoint cycles in its decomposition. After relabelling the edges $L_j = [a_j, b_j]$, we may assume that

$$\pi_0 = (1 \cdots n_1)(n_1 + 1 \cdots n_2) \cdots (n_k + 1 \cdots r).$$

Suppose the path defined by the cycle $(1 \cdots n_1)$ is not simple. Then – again modulo relabelling – $a_1 = a_s$ for some $s < n_1$. But then

$$\pi' = (1 \cdots s - 1)(s \cdots n_1)(n_1 + 1 \cdots n_2) \cdots (n_k + 1 \cdots r),$$

is a new permutation satisfying that $b_j = a_{\pi'(j)}$ for all j and π' has more disjoint cycles in its decomposition than π_0.

Lemma 9.6. *Let Ω be open, let $F \subseteq \Omega$ be compact and let f be meromorphic in Ω. Suppose all the zeros and poles of f are in F. Then if the γ_j are as in Lemma 9.4 we have*

$$\frac{1}{2\pi i} \sum_j \int_{\gamma_j} \frac{f'(z)}{f(z)} \, dz = Z - P$$

where Z and P are the number of zeros and poles of f, counted with multiplicities.

Proof. The proof is similar to that of Theorem 9.2. Represent f as there with the z_i and p_j as the zeros and poles of f with the corresponding multiplicities n_i and m_j. Differentiating we find

$$\frac{f'}{f} = \frac{g'}{g} + \sum_i \frac{n_i}{z - z_i} - \sum_j \frac{m_j}{z - p_j}.$$

Lemma 9.4 applied to the holomorphic function $f = 1$ gives

$$\frac{1}{2\pi i} \sum_j \int_{\gamma_j} \frac{dz}{z - a} = 1 \qquad \text{for all } a \in F.$$

Since g'/g is holomorphic in Ω we get the desired result from Lemma 9.4.   □

The following result (Rouché's theorem) has many applications as is apparent from the exercises to this chapter. Another version of it was proved in Chapter 4 (Theorem 4.13).

Theorem 9.7 (Rouché's Theorem, version 2). *Let K be a compact subset of $\mathbb{C}$ with interior $\mathrm{int}(K)$. Let the two mappings $f, g \colon K \to \mathbb{C}_\infty$ be meromorphic in $\mathrm{int}(K)$, continuous on K, finite-valued on ∂K and there satisfying*

$$|f(z) - g(z)| < |f(z)| + |g(z)| \qquad \text{for all } z \in \partial K. \tag{9.6}$$

Let $Z(f)$ and $P(f)$ (resp. $Z(g)$ and $P(g)$) denote the number of zeros and poles of f (resp. g) in $\mathrm{int}(K)$, counted with multiplicity.

Then $Z(f)$, $P(f)$, $Z(g)$ and $P(g)$ are finite, and

$$Z(f) - P(f) = Z(g) - P(g).$$

The crucial condition (9.6) expresses the natural condition that f and g should be "close" to one another, if only on ∂K.

Proof. By (9.6) neither f nor g has zeros on ∂K. Since they are finite-valued there, we see that there is a neighborhood $U \subseteq \mathbb{C}$ of ∂K such that f and g are finite-valued on $U \cap K$ and the inequality (9.6) is still true for all $z \in U \cap K$. In particular all the poles and zeros of f and g belong to the compact subset $F := K \setminus U$ of $\mathrm{int}(K)$. Hence their numbers are finite (Observation 9.1 above).

Let us for $\lambda \in [0, 1]$ consider the meromorphic function $h_\lambda := \lambda f + (1-\lambda)g$, which has all its poles and zeros off U, i.e., in F. By Lemma 9.6 we see that the quantity

$$\frac{1}{2\pi i} \sum_j \int_{\gamma_j} \frac{h_\lambda'(z)}{h_\lambda(z)}\, dz$$

is integer for each $\lambda \in [0, 1]$. Being continuous as a function of λ it takes the same value at $\lambda = 0$ and at $\lambda = 1$. □

A special case of Rouché's theorem is the following:

Theorem 9.8 (Rouché's Theorem, version 3). *Let f and g be holomorphic in an open set containing a compact set K. If*

$$|f(z) - g(z)| < |f(z)| + |g(z)| \qquad \text{for all } z \in \partial K$$

then f and g have the same (finite) number of zeros in K.

9.3   Runge's theorems

We know that a function which is holomorphic in a disc, can be expanded
in a power series, and so be approximated uniformly by polynomials on
any compact subdisc. If a function is holomorphic in an annulus, say in
$B(0,1) \setminus \{0\}$, then it cannot necessarily be approximated by polynomials.
The obvious counter example is $z \mapsto z^{-1}$. But we can in an annulus resort
to a Laurent expansion and hence approximate it by rational functions.
Runge's theorems deal with approximation of holomorphic functions by
polynomials or rational functions on more general sets than discs and annuli.

We remind the reader that we have introduced the concept of a pole at ∞
(Definition 8.4). To set the stage for Runge's theorem on approximation by
rational functions we note that a polynomial (non-constant to be pedantic)
is a rational function with a pole at ∞. And that a rational function whose
only pole is at ∞, is a polynomial.

**Theorem 9.9 (Runge's Theorem on Approximation by Rational
Functions).** *Let F be a compact subset of $\mathbb{C}$. Let $S \subseteq \mathbb{C}_\infty$ be a set which
meets each connected component of $\mathbb{C}_\infty \setminus F$. Let f be holomorphic in an
open set containing F.*

*Then to each $\varepsilon > 0$ there exists a rational function P/Q with poles only
in S such that*

$$\left| f(z) - \frac{P(z)}{Q(z)} \right| < \varepsilon \qquad \text{for all } z \in F.$$

Proof. In the proof it will be convenient to interpret expressions of the form
$P(\frac{1}{z-\infty})$, where P is a polynomial, as polynomials in z.

Letting Ω denote the open set in question there is according to Lemma 9.4
a finite set $\gamma_1, \gamma_2, \ldots, \gamma_n$ of oriented segments in $\Omega \setminus F$ such that

$$f(z) = \frac{1}{2\pi i} \sum_{j=1}^n \int_{\gamma_j} \frac{f(w)}{w - z}\, dw \qquad \text{for all } z \in F, \tag{9.7}$$

so the theorem will be proved once we show that each summand on the
right hand side of (9.7) can be approximated uniformly on F by finite sums
of the type $P(\frac{1}{z-c})$, where the P's are polynomials and the c's belong to S.
Let us just treat the first summand; the other terms may be handled in
exactly the same way.

The line integral

$$\int_{\gamma_1} \frac{f(w)}{w - z}\, dw$$

can be uniformly approximated by Riemann sums

$$\sum_j \frac{A_j}{b_j - z} \qquad \text{where } b_j \in \gamma^*.$$

Again let us just treat the first term $(b_1 - z)^{-1}$, the rest being handled similarly.

We drop the subscript and write b for b_1. γ_1 does not meet F so $b \notin F$. Let $c \in S$ be in the same connected component of $\mathbb{C}_\infty \setminus F$ as b, and assume first that $c \neq \infty$. Join b and c by a curve contained entirely in this component. Since this is disjoint from F, there is a positive distance between this curve and F. We can find points $b = c_0, c_1, \ldots, c_m = c$ on the curve such that

$$2|c_i - c_{i+1}| < \text{dist}(\text{curve}, F). \tag{9.8}$$

Then writing $z - b = z - c_1 + c_1 - b$ we get

$$\frac{1}{z - b} = (z - c_1)^{-1} \sum_{k=0}^\infty (b - c_1)^k (z - c_1)^{-k}$$

the series converging uniformly for $z \in F$ by (9.8). So we may choose N so that

$$\left| (z - b)^{-1} - \sum_{k=0}^N (b - c_1)^k (z - c_1)^{-k-1} \right| \tag{9.9}$$

is uniformly small for $z \in F$. We can use the same argument on each term of the sum in (9.9) to replace c_1 by c_2. Namely

$$(z - c_1)^{-k} = (z - c_2 + c_2 - c_1)^{-k}$$

which can be expanded in a series that – because of (9.8) – converges uniformly for $z \in F$. In other words we can find a polynomial in $(z - c_2)^{-1}$ which approximates $(z - b)^{-1}$ uniformly on F. This procedure can be repeated until we reach $c = c_m$.

The argument is thus complete except when the point c chosen in S is ∞. To treat this case consider $(z - b)^{-1}$ where b belongs to the unbounded component of $\mathbb{C} \setminus F$. Choose d in the same component such that $|\frac{z}{d}| < \frac{1}{2}$ for all $z \in F$. From what we have done above there is a polynomial P such that $P(\frac{1}{z-d})$ approximates $(z - b)^{-1}$ uniformly on F. Since $|\frac{z}{d}| < \frac{1}{2}$ each $(z - d)^{-1}$ occurring in the polynomial can be expanded in powers of z/d, i.e., there is a polynomial which approximates $(z - b)^{-1}$ uniformly on F.

The argument is thus complete. $\qquad \square$

For the next result (Runge's polynomial approximation theorem) and for the discussion of the inhomogeneous Cauchy-Riemann equation we need a fact from point set topology:

Proposition 9.10. *Let Ω be an open subset of the complex plane. Then there exists an increasing sequence $K_1, K_2, \ldots$ of compact subsets of Ω such that*

(a) $K_1 \cup K_2 \cup \cdots = \Omega$.

(b) $K_n \subseteq \operatorname{int} K_{n+1}$ *for all* $n = 1, 2, \ldots$

(c) *Every compact subset of Ω is contained in K_n for some n.*

(d) *Every connected component of $\mathbb{C}_\infty \setminus K_n$ contains a connected component of $\mathbb{C}_\infty \setminus \Omega$ for $n = 1, 2, \ldots$*

Statement (d) says that K_n has no holes except the ones coming from the holes of Ω (make a drawing!).

Proof. The cases $\Omega = \emptyset$ and $\Omega = \mathbb{C}$ are trivial, so we shall concentrate on the remaining case in which $\partial\Omega \neq \emptyset$.

We will show that we for $n = 1, 2, \ldots$ may take

$$K_n := \{z \in \Omega \mid \operatorname{dist}(z, \partial\Omega) \geq \tfrac{1}{n}\} \cap B[0, n].$$

This K_n is a closed and bounded subset of $\mathbb{C}$, so it is compact. It is now almost obvious that (a), (b) and (c) are satisfied, so left is only point (d).

Here we begin by noting that if a connected component of $\mathbb{C}_\infty \setminus K_n$, say U, intersects a connected component of $\mathbb{C}_\infty \setminus \Omega$, say V, then U contains V: Indeed, since $\mathbb{C}_\infty \setminus K_n \supseteq \mathbb{C}_\infty \setminus \Omega$ the connected components of $\mathbb{C}_\infty \setminus K_n$ cover $\mathbb{C}_\infty \setminus \Omega$; in particular they cover V. They are open and disjoint, so by the connectedness of V we get $U \supseteq V$.

Hence (d) boils down to proving that each connected component U of $\mathbb{C}_\infty \setminus K_n$ intersects $\mathbb{C}_\infty \setminus \Omega$. To do so we notice that

$$\mathbb{C}_\infty \setminus K_n = \bigcup_{z \notin \Omega} B(z, \tfrac{1}{n}) \cup B(\infty, n)$$

Let $w \in U$. Then w lies in one of the balls, say $w \in B(z_0, \tfrac{1}{n})$ where $z_0 \notin \Omega$. The ball is a connected subset of $\mathbb{C}_\infty \setminus K_n$, so $B(z_0, \tfrac{1}{n}) \subseteq U$, U being a connected component. Now, $z_0 \in U \cap (\mathbb{C}_\infty \setminus \Omega)$, so U intersects $\mathbb{C}_\infty \setminus \Omega$. $\square$

Theorem 9.11 (Runge's Polynomial Approximation Theorem). *Let Ω be an open subset of $\mathbb{C}$. Every holomorphic function on Ω can be approximated uniformly on compacts by polynomials iff $\mathbb{C}_\infty \setminus \Omega$ is connected.*

Proof. We will first show that the condition is sufficient. Proposition 9.10 tells (c) that it suffices to approximate the given holomorphic function f on any given K_n, (d) that $\mathbb{C}_\infty \setminus K_n$ has only one connected component. That one contains $\mathbb{C}_\infty \setminus \Omega$ and so it contains ∞. Take now $S = \{\infty\}$ and $F = K_n$ in Runge's theorem on approximation by rational functions (Theorem 9.9) to get polynomials to approximate f uniformly on K_n.

Now for the necessity: Suppose that $\mathbb{C}_\infty \setminus \Omega$ is not connected. Then there exist two closed, disjoint, non-empty sets F and F_1 with union $\mathbb{C}_\infty \setminus \Omega$. We may suppose that $\infty \in F_1$ so that F is a compact subset of $\mathbb{C}$. Clearly

$$\mathbb{C}_\infty \setminus F_1 = \Omega \cup F$$

is open. Put $O := \Omega \cup F$. F is a compact subset of the open set O. Let $\gamma_1, \gamma_2, \ldots, \gamma_n$ be paths in $O \setminus F = \Omega$ as in Lemma 9.4. Let $a \in F$. The function $z \to (z-a)^{-1}$ is holomorphic on Ω.

Assume now that all holomorphic functions on Ω are uniform limits on compacta of polynomials. Then we can in particular approximate the function $z \to (z-a)^{-1}$ by a sequence $\{P_k\}$ of polynomials on the union of the paths $\gamma_1, \gamma_2, \ldots, \gamma_n$. These polynomials are of course holomorphic in O and hence by the first part of the conclusion of Lemma 9.4 we have

$$\sum_j \int_{\gamma_j} P_k(z)\, dz = 0$$

which in the limit as $k \to \infty$ becomes

$$\sum_j \int_{\gamma_j} \frac{1}{z-a}\, dz = 0$$

But that contradicts the second part of Lemma 9.4. $\qquad\square$

Corollary 9.12. *If Ω be an open subset of $\mathbb{C}$, then $\mathbb{C}_\infty \setminus \Omega$ is connected iff Ω is simply connected.*

Proof. Suppose first that $\mathbb{C}_\infty \setminus \Omega$ is connected. Any function f which is holomorphic on Ω may by Runge's polynomial approximation theorem be approximated uniformly by polynomials on compact subsets of Ω. So if γ is a closed path in Ω we conclude that $\int_\gamma f\, dz = 0$ which implies that f

has a primitive (use the proof of Lemma 2.13) and hence that Ω is simply connected (Theorem 8.22).

Assume conversely that $\mathbb{C}_\infty \setminus \Omega$ is not connected. We then want to show that Ω is not simply connected. We assume it is and arrive at a contradiction as follows: Let F and F_1 be closed, disjoint, non-empty sets with union $\mathbb{C}_\infty \setminus \Omega$. Without loss of generality we may suppose that $\infty \in F_1$ so that F is a compact subset of $\mathbb{C}$. Clearly

$$\mathbb{C}_\infty \setminus F_1 = \Omega \cup F$$

is open. Put $O := \Omega \cup F$. F is a compact subset of the open set O. Let $\gamma_1, \gamma_2, \ldots, \gamma_n$ be the closed paths in $O \setminus F = \Omega$ from Lemma 9.4. Let $a \in F$. Since $z \mapsto (z - a)^{-1}$ is holomorphic in Ω we get according to the Global Cauchy Theorem (Theorem 5.7) that

$$\sum_j \int_{\gamma_j} \frac{1}{z - a}\, dz = 0$$

which contradicts the second part of Lemma 9.4. $\qquad\qquad\square$

Remark 9.13. Let K be a compact set. Then the unbounded component of $\mathbb{C} \setminus K$ is precisely those $a \in \mathbb{C}$ for which the function $z \mapsto (z - a)^{-1}$ is a uniform limit on K of polynomials (this follows from the proof of Theorem 9.9).

Runge's theorem on polynomial approximation is not the last word in these matters. Clearly, if a function f on a compact set K is the uniform limit of a sequence of polynomials then $f \in C(K) \cap \mathrm{Hol}(\mathrm{int}(K))$, so this is a necessary condition for approximation by polynomials. Mergelyan's theorem states that it is also sufficient:

Theorem 9.14 (Mergelyan's Theorem, 1952). *Let K be a compact subset of $\mathbb{C}$ with $\mathbb{C} \setminus K$ connected. Let $f \in C(K) \cap \mathrm{Hol}(\mathrm{int}(K))$*

Then there exists a sequence of polynomials converging to f uniformly on K.

Note that when K has empty interior then the only condition remaining on f is that $f \in C(K)$. So the classical approximation theorem of Weierstrass for an interval is a particular case of Mergelyan's theorem. For a comprehensible proof of Mergelyan's theorem the reader can consult Chapter 20 of Rudin (1974).

Another extension of Weierstrass' approximation theorem is provided by

Theorem 9.15 (Carleman's Theorem, 1927). *If $f \in C(\mathbb{R})$ then there exists to every continuous function $\varepsilon \colon \mathbb{R} \to \,]0, \infty[$ an entire function F with the property that*

$$|f(x) - F(x)| < \varepsilon(x) \qquad \text{for all } x \in \mathbb{R}.$$

A proof can be found in Burckel (1979, Chap. VIII, Section 5). For further generalizations (Arakelian's theorem, Keldysh-Lavrent'ev's theorem) see the references in the notes to Chapter VIII of Burckel (1979), the note Gauthier (1985), Rosay and Rudin (1989) and the survey article Vitushkin (1987).

9.4 The inhomogeneous Cauchy-Riemann equation

In this section we shall use Runge's theorem to show that the inhomogeneous Cauchy-Riemann equation has a solution on any open subset Ω of $\mathbb{C}$. We recall from Chapter 2 that the Cauchy-Riemann operator is the first order differential operator

$$\frac{\partial}{\partial \overline{z}} := \frac{1}{2}\left(\frac{\partial}{\partial x} + i\frac{\partial}{\partial y}\right) \qquad \text{on } \mathbb{C} = \mathbb{R}^2$$

and that a function $u \in C^1(\Omega)$ is holomorphic on Ω iff it satisfies the homogeneous Cauchy-Riemann equation

$$\frac{\partial u}{\partial \overline{z}} = 0$$

on Ω. In the case of $\Omega = \mathbb{C}$ an explicit formula for a solution can be displayed:

Proposition 9.16. *If $f \in C_0^\infty(\mathbb{R}^2)$ and*

$$u(z) := \frac{1}{\pi} \int_{\mathbb{R}^2} \frac{f(z - w)}{w}\, d\mu(w) \qquad \text{for } z \in \mathbb{C}$$

where $d\mu$ denotes the Lebesgue measure on $\mathbb{R}^2$, then $u \in C^\infty(\mathbb{R}^2)$ and $\partial u/\partial \overline{z} = f$.

Although the function $w \mapsto w^{-1}$ has a pole at $w = 0$ it is nevertheless integrable over any ball, so that the integral in the proposition makes sense:

Indeed, using polar coordinates we get

$$\int_{B(0,1)} \frac{d\mu(w)}{|w|} = \int_0^1 \int_0^{2\pi} \frac{1}{r} r \, d\phi dr = 2\pi < \infty.$$

For any constant coefficient differential operator $D \neq 0$ on $\mathbb{R}^n$ the equation $Du = f \in C_0^\infty(\mathbb{R}^n)$ has a solution $u \in C^\infty(\mathbb{R}^n)$: That is known from the theory of partial differential equations. We may for example choose $u = f * E$, where E is a fundamental solution of D. A fundamental solution of the Cauchy-Riemann operator $\partial/\partial \bar{z}$ is $E(z) = 1/(\pi z)$; that produces the formula of the proposition. We will, however, not appeal to the theory of partial differential equations, but prove directly that the formula of the proposition works.

Proof of Proposition 9.16. Since the function $w \mapsto w^{-1}$ is locally integrable we see that u is continuous and that we may differentiate u by doing so under the integral sign. So for any differential operator D with constant coefficients we get

$$(Du)(z) = \frac{1}{\pi} \int_{\mathbb{R}^2} \frac{(Df)(z-w)}{w} \, d\mu(w)$$

which shows that $u \in C^\infty(\mathbb{R}^2)$.

To prove the other statement we fix $z \in \mathbb{C}$. For simplicity in writing we delete the subscript $\mathbb{R}^2$ on the integrals and put $F(w) := f(z_0 - w)$. Now,

$$\frac{\partial u}{\partial \bar{z}}(z_0) = \frac{1}{\pi} \int \frac{\partial f}{\partial \bar{z}}(z_0 - w) \frac{d\mu(w)}{w}$$

$$= -\frac{1}{\pi} \int \frac{\partial}{\partial \bar{w}} \{f(z_0 - w)\} \frac{d\mu(w)}{w} = -\frac{1}{\pi} \int \frac{\partial F}{\partial \bar{w}}(w) \frac{d\mu(w)}{w}.$$

Using that $\partial/\partial \bar{w}$ is of first order (so that the usual formula for differentiating a product holds) and that $w \mapsto w^{-1}$ is holomorphic, we get

$$\frac{\partial}{\partial \bar{w}} \left\{ \frac{F(w)}{w} \right\} = \frac{\partial F}{\partial \bar{w}}(w) \frac{1}{w} + F(w) \frac{\partial(w^{-1})}{\partial \bar{w}} = \frac{\partial F}{\partial \bar{w}}(w) \frac{1}{w}$$

so

$$\frac{\partial u}{\partial \bar{z}}(z_0) = -\frac{1}{\pi} \int \frac{\partial}{\partial \bar{w}} \left\{ \frac{F(w)}{w} \right\} d\mu(w) = -\frac{1}{\pi} \lim_{\varepsilon \to 0} \int_{|w| > \varepsilon} \frac{\partial G}{\partial \bar{w}}(w) \, d\mu(w)$$

where we, again for the sake of simplicity, use the notation $G(w) := F(w)/w$. Introducing the definition of the Cauchy-Riemann operator we find by a small computation that

$$2\frac{\partial G}{\partial \overline{w}} = \text{div}(G, iG)$$

so Green's theorem for the annulus $\varepsilon < |w| < R$, where R is chosen so large that G vanishes out there, implies that

$$\frac{\partial u}{\partial \overline{z}} = \frac{1}{2\pi} \lim_{\varepsilon \to 0} \int_{|w|=\varepsilon} (G(w), iG(w)) \cdot \left(\frac{\text{Re}\, w}{|w|}, \frac{\text{Im}\, w}{|w|}\right) d\lambda(w)$$

where $d\lambda$ denotes the usual arc length measure on the circle $|w| = \varepsilon$. Now,

$$\begin{aligned}
\frac{\partial u}{\partial \overline{z}} &= \frac{1}{2\pi} \lim_{\varepsilon \to 0} \left\{ \frac{1}{\varepsilon} \int_{|w|=\varepsilon} [G(w)\,\text{Re}\,w + iG(w)\,\text{Im}\,w]\, d\lambda(w) \right\} \\
&= \frac{1}{2\pi} \lim_{\varepsilon \to 0} \left\{ \frac{1}{\varepsilon} \int_{|w|=\varepsilon} G(w)w\, d\lambda(w) \right\} \\
&= \frac{1}{2\pi} \lim_{\varepsilon \to 0} \left\{ \frac{1}{\varepsilon} \int_{|w|=\varepsilon} F(w)\, d\lambda(w) \right\} \\
&= \lim_{\varepsilon \to 0} \left\{ \frac{1}{2\varepsilon\pi} \int_{|w|=\varepsilon} [F(w) - F(0)]\, d\lambda(w) + F(0) \right\} \\
&= 0 + F(0) = F(0) = f(z_0).
\end{aligned}$$

$\square$

Theorem 9.17. *Let Ω be an open subset of the complex plane. The inhomogeneous Cauchy-Riemann equation $\partial u/\partial \overline{z} = f$ has a solution $u \in C^\infty(\Omega)$ for any $f \in C^\infty(\Omega)$.*

Proof. Let $K_0 = \emptyset \subseteq K_1 \subseteq K_2 \subseteq \cdots$ be an increasing sequence of compact subsets of Ω as in Proposition 9.10.

Note: Any function which is holomorphic on an open set containing K_n, can be approximated uniformly on K_n by elements from $\text{Hol}(\Omega)$. (Combine Runge's theorem (Theorem 9.9) with the property (d) of Proposition 9.10.)

Choose $\phi_1, \phi_2, \ldots \in C_0^\infty(\Omega)$ such that $\phi_n = 1$ on K_n.

By Proposition 9.16 there exists $u_1 \in C^\infty(\Omega)$ (even in $C^\infty(\mathbb{R}^2)$) such that $\partial u_1/\partial \overline{z} = \phi_1 f$, and $u_2 \in C^\infty(\Omega)$ such that $\partial u_2/\partial \overline{z} = \phi_2 f$. Since $\phi_1 = \phi_2(= 1)$ on K_1 we get

$$\frac{\partial(u_2 - u_1)}{\partial \overline{z}} = 0 \qquad \text{on } K_1$$

so $u_2 - u_1$ is holomorphic on $\text{int}(K_1)$. Trivially

$$|u_2 - u_1| < \tfrac{1}{2} \qquad \text{on } K_0$$

because $K_0 = \emptyset$.

Assume now that we have produced $u_1, u_2, \ldots, u_n \in C^\infty(\Omega)$ such that

$$\frac{\partial u_j}{\partial \bar{z}} = \phi_j f,$$

$$u_{j+1} - u_j \in \text{Hol}(\text{int } K_j)$$

and

$$|u_{j+1} - u_j| < \tfrac{1}{2^j} \qquad \text{on } K_{j-1} \text{ for } j = 1, 2, \ldots, n-1.$$

We will extend the system $\{u_1, u_2, \ldots, u_n\}$ by adding one function more: Choose $v \in C^\infty(\Omega)$ such that

$$\frac{\partial v}{\partial \bar{z}} = \phi_{n+1} f.$$

Since $\phi_n = \phi_{n+1}$ on K_n we get

$$\frac{\partial (v - u_n)}{\partial \bar{z}} = 0 \quad \text{and so} \quad v - u_n \in \text{Hol}(\text{int } K_n)$$

By the note we may find a $w \in \text{Hol}(\Omega)$ such that $|v - u_n - w| < 2^{-n}$ on K_{n-1}. We add $u_{n+1} := v - w$ to the system.

In this way we produce a sequence $u_1, u_2, \ldots$ from $C^\infty(\Omega)$ such that

$$\frac{\partial u_n}{\partial \bar{z}} = \phi_n f,$$

$$u_{n+1} - u_n \in \text{Hol}(\text{int } K_n)$$

and

$$|u_{n+1} - u_n| < \tfrac{1}{2^n} \qquad \text{on } K_{n-1}.$$

Any compact subset of Ω is contained in some K_N (by property (c) of Proposition 9.10), so the sum

$$\sum_{n=1}^{\infty} (u_{n+1} - u_n)$$

converges uniformly on any compact subset of Ω. Thus

$$u := \sum_{n=1}^{\infty} (u_{n+1} - u_n) + u_1$$

is a well defined function on Ω. For any fixed N we have $u_{n+1} - u_n \in$ Hol(int K_N) whenever $n \geq N$, so the right hand side of

$$u - u_N = \sum_{n=N}^{\infty} (u_{n+1} - u_n) \tag{9.10}$$

is on int K_N the limit of a uniformly convergent sequence of holomorphic functions, so it is itself holomorphic on int K_N (Weierstrass' theorem). In particular $u \in C^\infty(\text{int } K_N)$, and since N is arbitrary, $u \in C^\infty(\Omega)$.

When we apply $\partial/\partial\bar{z}$ to (9.10) on int K_N the right hand side vanishes (it is holomorphic), so

$$\frac{\partial(u - u_N)}{\partial\bar{z}} = 0 \qquad \text{on int } K_N$$

so

$$\frac{\partial u}{\partial\bar{z}} = \frac{\partial u_N}{\partial\bar{z}} = \phi_N f = f \qquad \text{on int } K_N.$$

Since N is arbitrary we read that $\partial u/\partial\bar{z} = f$ throughout Ω. $\qquad\square$

Exercises

9.1. Generalize the Argument Principle as follows: Replace the condition about simple connectivity by the condition that γ is a closed path in Ω with the property that

$$\text{Ind}_\gamma(z) = 0 \qquad \text{for all } z \in \mathbb{C} \setminus \Omega.$$

9.2. Show that the polynomial $z \mapsto z^4 + z^3 + 5z^2 + 2z + 4$ has no zeros in the first quadrant.

9.3. Let f be meromorphic in a neighborhood of $B[0, 1]$, and assume that there are no poles on the unit circle $|z| = 1$. Let $a \in \mathbb{C}$ satisfy

$$|a| > \sup_{|z|=1} |f(z)|.$$

Show that f has the same number of poles and a-values (both counted with multiplicity) in $B(0, 1)$.

9.4. Consider the holomorphic function $f(z) := \sinh(\pi z)$ in the half strip $\{x + iy \mid x > 0, -\frac{1}{2} < y < \frac{1}{2}\}$. Show that f takes each value in the upper half plane $\{w \in \mathbb{C} \mid \text{Im } z > 0\}$ exactly once.

9.5. Find the number of zeros of the polynomial $z \mapsto z^7 - 5z^4 + z^2 - 2$ in the annulus $\{z \in \mathbb{C} \mid 1 < |z| < 2\}$.

9.6. Show that the polynomial

$$z \mapsto 3z^{15} + 4z^8 + 6z^5 + 19z^4 + 3z + 1$$

has 4 zeros in $|z| < 1$ and 11 zeros in $1 < |z| < 2$.

9.7. Find a ball $B(0, R)$ such that the polynomial $z^3 - 4z^2 + z - 4$ has exactly two roots in it.

9.8. Let $a \in \mathbb{C}$, $|a| > e$ and let $n \in \{1, 2, 3, \ldots\}$. Show that the function $f(z) := e^z - az^n$ has exactly n zeros in $B(0, 1)$, and that they are different from one another.

9.9. Let $\lambda > 1$. Show that the equation $ze^{\lambda - z} = 1$ has exactly one root z in $B[0, 1]$, and that the root belongs to the open interval $]0, 1[$.

9.10. Let $\lambda > 1$. Show that the equation $\lambda - z - e^{-z} = 0$ has exactly one root z_0 in the half plane $\{z \in \mathbb{C} \mid \operatorname{Re} z \geq 0\}$, and that $z_0 > 0$. What happens as $\lambda \to 1$?

9.11. Show that all four roots of the polynomial $z^4 - z + 5$ lie in the annulus $\{z \in \mathbb{C} \mid 1.35 < |z| < 1.65\}$, and that there is exactly one root in each open quadrant.

9.12. Consider the polynomial $P(z) = a_0 + a_1 z + \cdots + a_n z^n$ where $0 < a_0 < a_1 < \cdots < a_n$. Show that P has n zeros in $B(0, 1)$.
Hint: Consider $(1 - z)P(z)$.

9.13. Use Rouché's theorem to prove that an n'th order polynomial has exactly n roots.

9.14. Use Rouché's theorem to prove that the positions of the roots of an n'th order polynomial depend continuously on the coefficients of the polynomial.

9.15. Prove the following lemma:

Lemma. *Suppose $a_1, a_2, \ldots, a_n$ are holomorphic functions on the disc $B(0, R)$, and suppose that $w_0 \in \mathbb{C}$ is a simple root of the polynomial*

$$w \mapsto w^n + a_1(0)w^{n-1} + \cdots + a_n(0)$$

Then there exist an $r \in {]0, R[}$, and a $\phi \in \mathrm{Hol}(B(0,r))$ such that $\phi(0) = w_0$ and

$$(\phi(z))^n + a_1(z)(\phi(z))^{n-1} + \cdots + a_n(z) = 0 \qquad \text{for all } z \in B(0,r).$$

9.16. Let f be analytic on a neighborhood of $B[0,1]$ and assume that $f(B[0,1]) \subseteq B(0,1)$. Show that f has exactly one fixed point.

9.17. Prove the following theorem:

Theorem. *Let $f_1, f_2, \ldots$ be a sequence of functions which are holomorphic on an open set Ω. Assume that $f_n \to f_0$ locally uniformly on Ω as $n \to \infty$. Assume furthermore that f_0 has a zero of order n_0 at $z_0 \in \Omega$.*

Then to every $r > 0$ there exists an $N = N(r)$ such that $n > N \Rightarrow$ the function f_n has at least n_0 zeros in the set $\Omega \cap B(z_0, r)$.

9.18. Let Ω be a connected open subset of $\mathbb{C}$. Assume that $f_n \in \mathrm{Hol}(\Omega)$ are univalent for $n = 1, 2, \ldots$ and that $f_n \to f$ uniformly on compacta as $n \to \infty$.

Show that f is either univalent or constant.

9.19. Show that the open set $\mathbb{C} \setminus S$ is simply connected, where S is the infinite spiral (Archimedes' spiral)

$$S := \{z = re^{ir} \mid 0 \le r < \infty\}.$$

9.20. Let Ω be an open connected non-empty subset of $\mathbb{C}$. Show that the following 8 statements are equivalent.

(α) Ω is simply connected.

(β) $\mathrm{Ind}_\gamma(z) = 0$ for all closed curves γ in Ω and all $z \in \mathbb{C} \setminus \Omega$.

(γ) $\mathbb{C}_\infty \setminus \Omega$ is connected.

(δ) $\int_\gamma f = 0$ for each $f \in \mathrm{Hol}(\Omega)$ and each closed path γ in Ω.

(ε) Every $f \in \mathrm{Hol}(\Omega)$ has a primitive.

(ζ) Every never-vanishing $f \in \mathrm{Hol}(\Omega)$ has a continuous logarithm.

(η) Every never-vanishing $f \in \mathrm{Hol}(\Omega)$ has a continuous square root.

(θ) Ω is homeomorphic to $B(0,1)$.

9.21. Prove the following theorem:

> **Theorem.** *Let K be a compact subset of $\mathbb{C}$ with the property that $\mathbb{C} \setminus K$ is connected. Let f be holomorphic in an open neighborhood of K. Then f is on K a uniform limit of polynomials.*

9.22. Let

$$\Omega := B(0,1) \setminus B[\tfrac{1}{2}, \tfrac{1}{2}]$$

and let $f \in \mathrm{Hol}(\Omega)$.

(α) Does there exist a sequence of polynomials converging uniformly to f on Ω ?

(β) Given a compact subset K of Ω, does there always exist a sequence of polynomials which converges uniformly to f on K?

(γ) Does there always exist a sequence of polynomials which converges uniformly to f on Ω, if f is even holomorphic on an open set containing the closure of Ω?

9.23. Does there exist a sequence $\{P_n\}$ of polynomials such that for all $z \in \mathbb{C}$ we have as $n \to \infty$:

$$P_n(z) \to \begin{cases} 1 & \text{if } \mathrm{Im}\, z > 0, \\ 0 & \text{if } \mathrm{Im}\, z = 0, \\ 1 & \text{if } \mathrm{Im}\, z < 0. \end{cases}$$

Chapter 10

Representations of Meromorphic Functions

Infinite products and construction of meromorphic functions with prescribed zeros and poles are the main themes of this chapter. The Euler product for sine, the partial fraction expansion of cotangent and the Γ-function will be the showpieces.

10.1 Infinite products

It is easy to construct a holomorphic function which has zeros exactly at finitely many given points $z_1, z_2, \ldots, z_N \in \mathbb{C}$. Simply take the function

$$z \mapsto (z - z_1)(z - z_2) \cdots (z - z_N) = \prod_{k=1}^{N} (z - z_k).$$

With an infinity of points $z_1, z_2, \ldots$ one is tempted to take $\prod_{k=1}^{\infty}(z - z_k)$. Although this product does not converge, a suitable modification (where we insert suitable convergence factors) does, as we shall see in Weierstrass' factorization theorem. So we are led to study infinite products.

An infinite sum behaves in most respects like a finite one. Analogously an infinite product should behave like a finite product; in particular $\prod_{k=1}^{\infty} c_k = 0$ should mean that at least one of the factors c_k vanishes. However, examples like

$$\lim_{n \to \infty} \underbrace{\frac{1}{2} \frac{1}{2} \cdots \frac{1}{2}}_{n \text{ factors}} = 0$$

demonstrate that we must be cautious, at least if the limit is zero. That is the background for the following rather involved definition of an infinite product.

Definition 10.1. *Let $c_1, c_2, \ldots$ be a sequence of complex numbers. The infinite product*

$$\prod_{k=1}^{\infty} c_k \tag{10.1}$$

*is said to **converge** if there to each $\varepsilon > 0$ exists an $N \in \mathbb{N}$ such that for all $n \geq N$ and $p \geq 1$:*

$$\left| \prod_{k=n+1}^{n+p} c_k - 1 \right| < \varepsilon. \tag{10.2}$$

If the infinite product (10.1) converges, then the numbers $C_n := \prod_{k=1}^{n} c_k$, $n = 1, 2, \ldots$ form a bounded sequence. Using (10.2) we find that $\{C_n\}$ is a Cauchy sequence and therefore has a limit in $\mathbb{C}$. We use the notation $\prod_{k=1}^{\infty} c_k$ also for this limit and say that it is the *value* of the infinite product (10.1).

If a finite number of factors is inserted in or removed from a convergent infinite product then the result will still be a convergent infinite product (and the value of the result will be the obvious one).

Taking $p = 1$ in (10.2) we see that $c_n \to 1$ as $n \to \infty$ if the product (10.1) converges. That is the reason why it is often convenient to write an infinite product in the form $\prod_{k=1}^{\infty} (1 + a_k)$.

Also note that if $\prod_{k=1}^{\infty} c_k$ is a convergent product, then $\prod_{k=1}^{\infty} c_k = 0$ if and only if at least one of the factors c_k is zero.

We leave it to the reader to verify the following two quite elementary properties of infinite products:

Properties. If $\prod_{k=1}^{\infty} c_k$ and $\prod_{k=1}^{\infty} d_k$ are both convergent. Then

$$\prod_{k=1}^{\infty} c_k d_k$$

converges and

$$\prod_{k=1}^{\infty} c_k d_k = \left(\prod_{k=1}^{\infty} c_k \right) \left(\prod_{k=1}^{\infty} d_k \right).$$

If the infinite product $\prod_{k=1}^{\infty} c_k$ is convergent and the limit $\prod_{k=1}^{\infty} c_k \neq 0$, then

$$\prod_{k=1}^{\infty} \frac{1}{c_k}$$

converges and

$$\prod_{k=1}^{\infty} \frac{1}{c_k} = \frac{1}{\prod_{k=1}^{\infty} c_k}.$$

The following result turns out to be very useful, because it connects infinite products with sums.

Proposition 10.2. *The infinite product* $\prod_{k=1}^{\infty}(1 + a_k)$ *converges if* $\sum_{k=1}^{\infty}|a_k| < \infty$, *and in that case*

$$\left|\prod_{k=1}^{\infty}(1 + a_k) - 1\right| \leq \exp\left(\sum_{k=1}^{\infty}|a_k|\right) - 1. \tag{10.3}$$

Proof. We first show the inequality

$$\left|\prod_{k=1}^{n}(1 + a_k) - 1\right| \leq \prod_{k=1}^{n}(1 + |a_k|) - 1. \tag{10.4}$$

The finite product minus 1, i.e., $\prod_{k=1}^{n}(1 + a_k) - 1$ is a polynomial in $a_1, a_2, \ldots, a_n$ with positive coefficients, say

$$p(a_1, a_2, \ldots, a_n) = \prod_{k=1}^{n}(1 + a_k) - 1.$$

p has positive coefficients so

$$|p(a_1, a_2, \ldots, a_n)| \leq p(|a_1|, |a_2|, \ldots, |a_n|)$$

which is the inequality (10.4).

Noting that $1 + x \leq e^x$ for any real x we put $x = |a_k|$ in (10.4) and deduce the finite version (10.3') of (10.3):

$$\left|\prod_{k=1}^{n}(1 + a_k) - 1\right| \leq \exp\left(\sum_{k=1}^{n}|a_k|\right) - 1. \tag{10.3'}$$

Coming now to the proof of the proposition we get from the inequality (10.3') applied to $a_{n+1}, a_{n+2}, \ldots, a_{n+p}$ that

$$\left|\prod_{k=n+1}^{n+p}(1 + a_k) - 1\right| \leq \exp\left(\sum_{k=n+1}^{n+p}|a_k|\right) - 1.$$

If $\sum_{k=1}^{\infty}|a_k| < \infty$ then the right hand side is small, independent of $p \geq 1$ if only n is large. $\qquad\square$

We next turn to infinite products where the factors are analytic functions. First a natural definition.

Definition 10.3. *Let $a_1, a_2, \ldots$ be complex valued functions defined in an open subset Ω of $\mathbb{C}$. We say that the infinite product $\prod_{k=1}^{\infty}(1+a_k)$ **converges locally uniformly in** Ω, if $\prod_{k=1}^{\infty}(1+a_k(z))$ converges at each $z \in \Omega$ and if furthermore to each compact subset K of Ω and each $\varepsilon > 0$ there exists an N such that for all $z \in K$ and $n \geq N$*

$$\left| \prod_{k=1}^{\infty}(1+a_k(z)) - \prod_{k=1}^{n}(1+a_k(z)) \right| < \varepsilon.$$

Lemma 10.4. *Let $a_1, a_2, \ldots$ be holomorphic on an open subset Ω of $\mathbb{C}$.*

(a) *If the infinite product $\prod_{k=1}^{\infty}(1+a_k)$ converges locally uniformly in Ω, then the limit function $F(z) := \prod_{k=1}^{\infty}(1+a_k(z))$ is holomorphic on Ω.*

   F vanishes precisely at those points where at least one of the factors $1+a_k$ vanishes. In particular F doesn't vanish identically on a connected open set unless at least one of the factors does so.

(b) *If as $N \to \infty$ the sum $\sum_{n=N}^{\infty}|a_n(z)|$ converges locally uniformly to 0, then the infinite product $\prod_{k=1}^{\infty}(1+a_k)$ converges locally uniformly in Ω.*

Proof. The last statement (b) of the lemma is a consequence of Proposition 10.2 and the inequality (10.3).

It has already been observed that a convergent infinite product vanishes at a point iff at least one of its factors does so. If F vanishes identically on a connected open subset Ω_0 of Ω then at least one of the factors vanishes at uncountably many points of Ω_0. In that case the zeros of that factor has a limit point in Ω_0. By the Unique Continuation Theorem (Theorem 4.11) the factor is identically 0 on Ω_0. $\qquad\square$

Definition 10.5. *If F is holomorphic then F'/F is called the **logarithmic derivative** of F.*

Placing ourselves in the situation of Lemma 10.4 where $F(z) = \prod_{k=1}^{\infty}(1+a_k(z))$ we compute formally to get $\log F(z) = \sum_{k=1}^{\infty}\log(1+a_k(z))$ which by formal differentiation yields

$$\frac{F'(z)}{F(z)} = \sum_{k=1}^{\infty}\frac{a_k'(z)}{1+a_k(z)} \qquad \text{for } z \in \Omega. \tag{10.5}$$

Even though it would be tricky to justify this way of deriving the formula (10.5) for the logarithmic derivative of an infinite product, the formula is true, and it is possible to prove it directly. That is done in the next proposition.

Proposition 10.6. *Let $a_1, a_2, \ldots$ be holomorphic in an open subset Ω of $\mathbb{C}$ and suppose that the sum $\sum_{n=N}^{\infty} |a_n(z)|$ as $N \to \infty$ converges locally uniformly in Ω to 0. Let F denote the value of the infinite product $F(z) = \prod_{k=1}^{\infty}(1 + a_k(z))$ and put $\Omega_0 = \{z \in \Omega \mid F(z) \neq 0\}$.*
   Then

$$\sum_{k=1}^{\infty} \frac{a_k'(z)}{1 + a_k(z)}$$

converges absolutely for any $z \in \Omega_0$, and it converges locally uniformly on Ω_0 towards the meromorphic function F'/F, so

$$\frac{F'(z)}{F(z)} = \sum_{k=1}^{\infty} \frac{a_k'(z)}{1 + a_k(z)} \qquad \text{for } z \in \Omega_0. \tag{10.6}$$

Proof. The finite product $F_N(z) = \prod_{k=1}^{N}(1 + a_k(z))$ converges as $N \to \infty$ locally uniformly on Ω to F (Lemma 10.4). By Weierstrass' theorem (Theorem 4.10) the sequence of derivatives $\{F_N'\}$ converges locally uniformly on Ω to F'.

Let K be a compact subset of Ω_0. Then F_N'/F_N converges uniformly on K to F'/F. Since

$$\frac{F_N'(z)}{F_N(z)} = \sum_{k=1}^{N} \frac{a_k'(z)}{1 + a_k(z)} \qquad \text{for } z \in \Omega_0$$

we have only left to show the claim on absolute convergence, i.e., that

$$\sum_{k=1}^{\infty} \frac{|a_k'(z_0)|}{|1 + a_k(z_0)|}$$

converges for each $z_0 \in \Omega_0$.

Given $z_0 \in \Omega_0$ there exists an N such that $|a_k(z_0)| \leq \frac{1}{2}$ for $k \geq N$, and hence that the denominator $|1 + a_k(z_0)| \geq \frac{1}{2}$ for $k \geq N$. Thus it suffices to verify that $\sum_{k=1}^{\infty} |a_k'(z_0)| < \infty$. For that purpose choose complex numbers ε_k for $k = 1, 2, \ldots$ such that $|\varepsilon_k| = 1$ and $\varepsilon_k a_k'(z_0) = |a_k'(z_0)|$. By assumption the series $\sum_{k=1}^{\infty} \varepsilon_k a_k$ converges locally uniformly on Ω, hence (Weierstrass' theorem) so does its derivative $\sum_{k=1}^{\infty} \varepsilon_k a_k'$. In particular it converges at $z = z_0$, which is the desired statement. $\qquad\square$

10.2   The Euler formula for sine

The infinite product

$$F(z) := z \prod_{n=1}^{\infty} \left(1 - \frac{z^2}{n^2}\right)$$

converges (by Lemma 10.4) locally uniformly in the entire z-plane, so that F is an entire function. The zero-set of F is $\mathbb{Z}$ and each zero has multiplicity one. The function $z \mapsto \sin \pi z$ has the same properties. Hence $z \mapsto \sin(\pi z)/F(z)$ is holomorphic and zero free in the entire z-plane. We can write

$$\sin \pi z = F(z)e^{A(z)} \tag{10.7}$$

where A is entire. The main difficulty is the determination of A. To find A we take logarithmic derivatives in (10.7):

$$\pi \cot \pi z = A'(z) + \frac{1}{z} + \sum_{n=1}^{\infty} \left(\frac{1}{z+n} + \frac{1}{z-n}\right) \tag{10.8}$$

The series on the right hand of (10.8) converges locally uniformly on the open set $\mathbb{C} \setminus \mathbb{Z}$. Differentiating (10.8) we get

$$\frac{-\pi^2}{\sin^2 \pi z} = A''(z) - \frac{1}{z^2} - \sum_{n=1}^{\infty} \left(\frac{1}{(z+n)^2} + \frac{1}{(z-n)^2}\right) \tag{10.9}$$

and since $\sum_{n=1}^{\infty} n^{-2}$ converges, we may rewrite (10.9) in the form

$$A''(z) = \sum_{n=-\infty}^{\infty} \frac{1}{(z+n)^2} - \frac{\pi^2}{\sin^2 \pi z}. \tag{10.10}$$

The right hand side in (10.10) is unaltered if we replace z by $z+1$. In other words, the entire function A'' has period 1. To show that A'' is a constant it is thus by Liouville's theorem sufficient to verify that A'' is bounded in the set $\{z = x + iy \mid 0 \le x \le 1, \ |y| \ge 1\}$. For such z we have

$$\left| \sum_{n=-\infty}^{\infty} \frac{1}{(z+n)^2} \right| \le \sum_{n=-\infty}^{\infty} \frac{1}{|z+n|^2}$$

$$= \sum_{n=0}^{\infty} \frac{1}{(x+n)^2 + y^2} + \sum_{n=-\infty}^{-1} \frac{1}{(x+n)^2 + y^2}$$

$$= \sum_{n=0}^{\infty} \frac{1}{(x+n)^2 + y^2} + \sum_{n=1}^{\infty} \frac{1}{(n-x)^2 + y^2}$$

$$= \sum_{n=0}^{\infty} \frac{1}{(x+n)^2 + y^2} + \sum_{n=0}^{\infty} \frac{1}{(n + (1-x))^2 + y^2}.$$

Since $0 \le x \le 1$ we estimate as follows

$$(x+n)^2 \ge n^2 \quad \text{and} \quad (n + (1-x))^2 \ge n^2$$

and get

$$\left| \sum_{n=-\infty}^{\infty} \frac{1}{(z+n)^2} \right| \le \sum_{n=0}^{\infty} \frac{1}{n^2 + y^2} + \sum_{n=0}^{\infty} \frac{1}{n^2 + y^2} = 2 \sum_{n=0}^{\infty} \frac{1}{n^2 + y^2}$$

If $y > N$ (N positive integer) then

$$2 \sum_{n=0}^{\infty} \frac{1}{n^2 + y^2} \le 2 \sum_{n=0}^{\infty} \frac{1}{n^2 + N^2} \le 2 \sum_{n=0}^{\infty} \frac{1}{\frac{1}{2}(n+N)^2} = 4 \sum_{n=N}^{\infty} \frac{1}{n^2}.$$

For the other term of the right hand side of (10.10) we estimate as follows for $y > 0$ (The case $y < 0$ can be treated similarly):

$$|\sin \pi z|^2 = \left| \frac{e^{\pi i z} - e^{-\pi i z}}{2} \right|^2 = \left| \frac{e^{\pi i x} e^{-\pi y} - e^{-\pi i x} e^{\pi y}}{2} \right|^2$$

$$= \left| \frac{e^{\pi y} - e^{2\pi i x} e^{-\pi y}}{2} \right|^2 \ge \left(\frac{e^{\pi y} - e^{-\pi y}}{2} \right)^2 = (\sinh \pi y)^2$$

so for any y we have

$$\left| \frac{\pi^2}{\sin^2 \pi z} \right| \le \frac{\pi^2}{\sinh^2 \pi y}.$$

These inequalities say that $A''(z)$ tends to 0 as z tends to infinity in the period strip $0 \le \operatorname{Re} z \le 1$, so we conclude that $A'' = 0$, i.e., that A' is a constant. By (10.8) A' is odd, so $A' = 0$, i.e., A is a constant.

Finally, dividing both sides of (10.7) by z and letting z tend to 0 we evaluate the constant e^A to be π, and arrive at the beautiful *Euler Formula for Sine*:

$$\sin \pi z = \pi z \prod_{n=1}^{\infty} \left(1 - \frac{z^2}{n^2} \right). \tag{10.11}$$

Let us note that we on the way also found (formula (10.8) the *partial fraction expansion of cotangent*:

$$\pi \cot \pi z = \frac{1}{z} + \sum_{n=1}^{\infty} \left(\frac{1}{z+n} + \frac{1}{z-n} \right) \qquad \text{for } z \in \mathbb{C} \setminus \mathbb{Z}. \qquad (10.12)$$

As an application of (10.11) take $z = 1/2$ to get

$$1 = \frac{\pi}{2} \prod_{n=1}^{\infty} \left(1 - \frac{1}{4n^2} \right) = \frac{\pi}{2} \prod_{n=1}^{\infty} \frac{4n^2 - 1}{4n^2}$$

from which we obtain *Wallis' product*:

$$\frac{\pi}{2} = \prod_{n=1}^{\infty} \frac{4n^2}{4n^2 - 1} = \prod_{n=1}^{\infty} \frac{(2n)^2}{(2n-1)(2n+1)} = \frac{2 \cdot 2}{1 \cdot 3} \cdot \frac{4 \cdot 4}{3 \cdot 5} \cdot \frac{6 \cdot 6}{5 \cdot 7} \cdots$$

As another illustration divide both sides of (10.11) by $1 - z$ and let $z \to 1$ to get the formula

$$\frac{1}{2} = \prod_{n=2}^{\infty} \left(1 - \frac{1}{n^2} \right).$$

The *Basel problem* was posed in 1644. It is to compute the value of the series $\sum_{n=1}^{\infty} 1/n^2$. Euler got instant fame in 1735 when he found the correct answer $\pi^2/6$. Today lots of ways of proceeding are in the literature, also elementary. For a discussion of Euler's contribution see the paper Ayoub (1974).

We shall derive the formula $\sum_{n=1}^{\infty} 1/n^2 = \pi^2/6$ by means of the results of the present section. Knowing that $A'' = 0$ we rewrite the formula (10.10) as

$$\sum_{n \neq 0} \frac{1}{(z+n)^2} = \frac{\pi^2}{\sin^2(\pi z)} - \frac{1}{z^2}. \qquad (10.13)$$

The left hand side of (10.13) converges and defines an analytic function around 0, which at $z = 0$ it assumes the value

$$\sum_{n \neq 0} \frac{1}{n^2} = 2 \sum_{n=1}^{\infty} \frac{1}{n^2}.$$

To find what this sum is we reformulate the right hand side of (10.13) as follows

$$\frac{\pi^2}{\sin^2(\pi z)} - \frac{1}{z^2} = \frac{\pi^2}{\sin^2(\pi z)}\left(1 - \frac{\sin^2(\pi z)}{\pi^2 z^2}\right)$$
$$= \frac{\pi^2}{\sin^2(\pi z)}\left(1 - \frac{\sin(\pi z)}{\pi z}\right)\left(1 + \frac{\sin(\pi z)}{\pi z}\right).$$

By the power series expansion of sin the factor in the middle is

$$1 - \frac{\sin(\pi z)}{\pi z} = (\pi z)^2 \sum_{n=0}^{\infty} \frac{(-1)^n}{(2n+3)!}(\pi z)^{2n},$$

which enables us to rewrite the right hand side as

$$\pi^2 \left(\frac{\pi z}{\sin(\pi z)}\right)^2 \sum_{n=0}^{\infty} \frac{(-1)^n}{(2n+3)!}(\pi z)^{2n}\left(1 + \frac{\sin(\pi z)}{\pi z}\right).$$

As $z \to 0$ it converges to $\pi^2 1^2 \frac{1}{3!}(1+1) = \pi^2/3$, giving us the desired value of the sum.

For another treatment of the topics of this section see Walter (1982).

10.3 Weierstrass' factorization theorem

In contrast to the example of the last section consider the problem of constructing an entire function with simple zeros exactly at the points $0, -1, -2, -3, \ldots$. The product $z \prod_{n=1}^{\infty}(1 + \frac{z}{n})$ will not do, simply because it does not converge. To treat such cases Weierstrass introduced certain explicitly given factors – the so-called primary factors – into the product to force convergence. Before we take up the general case let us illustrate the idea by the following lemma:

Lemma 10.7. *Let $a_1, a_2, \ldots$ be different complex numbers with the property that*

$$\sum_{n=1}^{\infty} \frac{1}{|a_n|^2} < \infty.$$

Then the infinite product

$$\prod_{k=1}^{\infty}\left\{\left(1 - \frac{z}{a_k}\right) e^{\frac{z}{a_k}}\right\} \tag{10.14}$$

converges locally uniformly on $\mathbb{C}$ and its value is an entire function with simple zeros precisely at $a_k, k = 1, 2, 3, \ldots$

Proof. We start with an estimate, valid for all $|a| < 1$:

$$|(1-a)e^a - 1| \le |a|^2. \tag{10.15}$$

To prove (10.15) note that we for $|a| < 1$ have

$$|(1-a)e^a - 1| = \left| \sum_{n=2}^{\infty} a^n \left(\frac{1}{n!} - \frac{1}{(n-1)!} \right) \right|$$

$$\le |a|^2 \sum_{n=2}^{\infty} |a|^{n-2} \left(\frac{1}{(n-1)!} - \frac{1}{n!} \right)$$

$$\le |a|^2 \sum_{n=2}^{\infty} \left(\frac{1}{(n-1)!} - \frac{1}{n!} \right) = |a|^2.$$

Now, $\sum_1^{\infty} |a_n|^{-2} < \infty$ so $|a_n|$ must tend to infinity as $n \to \infty$. For z in a fixed compact set K we have $|z/a_n| < 1$ uniformly in z for all large n. Using (10.15) we realize that the series

$$\sum_{k=N}^{\infty} \left| \left(1 - \frac{z}{a_k} \right) e^{\frac{z}{a_k}} - 1 \right|$$

as $N \to \infty$ converges to 0 locally uniformly on $\mathbb{C}$. Lemma 10.4 now takes over. $\square$

In the general case we will modify the infinite product $\prod_1^{\infty}(1 - z/a_n)$ by replacing each factor $(1 - z/a_n)$ by a factor of the form

$$\left(1 - \frac{z}{a_n} \right) \exp \lambda \left(\frac{z}{a_n} \right). \tag{10.16}$$

We choose the functions λ in such a way that the new factors are so close to 1 that the product converges. That $(1 - z) \exp(\lambda(z))$ is close to 1 means that $\lambda(z)$ approximates $\log((1-z)^1)$. The technical details are as follows:

Definition 10.8. *The **primary factors of Weierstrass** are the entire functions*

$$E(z, 0) = 1 - z,$$

$$E(z, n) = (1 - z) \exp \left(z + \frac{z^2}{2} + \cdots + \frac{z^n}{n} \right) \qquad \text{for } n = 1, 2, \ldots$$

Note that the primary factors vanish only at $z = 1$.

We will need the estimate

$$|E(z, n) - 1| \leq |z|^{n+1} \qquad \text{for } n = 0, 1, \ldots \text{ and } |z| \leq 1 \qquad (10.17)$$

which shows that $E(z, n)$ is close to 1 for large n, even though it is zero at $z = 1$.

To prove (10.17) we find by a simple calculation that

$$-E'(z, n) = z^n \exp\left(z + \frac{z^2}{2} + \cdots + \frac{z^n}{n}\right)$$

so the power series expansion of $-E'(z, n)$ around $z = 0$ starts from z^n and all its coefficients are nonnegative. From

$$1 - E(z, n) = \int_{[0,z]} (-E'(w, n)) \, dw$$

we then see that the power series expansion of $1 - E(z, n)$ starts from z^{n+1} and has nonnegative coefficients as well, i.e.,

$$1 - E(z, n) = \sum_{k=n+1}^{\infty} a_k z^k \qquad \text{where all } a_k \geq 0.$$

So for $|z| \leq 1$ we get

$$|1 - E(z, n)| \leq \sum_{k=n+1}^{\infty} a_k |z|^k \leq |z|^{n+1} \sum_{k=n+1}^{\infty} a_k 1 = |z|^{n+1}(1 - E(1, n))$$

$$= |z|^{n+1}(1 - 0) = |z|^{n+1}$$

which is (10.17).

And then for any $m \geq 0$, $n \geq 0$ and $|z| \leq 1$:

$$|E(z, n)^m - 1| \leq 2^m |E(z, n) - 1| \leq 2^m |z|^{n+1} \qquad (10.18)$$

where we have used the identity

$$|a^m - 1| = |a - 1||a^{m-1} + a^{m-2} + \cdots + a + 1|$$

and the fact (evident from (10.17)) that $|E(z, n)| \leq 2$ for $|z| \leq 1$.

In particular if $n = m + k$ then for any $m \geq 0$ and $k \geq 0$:

$$|E(z, m + k)1| \leq 2^{-k} \qquad \text{for } |z| \leq \tfrac{1}{2}. \qquad (10.19)$$

With these preliminaries to a side we can prove *Weierstrass' factorization theorem*:

Theorem 10.9 (Weierstrass' Factorization Theorem). *Let Ω be an open subset of $\mathbb{C}$, let $z_1, z_2, \ldots$ be a sequence of different points from Ω with no cluster point in Ω, and let finally $m_1, m_2, \ldots$ be a sequence of positive integers.*

There exists a function $g \in \mathrm{Hol}(\Omega)$ such that g has a zero of order m_k at z_k for $k = 1, 2, \ldots$, and such that g vanishes nowhere else in Ω.

Proof. Since the sequence $\{z_k\}$ does not cluster in Ω, there exist $a \in \Omega$ and $r > 0$ such that $B(a, r) \subseteq \Omega$ and such that $B(a, r)$ does not intersect the set $\{z_k\}$. Let us for convenience assume that $a = 0$ and $r = 1$. Under the transformation $z \mapsto z^{-1}$ the open set $\Omega \setminus \{0\}$ goes into an open set V. If $c_k := z_k^{-1}$, then $|c_k| \leq 1$, $c_k \in V$ and $\{c_k\}$ does not cluster in V (Recall that 0 does not belong to V). Choose for each k a point $a_k \in \mathbb{C} \setminus V$ closest to c_k. Since $\{c_k\}$ does not cluster in V we get

$$\lim_{k \to \infty} |a_k - c_k| = 0 \qquad (10.20)$$

[Indeed, if for some $\delta > 0$ and subsequence $\{k_i\}$ we have that $|a_{k_i} - c_{k_i}| \geq \delta$, where c_{k_i} tends to c (recall that $\{c_k\}$ is bounded), then c must belong to $\mathbb{C} \setminus V$. But by choice of a_k we have

$$|c_{k_i} - c| \geq |c_{k_i} - a_{k_i}| \geq \delta$$

which is a contradiction].

Now put

$$f(z) := \prod_{k=1}^{\infty} E\left(\frac{c_k - a_k}{z - a_k}, m_k + k\right)^{m_k} \qquad \text{for } z \in V. \qquad (10.21)$$

If F is any compact subset of V, then by (10.20) for all large k, $2|c_k - a_k| < |z - a_k|$, uniformly for $z \in F$.

The local uniform convergence of the infinite product in (10.21) now follows from (10.19) and Lemma 10.4.

By assumption $B(0, 1) \subseteq \Omega$, so $\{z \mid |z| > 1\} \subseteq V$. Since $|a_k| \leq 1$ and $|c_k| \leq 1$, we have

$$\left|\frac{c_k - a_k}{z - a_k}\right| \leq \frac{1}{2} \qquad \text{whenever } |z| \geq 3.$$

Using (10.19) and (10.21) we see that

$$|f(z)| \qquad \text{is bounded for } |z| \geq 3. \qquad (10.22)$$

Now define $g(z) := f(1/z)$ for $z \neq 0$. Then g has a zero of order m_k at z_k and vanishes nowhere else in $\Omega \setminus \{0\}$. (10.22) says that g is bounded in $B(0, \frac{1}{3}) \setminus \{0\}$. 0 is therefore a removable singularity of g and g can hence be defined to be analytic at 0 as well. If $g(0) \neq 0$, g is the required function. If 0 is a zero of g of order n, then $z^{-n}g$ fulfills all the requirements. □

10.4 The Γ-function

Taking $a_n = -n$ in (10.14) we find that

$$g(z) := e^{\gamma z} z \prod_{n=1}^{\infty} \left(1 + \frac{z}{n}\right) e^{-\frac{z}{n}} \tag{10.23}$$

is entire with simple zeros precisely at $z = 0, -1, -2, \ldots$ Here γ is a constant chosen so that $g(1) = 1$. γ is called the *Euler constant*. The reciprocal of g is meromorphic in $\mathbb{C}$ with simple poles at $0, -1, -2, \ldots$ This is the famous Γ-*function* (gamma-function):

$$\frac{1}{\Gamma(z)} = g(z) \tag{10.24}$$

and it is of utmost importance. The Γ-function was introduced by Euler in 1729 (See Davis (1959) for a readable account of its history).

Let us derive some of the most important properties of the Γ-function. Since g is entire, Γ has no zeros. Writing

$$g(z) = e^{\gamma z} z(z+1) e^{-z} \prod_{n=1}^{\infty} \left\{ \left(1 + \frac{z}{n+1}\right) \exp\left(-\frac{z}{n+1}\right) \right\} \tag{10.25}$$

and replacing z by $z + 1$ in (10.23), the following computation is valid for any $z \neq 0, -1, -2, \ldots$

$$\frac{g(z)}{zg(z+1)} = \frac{e^{-z}}{e^{\gamma}} \prod_{n=1}^{\infty} \left\{ \frac{1 + \frac{z}{n+1} \exp(-\frac{z}{n+1})}{1 + \frac{z+1}{n} \exp(-\frac{z+1}{n})} \right\}$$

$$= e^{-z-\gamma} \prod_{n=1}^{\infty} \left\{ \frac{n}{n+1} \exp\left(\frac{z}{n(n+1)} + \frac{1}{n}\right) \right\}$$

$$= e^{-z-\gamma} \exp\left\{ z \sum_{n=1}^{\infty} \frac{1}{n(n+1)} \right\} \prod_{n=1}^{\infty} \left\{ \frac{n}{n+1} \exp\left(\frac{1}{n}\right) \right\}.$$

Using that

$$\sum_{n=1}^{\infty} \frac{1}{n(n+1)} = \sum_{n=1}^{\infty} \left(\frac{1}{n} - \frac{1}{n+1} \right) = 1$$

we find further that

$$\frac{g(z)}{zg(z+1)} = e^{-\gamma} \prod_{n=1}^{\infty} \left\{ \frac{n}{n+1} \exp\left(\frac{1}{n}\right) \right\}. \tag{10.26}$$

Thus $g(z) = czg(z+1)$ for $z \neq 0, -1, -2, \ldots$, where c is a constant. By continuity this holds for all $z \in \mathbb{C}$. To find the constant c we note from (10.23) that

$$\lim_{z \to 0} \frac{g(z)}{z} = 1.$$

Since also $g(1) = 1$ we see that the constant is 1. Translating from g to Γ we find the *functional equation for* Γ:

$$\Gamma(z+1) = z\Gamma(z) \qquad \text{for } z \neq 0, -1, -2, \ldots \tag{10.27}$$

Since $\Gamma(1) = 1$, (10.27) leads to

$$\Gamma(n+1) = n! \qquad \text{for } n = 0, 1, 2, \ldots \tag{10.28}$$

which explains why the Γ-function is also called the *factorial function*.

We proceed by deriving another fundamental relation for the Γ-function, called the Formula of Complementary Arguments. Use (10.23) with z and $-z$ and multiply to get

$$g(z)g(-z) = -z^2 \prod_{n=1}^{\infty} \left(1 - \frac{z^2}{n^2} \right).$$

Appealing to the Euler Formula (10.11) for sine and to (10.27) the above relation reads in terms of Γ:

$$\Gamma(z)\Gamma(1-z) = \frac{\pi}{\sin \pi z} \qquad \text{for } z \notin \mathbb{Z}. \tag{10.29}$$

(10.29) is the *Formula of Complementary Arguments*.

In particular since $\Gamma(x) > 0$ for $x > 0$, we get taking $z = 1/2$ in (10.29) that

$$\Gamma(\tfrac{1}{2}) = \sqrt{\pi}. \tag{10.30}$$

The reader might in other connections have encountered the Mellin transform (= the Euler integral) of a complex valued function g on $\mathbb{R}^+$, which is the function Γ_g defined by

$$\Gamma_g(x) := \int_0^\infty t^{x-1} g(t) \, dt \qquad \text{wherever the integral converges.}$$

The particular case of $g(t) = e^{-t}$ is of special interest for us here because this Γ_g equals the Γ-function. This result is called *Euler's integral formula*

$$\Gamma(z) = \int_0^\infty t^{z-1} e^{-t} \, dt \qquad \text{for } \operatorname{Re} z > 0. \tag{10.31}$$

Proof of (10.31). To derive it we use Prym's decomposition

$$\int_0^\infty t^{z-1} e^{-t} \, dt = \int_0^1 t^{z-1} e^{-t} \, dt + \int_1^\infty e^{-t} t^{z-1} \, dt = P(z) + Q(z). \tag{10.32}$$

The second integral on the right, i.e., $Q(z)$ is entire. Expanding e^{-t} and integrating term by term we find that the other integral equals

$$P(z) = \sum_{n=0}^\infty \frac{(-1)^n}{n!(z+n)}. \tag{10.33}$$

The right hand side of (10.33) converges uniformly in any compact set not containing the points $z = 0, -1, -2, \ldots$, and we see that

$$F(z) := \sum_{n=0}^\infty \frac{(-1)^n}{n!(z+n)} + Q(z) \tag{10.34}$$

is a meromorphic function in the entire complex plane with simple poles at $z = 0, -1, -2, \ldots$ It also equals the integral in (10.31) for $\operatorname{Re} z > 0$. Using this, routine integration shows that

$$zF(z) = F(z+1) \qquad \text{when } \operatorname{Re} z > 0. \tag{10.35}$$

But then (10.35) must hold in the entire z-plane minus the points $0, -1, -2, \ldots$ (the Unique Continuation Theorem). We know that Γ satisfies the identity (10.35) in form of (10.27) and that Γ never vanishes. Thus from (10.24) we see that

$$G := \frac{F}{\Gamma} = Fg \tag{10.36}$$

is entire and has period 1, i.e., $G(z + 1) = G(z)$. Since $F(1) = \Gamma(1) = 1$, $G - 1$ vanishes at the (real) integers. Now, $\sin \pi z$ has period 2, has simple poles at the integers and vanishes nowhere else. Comparing G with $\sin \pi z$ we see that

$$H(z) := \frac{G(z) - 1}{\sin \pi z} \tag{10.37}$$

is entire and has period 2. We will prove that H is zero by showing that H is bounded in the period strip $1 \le \operatorname{Re} z \le 3$, and then applying Liouville's theorem.

F is bounded in this strip, as is evident from its integral representation (10.32). Let us estimate g in the strip. If $z = x + iy$ is in the strip under consideration then

$$\left| \left(1 + \frac{z}{n}\right) e^{-\frac{z}{n}} \right| = \sqrt{\left(1 + \frac{x}{n}\right)^2 + \frac{y^2}{n^2}}\, e^{-\frac{x}{n}} \le \left(1 + \frac{x}{n}\right) e^{-\frac{x}{n}} \sqrt{1 + \frac{y^2}{n^2}}\, .$$

Using this estimate in (10.23) and separating products we get:

$$|g(z)| \le |z| e^{\gamma x} \left\{ \prod_{n=1}^{\infty} \left(1 + \frac{x}{n}\right) e^{-\frac{x}{n}} \right\} \sqrt{\prod_{n=1}^{\infty} \left(1 + \frac{y^2}{n^2}\right)}$$

$$= |z| \frac{g(x)}{x} \sqrt{\prod_{n=1}^{\infty} \left(1 + \frac{y^2}{n^2}\right)} = |z| \frac{g(x)}{x} \sqrt{\frac{\sin \pi i y}{\pi i y}}$$

where the last equality comes from the Euler Sine Formula (10.11). Using this estimate it is elementary to show that (remember that $g(x)$ is bounded for $1 \le x \le 3$ and that $\sin(\pi i y) = i \sinh(\pi y)$)

$$\frac{g(z)}{\sin \pi z} \to 0 \qquad \text{as } z \to \infty \text{ in } 1 \le \operatorname{Re} z \le 3. \tag{10.38}$$

From (10.36) and (10.37)

$$|H(z)| \le |F(z)| \left| \frac{g(z)}{\sin \pi z} \right| + \frac{1}{|\sin \pi z|}.$$

Using (10.38) we conclude that H is bounded in the period strip and hence everywhere. By Liouville's theorem H is a constant. Since $H(z)$ tends to 0 as z tends to ∞ in the period strip, the constant is 0, so H is identically 0, i.e., $F = \Gamma$. We have thus derived (10.31). $\square$

From (10.30) we then get by a change of variables the important formula

$$\int_{-\infty}^{\infty} e^{-x^2} \, dx = \int_0^{\infty} t^{-\frac{1}{2}} e^{-t} \, dt = \sqrt{\pi}. \tag{10.39}$$

As an application of the Euler Integral Formula (10.31) we shall derive Stirling's asymptotic approximation of the factorial function:

Theorem 10.10 (Stirling's Approximation Formula).

$$\frac{\Gamma(n) e^n \sqrt{n}}{n^n} \to \sqrt{2\pi} \qquad \text{as } n \to \infty.$$

Proof. The proof is adapted from the one in Patin (1989).

Introducing $x = \sqrt{t} - \sqrt{n}$ as new variable in (10.31) we find

$$\frac{\Gamma(n) e^n \sqrt{n}}{n^n} = 2 \int_{-\sqrt{n}}^{\infty} \left(1 + \frac{x}{\sqrt{n}} \right)^{2n+1} e^{-2\sqrt{n}x} e^{-x^2} \, dx.$$

Since $1 + y \le \exp y$ we get for $x > -\sqrt{n}$ the following estimate of the integrand (except for the factor $\exp(-x^2)$)

$$\left(1 + \frac{x}{\sqrt{n}} \right)^{2n-1} e^{-2\sqrt{n}x} \le \exp\left\{ \frac{x}{\sqrt{n}} (2n - 1) \right\} \exp(-2\sqrt{n}x)$$

$$= \exp\left(-\frac{x}{\sqrt{n}} \right) \le e$$

so the integrand above is bounded uniformly in n by the integrable function $e \exp(-x^2)$.

We next show that the integrand converges pointwise, so x is fixed for a moment. Using the power series expansion of the logarithm we find for $n \to \infty$ that

$$\text{Log} \left(1 + \frac{x}{\sqrt{n}} \right) = \frac{x}{\sqrt{n}} - \frac{x^2}{2n} + O\left(\frac{1}{n^{3/2}} \right)$$

and so

$$\log\left\{ \left(1 + \frac{x}{\sqrt{n}} \right)^{2n-1} e^{-2\sqrt{n}x} \right\} = (2n - 1) \log\left(1 + \frac{x}{\sqrt{n}} \right) - 2\sqrt{n}x$$

$$= (2n - 1) \left(\frac{x}{\sqrt{n}} - \frac{x^2}{2n} \right) + O\left(\frac{1}{\sqrt{n}} \right) - 2\sqrt{n}x$$

$$= -x^2 + O\left(\frac{1}{\sqrt{n}} \right) \to -x^2.$$

Finally, by the Dominated Convergence Theorem

$$\frac{\Gamma(n)e^n \sqrt{n}}{n^n} \to 2 \int_{-\infty}^{\infty} e^{-x^2} e^{-x^2} \, dx = \sqrt{2\pi} \qquad \text{as } n \to \infty.$$

$\square$

As the end of this paragraph we mention the famous Duplication Formula of Legendre (see the exercises for a proof):

$$\sqrt{\pi}\Gamma(2z) = 2^{2z-1}\Gamma(z)\Gamma\left(z + \tfrac{1}{2}\right). \tag{10.40}$$

10.5   The Mittag-Leffler expansion

The theorem of this paragraph is analogous to the Weierstrass Theorem on zero sets in Section 10.4. Instead of zeros we deal with poles. Before we state and prove the theorem we remind the reader of the following fact from the theory of Laurent expansions:

Theorem 10.11. *Let f be holomorphic on the punctured disc $B(0,R) \setminus \{0\}$. Then f may in exactly one way be written in the form*

$$f(z) = g(z) + h(z^{-1}) \qquad \text{for } z \in B(0,R) \setminus \{0\}$$

where $g \in \mathrm{Hol}(B(0,R))$ and where h is an entire function such that $h(0) = 0$.

Proof. This was seen during the proof of Theorem 6.3. $\square$

More generally, if $a \in \mathbb{C}$ is an isolated singularity of a holomorphic function f then f can be written in the form

$$f(z) = h\left(\frac{1}{z-a}\right) + g(z) \tag{10.41}$$

where h is an entire function with $h(0) = 0$, and where g is a holomorphic function on the domain of definition of f and a is a removable singularity for g. The decomposition (10.41) is unique. The first term, i.e., the function $z \mapsto h((z-a)^{-1})$ is called the *principal part of f at a*.

Example. The function (cf. formula (10.10) above)

$$\frac{\pi^2}{\sin^2(\pi z)} = \sum_{n=-\infty}^{\infty} \frac{1}{(z-n)^2}$$

has for each $n \in \mathbb{N}$ the principal part $(z-n)^{-2}$ at $z = n$.

However, if we are given a sequence of principal parts, then their sum will normally not converge. Mittag-Leffler's theorem says that a meromorphic function with the prescribed principal parts nevertheless exists:

Theorem 10.12 (Mittag-Leffler's Theorem, 1884). *Let Ω be an open subset of $\mathbb{C}$. Let $p_1, p_2, \ldots$ be a sequence of different points from Ω without cluster point in Ω, and let $P_1, P_2, \ldots$ be a sequence of polynomials without constant term.*

Then there exists a function which is meromorphic on Ω, has $\{p_1, p_2, \ldots\}$ as its set of poles in Ω and whose principal part at p_k is $P_k((z - p_k)^{-1})$ for all $k = 1, 2, \ldots$

Remark. If we only specify that the desired function should have poles at $p_1, p_2, \ldots$ of given orders $m_1, m_2, \ldots$, then it can be found by help of Weierstrass' theorem on zero sets from Section 10.4: Indeed, if f is a holomorphic function with zeros at $p_1, p_2, \ldots$ of orders $m_1, m_2, \ldots$ then $1/f$ satisfies the requirements.

Proof of Theorem 10.12. Let us write $f_k(z) = P_k((z - p_k)^{-1})$ for $k = 1, 2, \ldots$ Let $B(p_k, r_k)$, $k = 1, 2, \ldots$ be disjoint balls in Ω and choose for each k a function $\phi_k \in C_0^\infty(B(p_k, r_k))$ such that ϕ_k is identically 1 on the smaller ball $B_k := B(p_k, r_k/2)$.

The function $f := \sum_{k=1}^\infty \phi_k f_k$ is smooth on $\Omega \setminus \{p_1, p_2, \ldots\}$ and reduces on B_k to f_k, so it satisfies the requirements of the theorem except that it is not holomorphic. To remedy that we consider the function

$$g(z) := \begin{cases} \frac{\partial f}{\partial \overline{z}}(z) & \text{for } z \in \Omega \setminus \{p_1, p_2, \ldots\}, \\ 0 & \text{for } z \in \{p_1, p_2, \ldots\}. \end{cases}$$

g is clearly C^∞ on $\Omega \setminus \{p_1, p_2, \ldots\}$. And on the punctured disc $B_k \setminus \{p_k\}$ the function f reduces to a holomorphic function (viz. f_k), so

$$g = \frac{\partial f}{\partial \overline{z}} = \frac{\partial f_k}{\partial \overline{z}} = 0$$

on $B_k \setminus \{p_k\}$. Thus $g \in C^\infty(\Omega)$.

The inhomogeneous Cauchy-Riemann equation $\partial u/\partial \overline{z} = g$ has a solution $u \in C^\infty(\Omega)$ (Theorem 9.17). Note that $\partial u/\partial \overline{z} = g = 0$ on B_k, so that $u \in \text{Hol}(B_k)$.

Now, $h := f - u$ is holomorphic in $\Omega \setminus \{p_1, p_2, \ldots\}$ by the very definition of g. On $B_k \setminus \{p_k\}$ we have $h = f_k - u$. As observed $u \in \text{Hol}(B_k)$, so h has the desired principal part f_k at p_k. $\square$

Weierstrass' factorization theorem (Theorem 10.9) told us how to construct a holomorphic function whose zero set is a prescribed sequence of points. The next result is that what can be done with zeros can be done with any sequence of values:

Corollary 10.13 (Germay's Interpolation Theorem). *Let Ω be an open subset of the complex plane. Let $\{z_1, z_2, \ldots\}$ be a subset of Ω without limit points in Ω, and let $w_1, w_2, \ldots$ be a sequence of complex numbers.*

Then there exists a function $f \in \text{Hol}(\Omega)$ such that $f(z_k) = w_k$ for $k = 1, 2, \ldots$

Proof. By Weierstrass' factorization theorem (Theorem 10.9) there is a holomorphic function f_0 with simple poles at the z_k, i.e., $f_0(z_k) = 0$ and $f_0'(z_k) \neq 0$ for $k = 1, 2, \ldots$ By the Mittag-Leffler Theorem there is a function $h \in \text{Hol}(\Omega \setminus \{z_1, z_2, \ldots\})$ such that

$$z \to h(z) - \frac{w_k}{f_0'(z_k)} \frac{1}{z - z_k}$$

is holomorphic in a ball around z_k for each k. Now, $f := f_0 h$ is holomorphic on Ω, because the zeros cancel the singularities, and for $z \to z_k$ we find that

$$f(z) = f_0(z)h(z) = f_0(z)\left\{h(z) - \frac{w_k}{f_0'(z_k)} \frac{1}{z - z_k}\right\} + f_0(z)\frac{w_k}{f_0'(z_k)} \frac{1}{z - z_k}$$
$$\to 0 + w_k = w_k$$

so $f(z_k) = w_k$ for $k = 1, 2, \ldots$ $\qquad\square$

For a generalization of Corollary 10.13 see Exercise 10.17.

10.6   The ζ- and $\wp$ -functions of Weierstrass

We have examples of periodic holomorphic functions, for instance: exp, cos and sin. Liouville's theorem rules out that a nonconstant holomorphic function can have two independent periods. We shall now give an example of a doubly periodic meromorphic function, viz. Weierstrass' $\wp$-function.

Let ω and ω' be non-zero complex numbers with non-real quotient, i.e., they are linearly independent over the reals. Let G be the group generated by ω and ω', i.e., G consists of the complex numbers of the form $g = m\omega + n\omega'$ where m and n are integers. If ω and ω' are periods of a given function then so is every $g \in G$.

By Mittag-Leffler's theorem there is a meromorphic function with principal part $(z-g)^{-1}$ at each element $g \in G$. However we do not have to appeal to that theorem, since an example is given by the so-called Weierstrass ζ-function (Weierstrass' zeta-function):

$$\zeta(z) := \frac{1}{z} + \sum_{0 \neq g \in G} \left\{ \frac{1}{z-g} + \frac{1}{g} + \frac{z}{g^2} \right\}. \tag{10.42}$$

To prove that ζ is such an example we prepare a lemma.

Lemma 10.14.

$$\sum_{0 \neq g \in G} \frac{1}{|g|^3} < \infty. \tag{10.43}$$

Proof. Each $g \in G$, $g \neq 0$ has the form $g = m\omega + n\omega'$ where $|m| + |n| \geq 1$. If $|m| + |n| = k$ either $|m|$ or $|n|$ is $\geq k/2$; if, say $|m| \geq k/2$ then

$$|m\omega + n\omega'| = |\omega'|\left|m\frac{\omega}{\omega'} + n\right| \geq |\omega'|\left|\text{Im}\,\frac{m\omega}{\omega'}\right| = |\omega'||m|\left|\text{Im}\,\frac{\omega}{\omega'}\right| \geq k\alpha$$

where $\alpha = 2\min\{|\omega'||\text{Im}\,\frac{\omega}{\omega'}|, |\omega||\text{Im}\,\frac{\omega'}{\omega}|\}$.

Now there are at most $4k$ pairs (m, n) such that $|m| + |n| = k$. Thus

$$\sum_{0 \neq g \in G} \frac{1}{|g|^3} \leq 4\sum_{k=1}^{\infty} \sum_{|m|+|n|=k} \frac{1}{\alpha^3 k^3} \leq 4\alpha^{-3} \sum_{k=1}^{\infty} \frac{1}{k^2} < \infty$$

which proves (10.43). $\qquad\square$

Let us now return to the right hand side of (10.42). Consider the disc $B(0, R)$. There are only finitely many $g \in G$ satisfying $|g| \leq R$, as is clear from f.ex. (10.43). Since

$$\frac{1}{z-g} + \frac{1}{g} + \frac{z}{g^2} = \frac{z^2}{(z-g)g^2},$$

and

$$|z - g| = |g|\left|1 - \frac{z}{g}\right| \geq \frac{|g|}{2} \qquad \text{if } |g| \geq 2R \text{ and } |z| \leq R$$

the series

$$\sum_{|g| \geq 2R} \left\{ \frac{1}{z-g} + \frac{1}{g} + \frac{z}{g^2} \right\}$$

converges uniformly in $B(0, R)$ and represents a holomorphic function in this disc. It follows that $\zeta(z)$, given by (10.42) is meromorphic in $\mathbb{C}$ with simple poles precisely at the points of G.

Differentiation of (10.42) termwise (which is permitted) leads to the Weierstrass $\wp$-function (Weierstrass' pe-function):

$$\wp(z) = -\zeta'(z) = \frac{1}{z^2} + \sum_{0 \neq g \in G} \left\{ \frac{1}{(z-g)^2} - \frac{1}{g^2} \right\}. \tag{10.44}$$

The $\wp$-function is meromorphic in $\mathbb{C}$ with double poles precisely at $g \in G$. Let us now show that the $\wp$-function is doubly periodic with periods ω and ω'. Differentiating (10.44) we get

$$\wp'(z) = -\frac{2}{z^3} - \sum_{0 \neq g \in G} \frac{2}{(z-g)^3} = -\sum_{g \in G} \frac{2}{(z-g)^3}. \tag{10.45}$$

For any $h \in G$ the series for $\wp'$ is unaltered by a change from z to $z + h$ because G is a group. In other words, $\wp'(z + h) = \wp'(z)$, implying that

$$\wp(z + h) - \wp(z) = C(h) \tag{10.46}$$

where $C(h)$ is a constant, perhaps depending on h. But the series for $\wp$ shows that $\wp$ is an even function: $\wp(z) = \wp(-z)$. Now it is immediate from (10.46) that

$$C(h + g) = C(h) + C(g) \qquad \text{for all } h, g \in G.$$

Using these facts we find that

$$C(-h) = \wp(z - h) - \wp(z) = \wp(-z - h) - \wp(-z)$$
$$= \wp(z + h) - \wp(z) = C(h).$$

But $C(h) + C(-h) = C(0) = 0$, so we must have $C(h) = 0$ and so $\wp(z + h) = \wp(z)$. We record this in the final statement of this section:

The functions $\wp$ and $\wp'$ are doubly periodic with periods ω and ω', and $\zeta' = -\wp$.

Exercises

10.1. Show that

$$\prod_{n=1}^{\infty} \left(1 - \frac{2}{(n+1)(n+2)} \right) = \frac{1}{3}.$$

10.2. For which $z \in \mathbb{C}$ will the infinite product

$$\prod_{n=0}^{\infty} (1 + z^{(2^n)})$$

converge? Show that the value of the product is $(1 - z)^{-1}$.

Hint:

$$(1 - z) \prod_{n=1}^{k-1} (1 + z^{(2^n)}) = 1 - z^{(2^k)}.$$

10.3. Assume that the limit $\lim_{N \to \infty} \prod_{n=1}^{N} c_n$ exists and is $\neq 0$. Show that the product $\prod_{n=1}^{\infty} c_n$ converges.

10.4. Assume that the infinite product $\prod_{n=1}^{\infty} (1 + a_n)$ converges. Discuss convergence of $\prod_{n=1}^{\infty} \sqrt{1 + a_n}$.

10.5. Let the sequence $\{a_n\}$ of complex numbers satisfy that $0 < |a_n| < 1$ for $n = 1, 2, \ldots$ and that $\sum_{n=1}^{\infty} (1 - |a_n|) < \infty$.

(α) Show that the infinite product (a so-called *Blaschke product*)

$$B(z) := \prod_{n=1}^{\infty} \frac{|a_n|}{a_n} \frac{a_n - z}{1 - \overline{a_n} z}$$

defines a function which is holomorphic in $B(0, 1)$. Find its zeros.

(β) Find a sequence $\{a_n\}$ as above with the property that every point on the unit circle $|z| = 1$ is a cluster point of $\{a_n\}$.

10.6. We shall in this exercise present another way of dealing with the entire function A from Section 10.2 (Herglotz' trick):

$$A'(z) = \pi \cot(\pi z) - \left(\frac{1}{z} + \sum_{n=1}^{\infty} \frac{2z}{z^2 - n^2} \right).$$

(α) Show that A' satisfies the functional equation

$$A'(z) = \frac{1}{2} \left\{ A' \left(\frac{z}{2} \right) + A' \left(\frac{z+1}{2} \right) \right\}.$$

(β) Use the functional equation to show that A' is a constant.
   Hint: The Maximum Modulus Principle.

10.7. Show that

$$\sum_{n=1}^{\infty} \frac{1}{n^2} = \frac{\pi^2}{6}$$

by help of the partial fraction expansion of the cotangent.

10.8. Show that

$$\cos \pi z = \prod_{n=1}^{\infty} \left\{ 1 - \frac{4z^2}{(2n-1)^2} \right\}.$$

10.9. Let f be meromorphic on an open subset Ω of $\mathbb{C}$. Show that f can be written as a quotient $f = g/h$, where g and h both are holomorphic in Ω.

10.10. Let $t \in \mathbb{R} \setminus \{0\}$. Show that

$$|\Gamma(it)| = \sqrt{\frac{\pi^2}{t \sinh(\pi t)}}.$$

10.11.

(α) Show that the Euler constant equals

$$\gamma = \lim_{N \to \infty} \left(1 + \tfrac{1}{2} + \cdots + \tfrac{1}{N} - \log N \right).$$

(β) Derive *Gauss' formula*

$$\Gamma(z) = \lim_{N \to \infty} \frac{N! N^z}{z(z+1)\cdots(z+N)} \qquad \text{for } z \neq 0, -1, -2, \ldots$$

and from it deduce Wallis' product

$$\frac{\pi}{2} = \frac{2 \cdot 2}{1 \cdot 3} \cdot \frac{4 \cdot 4}{3 \cdot 5} \cdot \frac{6 \cdot 6}{5 \cdot 7} \cdots$$

(γ) Prove the *Legendre Duplication Formula*

$$\sqrt{\pi}\,\Gamma(2z) = 2^{2z-1}\Gamma(z)\Gamma\left(z + \tfrac{1}{2}\right).$$

10.12. Find the residue of Γ at $z = 0$. More generally at $z = -n$ for $n = 0, 1, 2, \ldots$

10.13. Let C be the contour indicated in Fig. 10.1. Prove *Hankel's formula*

$$\Gamma(z) = \frac{i}{2\sin(\pi z)} \int_C (-t)^{z-1} e^{-t}\, dt.$$

For which $z \in \mathbb{C}$ is Hankel's formula valid?

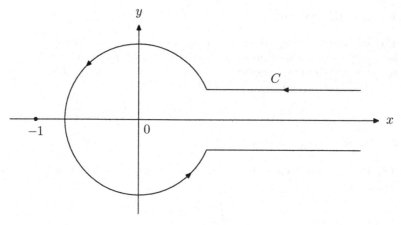

Fig. 10.1

10.14. Let Ω and $B = \{b_1, b_2, \ldots\}$ be as in the Mittag-Leffler Theorem. Let f_k for $k = 1, 2, \ldots$ be entire functions with $f_k(0) = 0$.

Show that there exists an $f \in \mathrm{Hol}(\Omega \setminus B)$ such that the principal part of f at b_k is $f_k((z - b_k)^{-1})$ for all $k = 1, 2, \ldots$

10.15. With the notation of Section 10.6 show that

$$\sigma(z) := z \prod_{0 \neq g \in G} \left\{ \left(1 - \frac{z}{g}\right) \exp\left(\frac{z}{g} + \frac{1}{2}\left(\frac{z}{g}\right)^2\right) \right\}$$

is entire with simple poles exactly at the $g \in G$.

The logarithmic derivative of σ is Weierstrass' ζ-function. σ is called the *sigma-function of Weierstrass.*

10.16. Show the following generalization of Proposition 10.2:

Let $\mu_1, \mu_2, \ldots$ be a sequence of reals with $\mu_k > 0$ for all k. Let $c_1, c_2, \ldots$ be a sequence of complex numbers from the open unit disc, and assume that

$$\sum_{k=1}^{\infty} |c_k| \mu_k < \infty.$$

Show that the product

$$\prod_{k=1}^{\infty} (1 + c_k)^{\mu_k}$$

converges (principal determinations).

10.17. Generalize Corollary 10.13 to the following theorem:

> **Theorem.** *Let Ω be an open subset of the complex plane and let $\{z_1, z_2, \ldots\}$ be a subset of Ω without limit points in Ω. Let there for each $k = 1, 2, \ldots$ be given finitely many complex numbers $w_{k,1}, w_{k,2}, \ldots, w_{k,N_k}$. Then there exists a function $f \in \mathrm{Hol}(\Omega)$ such that*
>
> $$f^{(n)}(z_k) = w_{k,n} \qquad for\ 0 \leq n \leq N_k\ and\ k = 1, 2, \ldots$$

Hint: Modify the proof of Corollary 10.13.

10.18. Derive the formula (10.39) directly without resort to the Γ-function as follows: Let

$$I := \int_{-\infty}^{\infty} e^{-x^2}\, dx,$$

show that

$$I^2 = \int_{\mathbb{R}^2} e^{-(x^2+y^2)}\, dxdy$$

and introduce polar coordinates.

Another elegant way of proving (10.39) can be found in Anonyme (1889).

Chapter 11

The Prime Number Theorem

11.1 The Riemann zeta function

The famous Prime Number Theorem states that $\pi(x)$, the number of primes less than or equal to x, is asymptotically the same as $x/\log x$. This was conjectured by Gauss and about a hundred years later proved by Hadamard and de la Vallée-Poussin. It is this proof that we present below. An "elementary" proof was given by A. Selberg Selberg (1949).

The proof uses the celebrated *Riemann zeta function* defined for z in the open half plane $\{w \in \mathbb{C} \mid \operatorname{Re} w > 1\}$ by:

$$\zeta(z) := \sum_{n=1}^{\infty} \frac{1}{n^z}. \tag{11.1}$$

The ζ-function was studied already by Euler, long before the time of Riemann (see Ayoub, 1974). Euler found e.g., formulas for $\zeta(2n), n = 1, 2, \ldots$ These formulas are derived in an elementary fashion in Berndt (1975). The special case $\zeta(2) = \pi^2/6$ has a short elegant proof Apostol (1983). Very little is known about the values of the ζ-function at other natural numbers, except that it has been established that $\zeta(3)$ is irrational. See the informal report van der Poorten (1978/79).

We shall presently show that the function given by (11.1) for $\operatorname{Re} z > 1$ is the restriction to this open set of a meromorphic function on $\mathbb{C}$ with a simple pole at $z = 1$. It is the extended function that is usually called the ζ-function. It is remarkable that the properties of this function should relate so decisively to the Prime Number Theorem. For more information on the ζ-function we refer to the monograph Edwards (1974).

Theorem 11.1. *The series* (11.1) *converges uniformly on every half plane of the form* $\{z \in \mathbb{C} \mid \operatorname{Re} z \geq a\}$ *where* $a > 1$.

There is a meromorphic function ζ *on* $\mathbb{C}$ *such that* $\zeta(z) - (z-1)^{-1}$ *is entire and such that* $\zeta(z)$ *is given by* (11.1) *for* $\operatorname{Re} z > 1$.

Proof. We concentrate on the second statement and leave the first one to the reader.

For $\operatorname{Re} z > 1$ we have

$$\Gamma(z) = \int_0^\infty t^{z-1} e^{-t}\, dt = n^z \int_0^\infty t^{z-1} e^{-nt}\, dt$$

so therefore

$$\zeta(z)\Gamma(z) = \sum_{n=1}^\infty \int_0^\infty t^{z-1} e^{-nt}\, dt = \int_0^\infty \frac{t^{z-1}}{e^t - 1}\, dt \qquad (11.2)$$

where interchange of sum and integral is justified by the Dominated Convergence Theorem. Write

$$\zeta(z)\Gamma(z) = \left(\int_0^1 + \int_1^\infty \right) \frac{t^{z-1}}{e^t - 1}\, dt. \qquad (11.3)$$

Noting that the second integral is entire in z, we concentrate on the first one. The function $z/(e^z - 1)$ is holomorphic in the ball $|z| < 2\pi$, so we may expand it

$$\frac{z}{e^z - 1} = \sum_{n=0}^\infty a_n z^n \qquad \text{for } |z| < 2\pi.$$

The series converges in particular for $z = 2$, so $a_n 2^n \to 0$. Thus there is a constant C such that $|a_n| \leq C 2^{-n}$ for $n = 0, 1, 2, \ldots$ Since $a_0 = 1$, we have

$$\frac{1}{e^z - 1} = \frac{1}{z} + \sum_{n=1}^\infty a_n z^{n-1}.$$

Using this expansion and term by term integration (justified by uniform convergence) the first integral in (11.3) can be written

$$\int_0^1 \frac{t^{z-1}}{e^t - 1}\, dt = \frac{1}{z-1} + \sum_{n=1}^\infty \frac{a_n}{z + n - 1}. \qquad (11.4)$$

Thus we may for $\operatorname{Re} z > 1$ write

$$\zeta(z)\Gamma(z) = \frac{1}{z-1} + \sum_{n=1}^\infty \frac{a_n}{z + n - 1} + \int_1^\infty \frac{t^{z-1}}{e^t - 1}\, dt. \qquad (11.5)$$

As said before the third term on the right in (11.5) is entire; the second term is by our estimate on a_n meromorphic in the complex plane with simple poles at $-(n-1)$ if $a_n \neq 0, n = 1, 2, \ldots$ And the first term has a simple pole at $z = 1$. Writing $g(z) = 1/\Gamma(z)$ we may define ζ on the complex plane by

$$\zeta(z) = \frac{g(z)}{z-1} + \Big(\sum_{n=1}^{\infty} \frac{a_n}{z+n-1}\Big) g(z) + g(z) \int_1^{\infty} \frac{t^{z-1}}{e^t - 1} \, dt. \qquad (11.6)$$

Since g is entire with zeros at $0, -1, -2, \ldots$ the right hand side of (11.6) is meromorphic on all of $\mathbb{C}$ with at most one pole, namely one at $z = 1$. Since $g(1) = 1$, the function $\zeta(z) - 1/(z-1)$ is entire. $\qquad\square$

Riemann discovered a remarkable functional relation for the ζ-function from which many of its properties can be deduced.

Theorem 11.2 (The Functional Equation for the Zeta-function).
The following relation is valid:

$$\zeta(1-z) = \frac{2}{(2\pi)^z}\zeta(z)\Gamma(z)\cos\frac{\pi z}{2} \qquad (11.7)$$

or written in a more symmetric form

$$\zeta(z)\Gamma\left(\frac{z}{2}\right)\pi^{-\frac{z}{2}} = \zeta(1-z)\Gamma\left(\frac{1-z}{2}\right)\pi^{-\frac{1-z}{2}}. \qquad (11.7')$$

Proof. Integrate by parts in (11.2) for $\operatorname{Re} z > 1$ to obtain

$$\zeta(z)\Gamma(z) = \frac{1}{z}\int_0^{\infty} \frac{t^z e^t}{(e^t - 1)^2} \, dt = -\frac{1}{4\pi}\int_0^{\infty} \frac{t^z}{\sin^2\left(\frac{it}{2}\right)} \, dt$$

so that

$$z\zeta(z)\Gamma(z) = -\int_0^1 \frac{t^z}{4\sin^2\left(\frac{it}{2}\right)} \, dt - \int_1^{\infty} \frac{t^z}{4\sin^2\left(\frac{it}{2}\right)} \, dt \qquad (11.8)$$

$$= -\int_0^1 t^z \left\{\frac{1}{4\sin^2\left(\frac{it}{2}\right)} + \frac{1}{t^2}\right\} dt + \frac{1}{z-1} - \int_1^{\infty} \frac{t^z}{4\sin^2\left(\frac{it}{2}\right)} \, dt.$$

Now from the formula (11.7) of Chapter 10 we get

$$\frac{1}{4\sin^2\left(\frac{it}{2}\right)} = -\frac{1}{t^2} + \sum_{n=1}^{\infty}\left\{\frac{1}{(it+2\pi n)^2} + \frac{1}{(it-2\pi n)^2}\right\}. \qquad (11.9)$$

The function $1 - \frac{\sin z}{z}$ behaves like $z^2/3!$ near $z = 0$, so

$$1 - \frac{\sin^2 z}{z^2} = \left(1 + \frac{\sin z}{z}\right)\left(1 - \frac{\sin z}{z}\right)$$

behaves like $z^2/3$ near $z = 0$. Hence the first integral in (11.8) is holomorphic for $\operatorname{Re} z > -1$. The last integral is in fact entire. Thus (11.8) holds for $\operatorname{Re} z > -1$. However, for $\operatorname{Re} z < 1$

$$\frac{1}{z-1} = -\int_1^\infty t^{z-2}\, dt$$

so we can certainly for $-1 < \operatorname{Re} z < 1$ write

$$z\zeta(z)\Gamma(z) = -\int_0^\infty t^z \left\{\frac{1}{4\sin^2\left(\frac{it}{2}\right)} + \frac{1}{t^2}\right\} dt. \tag{11.10}$$

And now we can use the expansion in (11.9). We may interchange the sum and the integral because of dominated convergence: Indeed, if $-1 < \operatorname{Re} z < 0$, then

$$\int_0^\infty t^{\operatorname{Re} z} \sum_{n=1}^\infty \frac{1}{|2\pi n + it|^2}\, dt = \sum_{n=1}^\infty \int_0^\infty \frac{t^{\operatorname{Re} z}}{4\pi^2 n^2 + t^2}\, dt$$

$$= \sum_{n=1}^\infty \int_0^\infty \frac{(2\pi n)^{\operatorname{Re} z} s^{\operatorname{Re} z}}{4\pi^2 n^2 (1 + s^2)}\, ds$$

$$= \sum_{n=1}^\infty \frac{(2\pi n)^{\operatorname{Re} z + 1}}{4\pi^2 n^2} \int_0^\infty \frac{s^{\operatorname{Re} z}}{1 + s^2}\, ds < \infty.$$

Thus using (11.9) in (11.10), changing the summation and integration around and using the substitution $t = 2\pi n u$ we get

$$z\zeta(z)\Gamma(z) = -\sum_{n=1}^\infty (2\pi n)^{z-1} \int_0^\infty \left\{\frac{1}{(1 + iu)^2} + \frac{1}{(1 - iu)^2}\right\} u^z\, du. \tag{11.11}$$

The sum on the right hand side is just $(2\pi)^{z-1}\zeta(1 - z)$. The integral – which is independent of n – can be evaluated as follows: The integrand is

$$2\frac{1 - u^2}{(1 + u^2)^2} u^z$$

so the integral equals

$$2\int_0^\infty \frac{1 - u^2}{(1 + u^2)^2} u^z\, du = 2\int_0^\infty \frac{u^z}{1 + u^2}\, du - 4\int_0^\infty \frac{u^{2+z}}{(1 + u^2)^2}\, du$$

$$= -2z\int_0^\infty \frac{u^z}{1 + u^2}\, du$$

where we have just integrated by parts in the second integral. The integral
on the right hand side is for $-1 < \operatorname{Re} z < 1$ evaluated below to be

$$2 \int_0^\infty \frac{u^z}{1+u^2} \, du = \frac{\pi}{\cos(\frac{\pi z}{2})}. \tag{11.12}$$

Using (11.12) in (11.11) we conclude that (11.7) is valid in the strip $-1 <$
$\operatorname{Re} z < 0$. Since both sides are meromorphic in all of the complex plane the
relation is an identity. □

Proof of (11.12). The substitution $v = u^2$ converts the above integral to

$$\int_0^\infty \frac{v^{\frac{z-1}{2}}}{1+v} \, dv = \int_0^\infty \frac{v^{\frac{z+1}{2}-1}}{1+v} \, dv$$

and the required result is a particular case of the formula

$$\int_0^\infty \frac{x^{p-1}}{(1+x)^{p+q}} \, dx = \frac{\Gamma(p)\Gamma(q)}{\Gamma(p+q)} \qquad \text{when } \operatorname{Re} p > 0, \operatorname{Re} q > 0$$

combined with the formula of complementary arguments (10.30). To derive
the formula we write

$$\Gamma(p+q)\frac{x^{p-1}}{(1+x)^{p+q}} = \int_0^\infty t^{p+q-1} e^{-t} \frac{x^{p-1}}{(1+x)^{p+q}} \, dt$$

$$= \int_0^\infty e^{-t(1+x)} t^{p+q-1} x^{p-1} \, dt.$$

Now we integrate both sides with respect to x from 0 to ∞ and change the
order of integration on the right hand side (which is justified by absolute
convergence):

$$\Gamma(p+q) \int_0^\infty \frac{x^{p-1}}{(1+x)^{p+q}} \, dx = \int_0^\infty e^{-t} t^{p+q-1} \left(\int_0^\infty e^{-tx} x^{p-1} dx \right) dt$$

$$= \int_0^\infty t^{q-1} e^{-t} \left(\int_0^\infty e^{-y} y^{p-1} dy \right) dt = \Gamma(p)\Gamma(q).$$

□

11.2   Euler's product formula and zeros of ζ

Euler's product formula relates the ζ-function directly to the prime numbers.
Simple consequences are some results on the zeros of the ζ-function.

Theorem 11.3 (Euler's Product Formula).

$$\zeta(z) = \prod_p \frac{1}{(1 - p^{-z})} \qquad when \ \operatorname{Re} z > 1 \qquad (11.13)$$

where p in the infinite product ranges over all primes $p > 1$.

Proof. Let us first prove that the infinite product converges: If we for each prime p write

$$\frac{1}{1 - p^{-z}} = 1 + \sum_{m=1}^{\infty} p^{-mz}$$

then it suffices by Proposition 10.2 to show that

$$\sum_p |p^{-z} + p^{-2z} + p^{-3z} + \cdots| < \infty.$$

Now

$$\sum_p |p^{-z} + p^{-2z} + p^{-3z} + \cdots| \le \sum_p \sum_{n=1}^{\infty} |p^{-nz}| \le \sum_p \sum_{n=1}^{\infty} \left(\frac{1}{p^n}\right)^{\operatorname{Re} z}.$$

The right hand side is a subset of the terms from the series

$$\sum_{m=1}^{\infty} \left(\frac{1}{m}\right)^{\operatorname{Re} z}$$

so

$$\sum_p |p^{-z} + p^{-2z} + p^{-3z} + \cdots| \le \sum_{m=1}^{\infty} \left(\frac{1}{m}\right)^{\operatorname{Re} z} = \zeta(\operatorname{Re} z) < \infty.$$

To prove that the value of the infinite product actually is $\sum_{n=1}^{\infty} n^{-z}$ we consider the finite product

$$P(N) := \prod_{p \le N} \{1 + p^{-z} + p^{-2z} + \cdots\} \qquad for \ N \in \mathbb{N}.$$

We can multiply this product out and rearrange the terms as we please without altering the result because it is just a product of (finitely many) absolutely convergent series. A typical term is of the form

$$p_1^{-\alpha_1 z} p_2^{-\alpha_2 z} \cdots p_k^{-\alpha_k z}$$

where $p_1, p_2, \ldots, p_k$ are different primes and $\alpha_1, \alpha_2, \cdots, \alpha_k$ are non-negative integers. By the Fundamental Theorem of Arithmetic $P(N) = \sum_{n \in A} n^{-z}$, where $A = A(N)$ consists of those $n \in \mathbb{N}$ whose prime factors all are $\leq N$. Thus

$$\left| \sum_{n=1}^{\infty} \frac{1}{n^z} - P(N) \right| \leq \sum_{n \in B} \frac{1}{|n^z|}$$

where B consists of those n having at least one prime factor $> N$. In particular

$$\left| \sum_{n=1}^{\infty} \frac{1}{n^z} - P(N) \right| \leq \sum_{n > N} \frac{1}{|n^z|} = \sum_{n > N} \frac{1}{n^{\mathrm{Re}\, z}}.$$

The right hand side converges to 0 as $N \to \infty$, so

$$P(N) \xrightarrow[N \to \infty]{} \sum_{n=1}^{\infty} \frac{1}{n^z}$$

as desired. $\qquad \square$

As an amusing commentary on (11.13) we may note that there are infinitely many primes (that fact was already known and proved by Euclid): Indeed, if there were but finitely many then the ζ-function would not diverge at $z = 1$. Not only are there infinitely many primes, but as observed by Euler (1737) there are so many that the series formed by the reciprocals of the prime numbers diverges, i.e.,

$$\sum_{p} \frac{1}{p} = \frac{1}{2} + \frac{1}{3} + \frac{1}{5} + \frac{1}{7} + \frac{1}{11} + \cdots = \infty.$$

(See Exercise 11.1.) For elementary derivations of this fact consult Artmann and Gerecke (1990) or Apostol (1976, Theorem 1.13).

Corollary 11.4. *The ζ-function does not vanish for* $\mathrm{Re}\, z > 1$. *For* $\mathrm{Re}\, z < 0$ *it vanishes only at the points* $-2m, m = 1, 2, \ldots$

Proof. The first assertion is immediate from (11.13). Since the Γ-function has no zeros, the second assertion follows from the first and the functional equation (11.7).

Taking logarithmic derivatives in (11.13) we obtain for $\mathrm{Re}\, z > 1$ that

$$-\frac{\zeta'(z)}{\zeta(z)} = \sum_{p} \frac{\mathrm{Log}\, p}{p^z} \frac{1}{1 - p^{-z}} = \sum_{p} \mathrm{Log}\, p \sum_{n=1}^{\infty} (p^n)^{-z} = \sum_{m=2}^{\infty} \Lambda(m) m^{-z}$$

$$(11.14)$$

where

$$\Lambda(m) = \begin{cases} \text{Log}\, p & \text{if } m \text{ is a power } (> 0) \text{ of } p, \\ 0 & \text{otherwise.} \end{cases}$$

It is permitted to rearrange terms because of the absolute convergence (cf. Proposition 10.6).

The fundamental link between the ζ-function and the distribution of primes is provided by the following function (called Chebyshev's function)

$$\psi(x) := \sum_{m \leq x} \Lambda(m) \qquad \text{for } x \in [1, \infty[. \tag{11.15}$$

The main result, known as the *Hadamard-de la Vallée Poussin Theorem* asserts that

$$\lim_{x \to \infty} \frac{\psi(x)}{x} = 1. \tag{11.16}$$

We will first show that the Prime Number Theorem is a consequence of (11.16):

Given a prime p the number of k such that $p^k \leq x$ is less than or equal to $(\text{Log}\, x)/(\text{Log}\, p)$, so recalling that $\Lambda(m) = \text{Log}\, p$ if m is a power of p we get from the definition (11.15) that

$$\psi(x) \leq \sum_{p \leq x} \frac{\text{Log}\, x}{\text{Log}\, p} \text{Log}\, p = \pi(x) \text{Log}\, x \tag{11.17}$$

where $\pi(x) =$ the number of primes less than or equal to x.

On the other hand, if $1 < y < x$ then

$$\pi(x) - \pi(y) = \sum_{y < p \leq x} 1 \leq \sum_{y < p \leq x} \frac{\text{Log}\, p}{\text{Log}\, y} \leq \frac{\psi(x)}{\text{Log}\, y}. \tag{11.18}$$

Taking $y = x/(\text{Log}\, x)^2$ and noting that $\pi(y) \leq y$ we obtain from (11.18) for any $x > e$ that

$$\pi(x) \frac{\log x}{x} \leq \frac{1}{\log x} + \frac{\psi(x)}{x} \frac{\log x}{\log x - 2\log\log x}. \tag{11.19}$$

That the Prime Number Theorem is equivalent to (11.16) is now evident from (11.17) and (11.19).   □

Before we embark on the proof of (11.16) we need one more fact, namely that ζ has no zeros on the line $\text{Re}\, z = 1$. That result is due to Hadamard and de la Vallée Poussin.

Theorem 11.5. *The* ζ-*function has no zeros on the line* $\operatorname{Re} z = 1$.

Proof. ζ is meromorphic, hence so is $\eta := \zeta'/\zeta$. Thus, for any complex number w, $\lim_{z \to w}(z - w)\eta(z)$ exists and is an integer. It is positive when w is a zero of ζ, negative when w is a pole of ζ, and zero otherwise.

If $z = 1 + \varepsilon + it$ where $\varepsilon > 0$ and $t \in \mathbb{R}$ we have from (11.14) that

$$\operatorname{Re}\{\eta(1 + \varepsilon + it)\} = -\sum_{m=2}^{\infty} \Lambda(m)m^{-1-\varepsilon}\cos(t\log m).$$

Using

$$0 \le (1 + \cos\theta)^2 = 1 + 2\cos\theta + \cos^2\theta = \tfrac{1}{2}(3 + 4\cos\theta + \cos 2\theta)$$

we then get

$$3\operatorname{Re}(\eta(1 + \varepsilon)) + 4\operatorname{Re}(\eta(1 + \varepsilon + it)) + \operatorname{Re}(\eta(1 + \varepsilon + 2it)) \qquad (11.20)$$

$$= -\sum_{m=2}^{\infty} \Lambda(m)m^{-1-\varepsilon}[1 + \cos(t\log m)]^2 \le 0. \qquad (11.21)$$

Now multiply the above by ε and let ε tend to 0. If $1 + it$ were a zero of ζ then we get that the limit of the left hand side of (11.20) is $8 - 3 + 4 + 0 = 1$.

This contradiction proves Theorem 11.5. $\square$

11.3   More about the zeros of ζ

In this section which is a digression, we mention a couple of results on the zeros of ζ. We have seen that ζ has no zeros in the closed half plane $\operatorname{Re} z \ge 1$. The functional equation (11.7) then tells us that the only zeros in the half plane $\operatorname{Re} z \le 0$ are $-2m, m = 1, 2, \ldots$ These latter zeros are called the trivial zeros. All the non-trivial zeros – if any – are thus in the open strip $0 < \operatorname{Re} z < 1$. This is called the *critical strip*.

Let us show that ζ has no *real* zeros in the critical strip:

Proof. We start with the formula (11.2) or rather the consequence

$$\zeta(z)\Gamma(z) = \int_0^1 \frac{t^{z-1}}{e^t - 1}\,dt + \int_1^\infty \frac{t^{z-1}}{e^t - 1}\,dt$$

which we for $\operatorname{Re} z > 1$ rewrite as

$$\zeta(z)\Gamma(z) = \int_0^1 t^{z-1} \left(\frac{1}{e^t - 1} - \frac{1}{t} \right) dt + \frac{1}{z-1} + \int_1^\infty \frac{t^{z-1}}{e^t - 1} \, dt.$$

The above formula is derived for $\operatorname{Re} z > 1$, but since the right hand side is meromorphic in $\operatorname{Re} z > 0$ it is valid in $\operatorname{Re} z > 0$. For $0 < t$ we have

$$\frac{1}{t} > \frac{1}{e^t - 1}$$

so the first integral above is negative when $z > 0$. For $t \geq 1$ we have $e^t - 1 \geq t^2$ so the third term is at most

$$\int_1^\infty \frac{dt}{t^2} = 1$$

Finally, for $0 < z < 1$ we have $1/(z-1) < -1$, so $\zeta(z)\Gamma(z) < 0$. $\qquad\square$

One of the most famous still unsolved problems in mathematics is the

Riemann Hypothesis. *All non-trivial zeros of ζ lie on the line* $\operatorname{Re} z = \frac{1}{2}$.

Numerical computations support the validity of the Riemann hypothesis. In van de Lune and te Riele (1983) it is shown that the ζ-function has exactly $300\,000\,001$ zeros whose imaginary parts lie between 0 and $119\,590\,809\,282$, and that all of them have real part $1/2$. See also Wagon (1986). On the theoretical side we will here just mention that ζ has infinitely many zeros on the line $\operatorname{Re} z = \frac{1}{2}$, a result due to Hardy (1914).

11.4 The prime number theorem

We now set out to prove the Prime Number Theorem. It is not easy to explain why or how the method came about.

Let us start with the right hand side of (11.14). Using Abel's partial summation formula (Proposition 1.10) we get

$$\sum_{m=2}^\infty \frac{\Lambda(m)}{m^z} = \sum_{m=2}^{N-1} \left\{ \frac{1}{m^z} - \frac{1}{(m+1)^z} \right\} \psi(m) + \frac{\psi(N)}{N^z}$$

where ψ is defined in (11.15). Since $\pi(x) \leq x$ the last term above tends to 0 as $N \to \infty$ by (11.17) when $\mathrm{Re}\, z > 1$. Thus

$$\sum_{m=2}^{\infty} \frac{\Lambda(m)}{m^z} = \sum_{m=2}^{\infty} \left\{ \frac{1}{m^z} - \frac{1}{(m+1)^z} \right\} \psi(m)$$

$$= \sum_{m=1}^{\infty} \left\{ \frac{1}{m^z} - \frac{1}{(m+1)^z} \right\} \psi(m).$$

Since $\psi(t)$ equals $\psi(m)$ for $m \leq t < m+1$ the last sum can be written

$$\sum_{m=1}^{\infty} z \int_{m}^{m+1} t^{-z-1}\, dt\, \psi(m) = z \int_{1}^{\infty} t^{-z-1} \psi(t)\, dt$$

$$= z \int_{0}^{\infty} e^{-(z-1)y} \left\{ e^{-y}\psi(e^y) \right\} dy$$

where we have just introduced $y = \log t$ as new variable – it is much more convenient to deal with the interval $[0, \infty[$. Thus from (11.14) we get

$$-\frac{1}{z}\frac{\zeta'(z)}{\zeta(z)} = \int_{0}^{\infty} e^{-(z-1)t} H(t)\, dt \qquad \text{when } \mathrm{Re}\, z > 1 \qquad (11.22)$$

where $H(t) := e^{-t}\psi(e^t)$.

Note that (11.16) is equivalent to $\lim_{t\to\infty} H(t) = 1$.

We can further rewrite (11.22) as

$$A(z) := -\frac{1}{z}\frac{\zeta'(z)}{\zeta(z)} - \frac{1}{z-1} = \int_{0}^{\infty} e^{-(z-1)t}(H(t) - 1)\, dt. \qquad (11.23)$$

The reason for doing so is: We know that $\zeta(z) - (z-1)^{-1}$ is entire (that is Theorem 11.1). A simple computation then shows that the function A is analytic wherever $z\zeta(z) \neq 0$; in particular (by Theorem 11.5) in a neighborhood of the line $\mathrm{Re}\, z = 1$.

The right hand side of (11.23) is the Laplace transform of the function $H - 1$. Referring back to the fact that (11.16) is equivalent to $H(t) - 1 \to 0$ as $t \to \infty$, we now realize that we need to prove a type of Tauberian result (a Tauberian result derives the asymptotic behavior of a function from the behavior of its averages).

We define $\rho \colon \mathbb{R} \to \mathbb{R}$ by

$$\rho(y) := \begin{cases} \frac{1 - |y|/2}{2} & \text{when } |y| \leq 2, \\ 0 & \text{when } |y| > 2, \end{cases}$$

and let $u > 0, \lambda > 0$ be real parameters that later on in the proof will converge to ∞.

With $z = 1 + \varepsilon + i\lambda y$, where $\varepsilon > 0$, we multiply both sides of (11.23) by $\rho(y)e^{iuy}$ and integrate with respect to y to get

$$\int_{-2}^{2} A(1 + \varepsilon + i\lambda y)\rho(y)e^{iuy}\, dy$$

$$= \int_{-2}^{2} \rho(y)e^{iuy}\left\{\int_{0}^{\infty} (H(t) - 1)e^{-\varepsilon t - i\lambda yt}\, dt\right\} dy. \tag{11.24}$$

Interchange of order of integration is permitted on the right hand side of (11.24) because

$$|H(t) - 1| \leq 1 + H(t) = 1 + e^{-t}\psi(e^t) \leq 1 + t$$

by (11.17), so the double integral is absolutely convergent. Effecting this interchange and evaluating the elementary inner integral we find

$$\int_{-2}^{2} A(1 + \varepsilon + i\lambda y)\rho(y)e^{iuy}\, dy$$

$$= \int_{0}^{\infty} (H(t) - 1)e^{-\varepsilon t}\frac{\sin^2(u - \lambda t)}{(u - \lambda t)^2}\, dt. \tag{11.25}$$

The left hand side of (11.25) is continuous in ε at $\varepsilon = 0$, because A is holomorphic on the line $\operatorname{Re} z = 1$, and ρ is bounded. The limit as $\varepsilon \searrow 0$ of the left hand side therefore exists. Also

$$\int_{0}^{\infty} \frac{\sin^2(u - \lambda t)}{(u - \lambda t)^2}\, dt < \infty$$

(because $\lambda > 0$ and u is real), so by the Monotone Convergence Theorem we get from (11.25) that

$$\int_{0}^{\infty} H(t)\frac{\sin^2(u - \lambda t)}{(u - \lambda t)^2}\, dt < \infty.$$

Going to the limit $\varepsilon = 0$ in (11.25) we get

$$\int_{-2}^{2} A(1 + i\lambda y)\rho(y)e^{iuy}\, dy = \int_{0}^{\infty} (H(t) - 1)\frac{\sin^2(u - \lambda t)}{(u - \lambda t)^2}\, dt.$$

The above holds for $\lambda > 0$ and any real u, so we may replace u by λu and change variable to get

$$\lambda \int_{-2}^{2} A(1 + i\lambda y)\rho(y)e^{i\lambda uy}\, dy = \int_{-\infty}^{\lambda u} \left(H\left(u - \frac{t}{\lambda}\right) - 1\right)\frac{\sin^2(t)}{t^2}\, dt.$$

Writing the integral on the left hand side above as a sum of the integral over $[-2, 0]$ and that over $[0, 2]$ and integrating by parts we see that the integral on the left hand side is bounded by $C/|u|$ for a suitable constant $C = C(\lambda)$. In particular the left hand side converges to 0 as $u \to \infty$ (the reader may possibly recognize the statement as a special case of the Riemann-Lebesgue lemma). Thus for $\lambda > 0$,

$$\lim_{u \to \infty} \int_{-\infty}^{\lambda u} H\left(u - \frac{t}{\lambda}\right) \frac{\sin^2 t}{t^2}\, dt = \lim_{u \to \infty} \int_{-\infty}^{\lambda u} \frac{\sin 2t}{t^2}\, dt = \int_{-\infty}^{\infty} \frac{\sin^2 t}{t^2}\, dt.$$
(11.26)

It is not necessary to know the value of the last integral, so we just call it B (actually $B = \pi$ by Exercise 6.16).

Observe that $e^t H(t) = \psi(e^t)$ is increasing in t, so that we for $b > 0$ have

$$e^t H(t) \le e^{t+b} H(t + b), \qquad \text{i.e., } H(t) \le H(t + b)e^b.$$
(11.27)

Now, for $|t| < \sqrt{\lambda}$,

$$H(u) \le H\left(u + \frac{1}{\sqrt{\lambda}} - \frac{t}{\lambda}\right) \exp\left(\frac{1}{\sqrt{\lambda}} - \frac{t}{\lambda}\right)$$

$$\le H\left(u + \frac{1}{\sqrt{\lambda}} - \frac{t}{\lambda}\right) \exp\left(\frac{2}{\sqrt{\lambda}}\right)$$

which implies that

$$\exp\left(-\frac{2}{\sqrt{\lambda}}\right) H(u) \int_{-\sqrt{\lambda}}^{\sqrt{\lambda}} \frac{\sin^2 t}{t^2}\, dt$$

$$\le \int_{-\sqrt{\lambda}}^{\sqrt{\lambda}} H\left(u + \frac{1}{\sqrt{\lambda}} - \frac{t}{\lambda}\right) \frac{\sin^2 t}{t^2}\, dt$$

$$\le \int_{-\infty}^{\lambda\left(u + \frac{1}{\sqrt{\lambda}}\right)} H\left(u + \frac{1}{\sqrt{\lambda}} - \frac{t}{\lambda}\right) \frac{\sin^2 t}{t^2}\, dt.$$

Using (11.26) and letting $u \nearrow \infty$ above we get

$$\exp\left(-\frac{2}{\sqrt{\lambda}}\right) \limsup_{u \to \infty} H(u) \int_{-\sqrt{\lambda}}^{\sqrt{\lambda}} \frac{\sin^2 t}{t^2}\, dt \le B.$$

This relation holds for all $\lambda > 0$, so letting $\lambda \to \infty$ we deduce that

$$\limsup_{u \to \infty} H(u) \le 1.$$
(11.28)

In particular H is bounded, say $H \le M$.

Let $\varepsilon > 0$ be arbitrary, but fixed. For sufficiently large λ, say $\lambda > \Lambda(\varepsilon)$ we have

$$\int_{|t|>\sqrt{\lambda}} M\left(\frac{\sin t}{t}\right)^2 dt < \varepsilon.$$

Replacing u by $u - \frac{1}{\sqrt{\lambda}}$ in (11.26) we get for all $\lambda > \Lambda(\varepsilon)$ that

$$B = \lim_{u \to \infty} \int_{-\infty}^{\lambda\left(u - \frac{1}{\sqrt{\lambda}}\right)} H\left(u - \frac{1}{\sqrt{\lambda}} - \frac{t}{\lambda}\right) \frac{\sin^2 t}{t^2} dt$$

$$\leq \varepsilon + \liminf_{u \to \infty} \int_{-\sqrt{\lambda}}^{\sqrt{\lambda}} H\left(u - \frac{1}{\sqrt{\lambda}} - \frac{t}{\lambda}\right) \frac{\sin^2 t}{t^2} dt.$$

From (11.27) we get for $|t| < \sqrt{\lambda}$ that

$$H\left(u - \frac{1}{\sqrt{\lambda}} - \frac{t}{\lambda}\right) \leq H(u)e^{\frac{1}{\sqrt{\lambda}} + \frac{|t|}{\lambda}} \leq H(u)e^{\frac{2}{\sqrt{\lambda}}}, \qquad (11.29)$$

so we may continue the estimates as follows

$$B \leq \varepsilon + \liminf_{u \to \infty} \int_{-\sqrt{\lambda}}^{\sqrt{\lambda}} H(u)e^{\frac{2}{\sqrt{\lambda}}} \frac{\sin^2 t}{t^2} dt$$

$$\leq \varepsilon + e^{\frac{2}{\sqrt{\lambda}}} \liminf_{u \to \infty} H(u) \int_{-\sqrt{\lambda}}^{\sqrt{\lambda}} \frac{\sin^2 t}{t^2} dt \leq \varepsilon + e^{\frac{2}{\sqrt{\lambda}}} \liminf_{u \to \infty} H(u)B.$$

Letting $\lambda \to \infty$ we find $B \leq \varepsilon + \liminf_{u \to \infty} H(u)B$, and so, $\varepsilon > 0$ being arbitrary, $1 \leq \liminf_{u \to \infty} H(u)$. Comparing with (11.28) we have finally arrived at

Theorem 11.6 (The Prime Number Theorem). *If $\pi(n)$ denotes the number of primes less than or equal to n, then*

$$\frac{\pi(n)\log n}{n} \to 1 \qquad \text{as } n \to \infty.$$

Exercises

11.1. Show that

$$\sum_p \frac{1}{p} = \infty$$

where the summation ranges over all primes p.

Hint: If not, then the infinite product

$$\prod_p \left(1 - \frac{1}{p}\right)$$

converges, and you may take $z = 1$ in the procedure of the proof of Theorem 11.3.

11.2. Let $\mu \colon \mathbb{N} \to \mathbb{R}$ be the *Möbius function*, i.e.,

$$\mu(1) = 1,$$
$$\mu(p_1^{\alpha_1} \cdots p_k^{\alpha_k}) = (-1)^k \quad \text{if } p_1, \ldots, p_k \text{ are distinct primes}$$
$$\text{and } \alpha_1 = \cdots = \alpha_k = 1,$$
$$\mu(n) = 0 \quad \text{otherwise.}$$

Show that the series (a so-called *Dirichlet series*)

$$\sum_{n=1}^{\infty} \frac{\mu(n)}{n^z}$$

converges absolutely when $\operatorname{Re} z > 1$ and that

$$\zeta(z) \sum_{n=1}^{\infty} \frac{\mu(n)}{n^z} = 1$$

when $\operatorname{Re} z > 1$.

Hint: Note first that

$$\sum_{d \mid n} \mu(d) = \begin{cases} 1 & \text{if } n = 1, \\ 0 & \text{if } n > 1. \end{cases}$$

11.3. Let $\phi \colon \mathbb{N} \to \mathbb{C}$ have the properties

(i) $\phi(nm) = \phi(n)\phi(m)$ whenever $(n, m) = 1$,

(ii) $\sum_{n=1}^{\infty} |\phi(n)| < \infty$.

Show that

$$\sum_{n=1}^{\infty} \phi(n) = \prod_p \{1 + \phi(p) + \phi(p^2) + \cdots\}$$

where p in the product ranges over all primes.

Apply this result to $\phi(n) := \mu(n)/n^z$, where $z \in \mathbb{C}$ is fixed with $\operatorname{Re} z > 0$ and μ is the Möbius function from the previous exercise, to get

$$\sum_{n=1}^{\infty} \frac{\mu(n)}{n^z} = \prod_p (1 - p^{-z}).$$

As another application take $\phi = 1$.

11.4. The result

$$2 \int_0^\infty \frac{u^z}{1+u^2} \, du = \frac{\pi}{\cos\left(\frac{\pi z}{2}\right)} \qquad \text{for } -1 < \operatorname{Re} z < 1$$

from the proof of Theorem 11.2 can also be verified by calculus of residues. Do it!

Chapter 12

Harmonic Functions

Harmonic functions are solutions to the Laplace equation $\Delta u = 0$, where

$$\Delta = \frac{\partial^2}{\partial x^2} + \frac{\partial^2}{\partial y^2}.$$

The Laplace operator Δ crops up in so many practical and theoretical situations, that it is hard to overestimate its importance. In the context of holomorphic functions every harmonic function is the real part of a holomorphic function in a simply connected domain.

Subharmonic functions, which are the topic for the next chapter, are very useful in the study of holomorphic functions. It may be said that this largely is due to the fact that when f is holomorphic then $\log|f(z)|$ is subharmonic. We will approach this via Jensen's formula. Harmonic and subharmonic functions are very closely related in much the same way as linear and convex functions.

12.1 Holomorphic and harmonic functions

Definition 12.1. *Let Ω be an open subset of $\mathbb{R}^n$. A real valued function $u \in C^2(\Omega)$ is said to **harmonic** if $\Delta u = 0$, where*

$$\Delta = \frac{\partial^2}{\partial x_1^2} + \frac{\partial^2}{\partial x_2^2} + \cdots + \frac{\partial^2}{\partial x_n^2}$$

*is the so-called **Laplace operator**.*

The Laplace equation $\Delta u = 0$ occurs not just in mathematics, but also in many different connections in physics and mechanics, in particular in descriptions of stationary situations and equilibrium states.

Theorem 12.2. *Let Ω be an open subset of $\mathbb{R}^2$.*

(a) *If $f \in \text{Hol}(\Omega)$ then $\text{Re}\, f$ and $\text{Im}\, f$ are harmonic in Ω.*

(b) *Assume that Ω is simply connected. If u is harmonic in Ω, then there exists an $f \in \text{Hol}(\Omega)$ such that $u = \text{Re}\, f$.*

The theorem yields a lot of interesting harmonic functions, e.g.,

$$u(x,y) = \text{Log}(x^2 + y^2) \qquad \text{defined for } (x,y) \in \mathbb{R}^2 \setminus \{0,0\}.$$

Proof of Theorem 12.2. Ad (a): The Cauchy-Riemann equations tell us that $\partial f / \partial \bar{z} = 0$, so the result is a consequence of the formula

$$\Delta f = 4 \frac{\partial}{\partial z} \left(\frac{\partial f}{\partial \bar{z}} \right).$$

Ad (b): The function

$$h := \frac{\partial u}{\partial x} - i \frac{\partial u}{\partial y} : \Omega \to \mathbb{C}$$

is C^1 in Ω and satisfies the Cauchy-Riemann equations, so it is holomorphic in Ω. Since Ω is simply connected h has a primitive f (Theorem 5.8), and (again by the Cauchy-Riemann equations)

$$h = f' = \frac{\partial(\text{Re}\, f)}{\partial x} - i \frac{\partial(\text{Re}\, f)}{\partial y}.$$

Comparing with the definition of h we get

$$\frac{\partial u}{\partial x} = \frac{\partial(\text{Re}\, f)}{\partial x} \quad \text{and} \quad \frac{\partial u}{\partial y} = \frac{\partial(\text{Re}\, f)}{\partial y}$$

so $u = \text{Re}\, f +$ some constant (on each connected component of Ω). When we absorb the constant in f the proof is complete. $\qquad\square$

Corollary 12.3. *A harmonic function is C^∞.*

The function $u(x,y) = \text{Log}(x^2 + y^2)$ is not the real part of an analytic function in all of $\mathbb{C} \setminus \{0\}$: The analytic function in question would be $2 \log z$, which, as we know, cannot be defined in all of $\mathbb{C} \setminus \{0\}$. However that is the only regrettable thing that may happen:

Theorem 12.4 (The Logarithmic Conjugation Theorem). *Let K be a connected compact subset of $B(0, R)$, where $R \in \,]0, \infty]$. Let $z_0 \in K$. If u is harmonic on $\Omega := B(0, R) \setminus K$ then there exist an analytic function f on Ω and a real constant c such that*

$$u(z) = \operatorname{Re} f(z) + c \operatorname{Log}|z - z_0| \qquad \text{for all } z \in \Omega.$$

The applications of the Logarithmic Conjugation Theorem are typically to ring shaped domains, i.e., the K of the theorem is a closed ball, possibly degenerated to a point.

Proof. As in the proof of Theorem 12.2(b) we note that

$$h := \frac{\partial u}{\partial x} - i\frac{\partial u}{\partial y} : \Omega \to \mathbb{C}$$

is holomorphic in Ω, because h satisfies the Cauchy-Riemann equations.

Choose $r \in \,]0, R[$ such that $K \subseteq B(0, r)$, and let C denote the circle $|z - z_0| = r$. Let us from now on take $z_0 = 0$ for simplicity.

Let γ be any closed path in Ω. If we put $m := \operatorname{Ind}_\gamma(z_0)$, then $\operatorname{Ind}_\gamma(z) = m$ for all $z \in K$, K being connected. Now,

$$\operatorname{Ind}_\gamma(z) = m \operatorname{Ind}_C(z) \qquad \text{for all } z \in \mathbb{C} \setminus \Omega$$

so by the Global Cauchy Theorem (Corollary 5.4) we get for any complex number c that

$$\int_\gamma \left(h(z) - \frac{c}{z} \right) dz = m \int_C \left(h(z) - \frac{c}{z} \right) dz = m\left(\int_C h(z)\, dz - 2\pi i c \right).$$

Choosing

$$c := \frac{1}{2\pi i} \int_C h$$

we get

$$\int_\gamma \left(h(z) - \frac{c}{z} \right) dz = 0 \qquad \text{for all closed paths } \gamma \text{ in } \Omega.$$

It follows that $z \mapsto h(z) - c/z$ has a primitive in each connected component of Ω, hence in all of Ω. Letting f denote such a primitive we have

$f'(z) = h(z) - c/z$. We shall soon need that c is real, so we prove that now:

$$c = \frac{1}{2\pi i} \int_C h(z)\, dz = \frac{1}{2\pi} \int_0^{2\pi} h(e^{i\phi}) e^{i\phi}\, d\phi$$

$$= \frac{1}{2\pi} \int_0^{2\pi} \left\{ \frac{\partial u}{\partial x}(e^{i\phi}) - i\frac{\partial u}{\partial y}(e^{i\phi}) \right\} e^{i\phi}\, d\phi$$

$$= \frac{1}{2\pi} \int_0^{2\pi} \left\{ \frac{\partial u}{\partial x}\cos\phi + \frac{\partial u}{\partial y}\sin\phi + i\frac{\partial u}{\partial x}\sin\phi - i\frac{\partial u}{\partial y}\cos\phi \right\} d\phi$$

so

$$\operatorname{Im} c = \frac{1}{2\pi} \int_0^{2\pi} \left\{ \frac{\partial u}{\partial x}\sin\phi - \frac{\partial u}{\partial y}\cos\phi \right\} d\phi$$

$$= -\frac{1}{2\pi} \int_0^{2\pi} \frac{d}{d\phi}\left\{ u(\cos\phi, \sin\phi) \right\} d\phi = 0$$

i.e., c is real.

Now,

$$\frac{\partial}{\partial x}\{\operatorname{Re} f + c\operatorname{Log}|z|\} = \operatorname{Re}\frac{\partial}{\partial x}\{f + c\operatorname{Log}|z|\}$$

$$= \operatorname{Re}\left\{ f' + \frac{c}{z} \right\} = \operatorname{Re} h = \frac{\partial u}{\partial x}$$

and similarly

$$\frac{\partial}{\partial y}\{\operatorname{Re} f + c\operatorname{Log}|z|\} = \frac{\partial u}{\partial y}$$

so that

$$\operatorname{Re} f + c\operatorname{Log}|z| = u + g$$

where g is constant on each connected component of Ω. Absorbing the constant in f the proof is finished. $\qquad\square$

In the above version of the Logarithmic Conjugation Theorem the complement of Ω has only one bounded connected component, viz. K. In general each bounded connected component contributes with a logarithmic term.

Let us use the Logarithmic Conjugation Theorem to prove the following classical result. A simple proof, that uses the Maximum Principle, but not the relationship between harmonic and analytic functions, can be found in Petrowski (1955, Satz III.30.6).

Theorem 12.5. *If a harmonic function is bounded in a neighborhood of an isolated singularity then that singularity is a removable singularity.*

Proof. We may assume that the harmonic function u is bounded in a punctured disc D' around 0. By the Logarithmic Conjugation Theorem u has the form

$$u(z) = \operatorname{Re} f(z) + c \operatorname{Log}|z| \qquad \text{where } f \in \operatorname{Hol}(D') \text{ and } c \in \mathbb{R}.$$

We will first show that $c = 0$. If not we may divide through by it and so assume that $c = 1$. Now, $\operatorname{Log}|z| \to -\infty$ as $z \to 0$, so to balance it $\operatorname{Re} f(z) \to \infty$ as $z \to 0$. In particular we see that $|f(z)| \to \infty$ as $z \to 0$, i.e., f has a pole at $z = 0$, and so f may be written in the form

$$f(z) = \frac{h(z)}{z^N}$$

where $N \geq 1$ and $h(0)$ is not 0.

Choosing z_n as an N'th root of $-h(0)/n$ for $n = 1, 2, \ldots$, we find

$$f(z_n) = -\frac{h(z_n)}{h(0)} n$$

so

$$\operatorname{Re} f(z_n) = -n \operatorname{Re} \frac{h(z_n)}{h(0)} \to -\infty, \qquad \text{for } n \to \infty.$$

But that contradicts $\operatorname{Re} f(z) \to \infty$ as $z \to 0$, so $c = 0$, and thus $u = \operatorname{Re} f$.

If we assume that f has a pole at 0 we may derive a contradiction in the same way as above. If we assume that 0 is an essential singularity then we get a contradiction by Picard's big theorem or just by Casorati-Weierstrass (remember $\operatorname{Re} f = u$ is bounded). So left is the desired case of 0 being a removable singularity for f. $\qquad \square$

For information on the Logarithmic Conjugation Theorem in finitely connected regions and its applications we recommend the paper Axler (1986) from where the above treatment is taken.

12.2   Poisson's formula

The important Poisson Formula expresses harmonic functions in terms of their boundary values:

Theorem 12.6 (Poisson's Formula). *If u is continuous in $B[0, R]$ and harmonic in $B(0, R)$ then*

$$u(z) = \frac{1}{2\pi} \int_0^{2\pi} u(Re^{i\phi}) \frac{R^2 - |z|^2}{|Re^{i\phi} - z|^2} \, d\phi \qquad for \ |z| < R.$$

*In particular u has the **Mean Value Property**, i.e.,*

$$u(0) = \frac{1}{2\pi} \int_0^{2\pi} u(Re^{i\phi}) \, d\phi.$$

Proof. The proof is from Åkerberg (1961). We will assume that u is harmonic in an open ball containing $B[0, R]$ and leave the derivation of the general case to the reader.

Let $|a| < R$. The Cauchy Integral Formula for the function

$$z \mapsto \frac{f(z)}{R^2 - z\bar{a}}, \qquad \text{where } u = \operatorname{Re} f,$$

in the disc $B[0, R]$ and where $u = \operatorname{Re} f$, is

$$\begin{aligned}
\frac{f(z)}{R^2 - z\bar{a}} &= \frac{1}{2\pi i} \int_{|w|=R} \frac{f(w)}{R^2 - w\bar{a}} \frac{1}{w - z} \, dw \\
&= \frac{1}{2\pi} \int_0^{2\pi} \frac{f(Re^{i\phi})}{R^2 - Re^{i\phi}\bar{a}} \frac{Re^{i\phi}}{Re^{i\phi} - z} \, d\phi \\
&= \frac{1}{2\pi} \int_0^{2\pi} \frac{f(Re^{i\phi})}{Re^{-i\phi} - \bar{a}} \frac{1}{Re^{i\phi} - z} \, d\phi.
\end{aligned}$$

Taking $z = a$ here we get Poisson's formula. $\square$

If we as center of the circle take z instead of 0 then the resulting formula is

$$u(z) = \frac{1}{2\pi} \int_0^{2\pi} u(z + Re^{i\phi}) \, d\phi$$

i.e., the value of u at the center of a ball is the mean value of the values of u at the boundary. We say that u has the *Mean Value Property*, and the last part of Theorem 12.6 is called the *Mean Value Theorem*.

The function

$$P_R(\phi, z) := \frac{1}{2\pi} \frac{R^2 - |z|^2}{|Re^{i\phi} - z|^2} \qquad for \ \phi \in \mathbb{R} \text{ and } |z| < R$$

is called the *Poisson kernel for $B(0, R)$* and is of great value in the study of harmonic functions. We list a couple of its properties that will be useful in the sequel:

Theorem 12.7 (Properties of the Poisson Kernel).
(a)
$$P_R > 0.$$

(b)
$$\int_0^{2\pi} P_R(\phi, z)\, d\phi = 1 \qquad \text{when } |z| < R.$$

(c)
$$P_R(\phi, z) = \operatorname{Re}\left\{\frac{1}{2\pi} \frac{Re^{i\phi} + z}{Re^{i\phi} - z}\right\} = \frac{1}{2\pi} \operatorname{Re}\left\{1 + 2\sum_{n=1}^{\infty} R^{-n} e^{-in\phi} z^n\right\}.$$

In particular $P_R(\phi, z)$ is harmonic in $B(0, R)$ for fixed ϕ.

(d)
$$\frac{1}{2\pi} \frac{R - |z|}{R + |z|} \leq P_R(\phi, z) \leq \frac{1}{2\pi} \frac{R + |z|}{R - |z|}.$$

(e) *If $0 < \delta < \pi/2$ then*

$$P_R(\phi, z) \leq \frac{1 - R^2}{\sin^2(\delta)} \qquad \text{for } \delta \leq \phi \leq 2\pi - \delta \text{ and } |z| < R.$$

Proof. The proofs of (a), (c), (d) and (e) are elementary estimates and computations left to the reader. You get (b) by putting $u = 1$ in Poisson's formula. $\qquad \square$

The following theorem is a consequence of the Mean Value Property. Maximum principles are among the most useful tools employed in differential equations. See Protter and Weinberger (1967).

Theorem 12.8 (The Strong Maximum Principle). *Let u be harmonic in an open connected subset Ω of $\mathbb{R}^2$, and assume that u takes its supremum, i.e., there exists a $z_0 \in \Omega$ such that $u(z_0) = \sup\{u(z) \mid z \in \Omega\}$.*
    Then u is constant.

Proof. It suffices to show that the set $\Omega_0 := \{z \in \Omega \mid u(z) = u(z_0)\}$ is both closed and open in Ω. It is closed because u is continuous. Let $z \in \Omega_0$; for the simplicity of writing, assume that $z = 0$. If $B[0, R] \subseteq \Omega$, then we get from the Mean Value Property for any $0 < r \leq R$ that

$$0 = u(0) - \frac{1}{2\pi} \int_0^{2\pi} u(re^{i\phi})\, d\phi = \frac{1}{2\pi} \int_0^{2\pi} \{u(0) - u(re^{i\phi})\}\, d\phi$$

and – since $u(0) \geq u(re^{i\phi})$ - we get $u(0) = u(re^{i\phi})$ for all ϕ. So $u(0) = u(w)$ for all $w \in B(0, R)$. This shows that the set Ω_0 is open. $\qquad \square$

We shall next formulate a converse to Poisson's formula:

Theorem 12.9. *Let f be continuous and real valued on the circle $|z| = R$. Then the function u, defined by*

$$u(z) := \begin{cases} \int_0^{2\pi} f(Re^{i\phi}) P_R(\phi, z)\, d\phi & \text{for } |z| < R, \\ f(z) & \text{for } |z| = R, \end{cases}$$

is continuous in the closed ball $B[0, R]$ and harmonic in the open ball $B(0, R)$.

Proof. The procedure of the proof is borrowed from Minda (1990).

We treat only the case $R = 1$ and leave the derivation of the general case to the reader. For $|z| < 1$ we have

$$u(z) = \int_0^{2\pi} f(e^{i\phi}) P_1(\phi, z)\, d\phi = \frac{1}{2\pi} \operatorname{Re} \int_0^{2\pi} f(e^{i\phi}) \frac{e^{i\phi} + z}{e^{i\phi} - z}\, d\phi.$$

The last integral can be expanded in a power series in z, so u is the real part of an analytic function and is hence harmonic.

Left is continuity at the boundary $|z| = 1$.

To motivate the procedure we recall the Möbius transformation $A(z; w)$ from Example 8.14. If our theorem is correct then the Mean Value Property for the harmonic function $u \circ A(-z, \cdot\,)$ implies that

$$\int_0^{2\pi} f(e^{it}) P_1(t, z)\, dt = u(z) = [u \circ A(-z, \cdot\,)](0)$$

$$= \frac{1}{2\pi} \int_0^{2\pi} u(A(-z, e^{i\theta}))\, d\theta = \frac{1}{2\pi} \int_0^{2\pi} f\left(\frac{e^{i\theta} + z}{1 + \bar{z}e^{i\theta}}\right) d\theta.$$

The crucial point of the proof is to establish the identity

$$\int_0^{2\pi} f(e^{it}) P_1(t, z)\, dt = \frac{1}{2\pi} \int_0^{2\pi} f\left(\frac{e^{i\theta} + z}{1 + \bar{z}e^{i\theta}}\right) d\theta \qquad \text{for } |z| < 1. \quad (12.1)$$

To do so we note that the map

$$w \mapsto A(z; e^{iw}) \frac{1 - \bar{z}}{1 - z}$$

is analytic and nowhere vanishing on a sufficiently small neighborhood of $[0, 2\pi]$, so it has a continuous and hence analytic logarithm there, say $i\Theta$. We may choose Θ so that $\Theta(0) = 0$. Letting $\theta := \Theta|_{[0, 2\pi]}$, we have the identity

$$e^{i\theta(t)} = \frac{e^{it} - z}{1 - \bar{z}e^{it}} \frac{1 - \bar{z}}{1 - z}.$$

A differentiation of this identity with respect to t shows that

$$\theta'(t) = \frac{1 - |z|^2}{|e^{it} - z|^2} = 2\pi P_1(t, z).$$

From property (b) of the Poisson kernel we infer that θ is an increasing bijection of $[0, 2\pi]$ onto $[0, 2\pi]$. And when we introduce t as new variable on the right hand side of (12.1) we get (12.1) by the change of variables formula.

Since the restriction of u to the boundary is continuous (being equal to f there), it suffices to finish the proof of the theorem to show that $u(z_n) \to f(e^{i\theta_0})$ for any sequence $\{z_n\}$ from the open ball $B(0, 1)$ converging to the arbitrary point $e^{i\theta_0}$ of the boundary. But f is continuous, hence bounded, so this is an immediate consequence of (12.1) and Lebesgue's dominated convergence theorem.   □

At this point we must mention the

Definition 12.10 (Dirichlet's Problem). *Given an open subset Ω of $\mathbb{R}^n$ and a continuous function f on the boundary $\partial\Omega$. Find a function u such that u is continuous on the closure $\overline{\Omega}$ of Ω, $\Delta u = 0$ in Ω and $u|_{\partial\Omega} = f$.*

The Dirichlet Problem in $\mathbb{R}^3$ may be interpreted as finding the electrostatic potential u inside a hollow cavity Ω, given the potential at the boundary $\partial\Omega$.

Proposition 12.11. *Let Ω be a bounded open subset of $\mathbb{R}^2$. Then the Dirichlet Problem has at most one solution, given the boundary values.*

Proof. We shall show that $u \in C(\overline{\Omega})$, u is harmonic in Ω and $u|_{\partial\Omega} = 0$ implies that $u = 0$.

Since $\overline{\Omega}$ is compact, u assumes its supremum somewhere in $\overline{\Omega}$, say at $z_0 \in \overline{\Omega}$. If $z_0 \in \partial\Omega$ then $u \leq 0(= u(z_0))$ everywhere. If $z_0 \in \Omega$ then u is by the Strong Maximum Principle constant throughout the connected component of z_0, so the value $u(z_0)$ is also taken at some boundary point. Thus $u \leq 0$ once again.

Replacing u by $-u$ we get $u = 0$.   □

Theorem 12.9 shows that the Dirichlet Problem for a disc has a solution. Even better, it provides us with an explicit formula for the (unique) solution.

We shall later need the following version of Proposition 12.11 in which we discard the boundedness assumption on the open set Ω.

Proposition 12.12. *Let $\Omega \neq \emptyset$ be an open subset of $\mathbb{R}^2$. Let $\overline{\Omega}$ denote its closure in the Riemann sphere. If $u \in C(\overline{\Omega})$ is harmonic in Ω, and $u(z) = 0$ for all $z \in \overline{\Omega} \setminus \Omega$, then $u = 0$.*

Proof. The Riemann sphere is compact, hence so is the closed subset $\overline{\Omega}$. We will first show that $u \leq 0$. By the compactness u assumes its maximum at some point $z_0 \in \overline{\Omega}$. If $z_0 \in \overline{\Omega} \setminus \Omega$, then $u(z_0) = 0$, and so $u \leq 0$ everywhere. The remaining possibility is $z_0 \in \Omega$. By the Strong Maximum Principle (Theorem 12.8) u is constant in the connected component of Ω that contains z_0. So the maximum value of u is attained throughout this component. The boundary of the component is not empty because the north pole is not in Ω (our hypothesis is $\Omega \subseteq \mathbb{R}^2$). So the maximum value of u is assumed at a boundary point. At that point $u = 0$, so $u \leq 0$ once again.

Replacing u by $-u$ we get $-u \leq 0$. We conclude that $u = 0$. $\qquad\square$

Theorem 12.13 (Harnack's Monotone Convergence Theorem). *Let Ω be an open connected subset of $\mathbb{R}^2$. Let $u_1 \leq u_2 \leq \cdots \leq u_n \leq \cdots$ be an increasing sequence of functions that are harmonic in Ω. Assume that there exists a point $z_0 \in \Omega$ such that $\sup\{u_n(z_0) \mid n = 1, 2, \ldots\} < \infty$.*

Then $\{u_n\}$ converges locally uniformly in Ω to a harmonic function.

Proof. Let $B[z_0, R] \subseteq \Omega$. We will for $n, p \in \mathbb{N}$ estimate the positive function $v_{n,p} := u_{n+p} - u_n$ in $B[z_0, R]$. Choose $\varepsilon > 0$ such that $B[z_0, R + \varepsilon] \subseteq \Omega$, and let us for the sake of simplicity assume that $z_0 = 0$. From the Poisson formula we get for any $z \in B(0, R + \varepsilon)$ that

$$v_{n,p}(z) = \int_0^{2\pi} v_{n,p}((R + \varepsilon)e^{i\phi}) P_{R+\varepsilon}(\phi, z) \, d\phi$$

and so, from Property (d) of the Poisson kernel, the estimates

$$\frac{1}{2\pi} \int_0^{2\pi} v_{n,p}([R + \varepsilon]e^{i\phi}) \frac{R + \varepsilon - |z|}{R + \varepsilon + |z|} \, d\phi \leq v_{n,p}(z)$$

$$\leq \frac{1}{2\pi} \int_0^{2\pi} v_{n,p}([R + \varepsilon]e^{i\phi}) \frac{R + \varepsilon + |z|}{R + \varepsilon - |z|} \, d\phi$$

which by the mean value property reduce to

$$\frac{R + \varepsilon - |z|}{R + \varepsilon + |z|} v_{n,p}(0) \leq v_{n,p}(z) \leq \frac{R + \varepsilon + |z|}{R + \varepsilon - |z|} v_{n,p}(0).$$

These inequalities show that $\{u_n\}$ converges uniformly in $B[z_0, R]$.

The argument actually shows that if $\{u_n\}$ converges in a point z, then it converges uniformly in any closed ball $B[z, R]$ in Ω. Thus the set $\{z \in \Omega \mid \{u_n(z)\}$ converges$\}\}$ is both open and closed in Ω, and so by connectedness it equals Ω or the empty set. But z_0 belongs to the set.

We have now shown that $\{u_n\}$ converges uniformly on any closed ball in Ω. Since any compact set in Ω can be covered by finitely many such balls the convergence is uniform on compact subsets of Ω.

Since u_n is continuous, so is the limit function u.

Let $z_1 \in \Omega$ and $B[z_1, R] \subseteq \Omega$. Letting $n \to \infty$ in the formula

$$u_n(z) = \int_0^{2\pi} u_n(z_1 + Re^{i\phi}) P_R(\phi, z - z_1) \, d\phi$$

we get for z in the ball $B(z_1, R)$ that

$$u(z) = \int_0^{2\pi} u(z_1 + Re^{i\phi}) P_R(\phi, z - z_1) \, d\phi.$$

It now follows from Theorem 12.9 that u is harmonic in $B(z_1, R)$. But z_1 was arbitrary in Ω. $\qquad\square$

12.3   On Liouville's theorem

The main topic of the present section is Liouville's theorem for harmonic functions, or rather various versions of it.

We shall see that Liouville's theorem (Theorem 1.11), which states that a bounded, entire function is constant, still holds, if the word *entire* is replaced by *harmonic* in $\mathbb{R}^n$. This result in turn reflects back onto entire functions, because the real and imaginary parts of a holomorphic function are harmonic (Theorem 12.2(a)). For instance it turns out that the boundedness assumption on the entire function can be relaxed: It suffices that its real or imaginary part is bounded above or below.

In the proofs we shall need the Mean Value Theorem for Balls:

Theorem 12.14. *Let u be harmonic in an open subset $\Omega \subseteq \mathbb{R}^2$. Let $B[z, R] \subseteq \Omega$. Then*

$$u(z) = \frac{1}{\pi R^2} \int_{B[z, R]} u(x, y) \, dx \, dy \qquad \text{for all } z \in \mathbb{R}^2.$$

Thus the value of u at any point is its average over any ball with the point as center.

Proof. The mean value theorem (Theorem 12.6) plus integration in polar coordinates.   □

Our first version (Theorem 12.15) of Liouville's theorem imposes no conditions at all on the growth at infinity of the harmonic function.

Theorem 12.15. *A positive, harmonic function on $\mathbb{R}^2$ is constant.*

The proof exploits the fact just proved that harmonic functions have the Mean Value Property over balls. It is adapted from Nelson's paper Nelson (1961) which is remarkable, because (1) it is very short (9 lines of text) and (2) it does not contain a single mathematical symbol!

Proof. First some notation. For any $R > 1$ let $A(z, R)$ for $z \in \mathbb{R}^2$ denote the annulus

$$A(z, R) := \{w \in \mathbb{R}^2 \mid R - 1 < |w - z| \leq R\} = B[z, R] \setminus B[z, R - 1].$$

Note for later use that

$$B[0, R] \setminus B[1, R] \subseteq A(0, R) \quad \text{and} \quad B[1, R] \setminus B[0, R] \subseteq A(1, R),$$

since $B[0, R - 1] \subseteq B[1, R]$ and $B[1, R - 1] \subseteq B[0, R]$. As a consequence we get for any positive function F that

$$\int_{B[0,R]\setminus B[1,R]} F(x, y)\, dxdy \leq \int_{A(0,R)} F(x, y)\, dxdy,$$

and

$$\int_{B[1,R]\setminus B[0,R]} F(x, y)\, dxdy \leq \int_{A(1,R)} F(x, y)\, dxdy.$$

We now start on the proof, where u denotes a positive, harmonic function on $\mathbb{R}^2$. We will show that u is constant, more precisely that $u(0) = u(z)$ for all $z \in \mathbb{R}^2$. To fix the ideas we take $z = 1$, because it is easy to generalize the arguments from $z = 1$ to any $z \neq 0$.

The Mean Value Theorem for Balls (Theorem 12.14) gives us the identity

$$\pi R^2[u(0) - u(1)] = \int_{B[0,R]} u(x, y)\, dxdy - \int_{B[1,R]} u(x, y)\, dxdy$$

$$= \int_{B[0,R]\setminus B[1,R]} u(x, y)\, dxdy - \int_{B[1,R]\setminus B[0,R]} u(x, y)\, dxdy, \quad (12.2)$$

We shall estimate the right hand side of the formula (12.2). Since u is positive we shall discard the second term on the right hand side and use that $B[0, R] \setminus B[1, R] \subseteq A(0, R)$ on the first term to get

$$\pi R^2 [u(0) - u(1)] \leq \int_{B[0,R] \setminus B[1,R]} u(x, y) \, dx \, dy \leq \int_{A(0,R)} u(x, y) \, dx \, dy.$$

Using the Mean Value Theorem on Balls on the right we get

$$\pi R^2 [u(0) - u(1)] \leq \pi [R^2 - (R - 1)^2] u(0) \leq 2\pi R u(0).$$

Letting $R \to \infty$ we get $u(0) - u(1) \leq 0$, i.e., $u(0) \leq u(1)$.

To get the converse inequality $u(1) \leq u(0)$ we proceed in the same way, using the formula that we get by multiplying (12.2) by minus one (equivalently by interchanging 0 and 1):

$$\pi R^2 [u(1) - u(0)]$$
$$= \int_{B[1,R] \setminus B[0,R]} u(x, y) \, dx \, dy - \int_{B[0,R] \setminus B[1,R]} u(x, y) \, dx \, dy.$$

$\square$

Corollary 12.16. *Any harmonic function on $\mathbb{R}^2$ which is bounded from below or from above, is constant.*

Proof. If the harmonic function is bounded below, say $u \geq C$, then Theorem 12.15 applies to the function $u - C$. If u is bounded above then consider the function $-u$, which is bounded below. $\square$

Exercise 12.10 presents a proof of Corollary 12.16 based on properties of the Poisson kernel.

We derive a beautiful extension of Liouville's classical theorem about bounded, entire functions from Corollary 12.16. In contrast to Theorem 1.11(a) the extension does not assume that the entire function in question is bounded.

Corollary 12.17. *An entire function whose real part is bounded from below or above, is constant.*

Proof. According to Corollary 12.16 the real part of the entire function is constant, so the result follows from Exercise 12.19(β). $\square$

Our next result (Theorem 12.18) is that all of Theorem 1.11 carries over to harmonic functions on $\mathbb{R}^n$.

Theorem 12.18. *Let u be a harmonic function on $\mathbb{R}^n$ with the following property: There exist positive constants A, B and $N \in \mathbb{N} \cup \{0\}$ such that $|u(x)| \leq A\|x\|^N + B$ for all $x \in \mathbb{R}^n$. Then u is a polynomial of degree at most N.*

We shall not need Theorem 12.18 later, so we refer to Exercise 12.18 for a proof. The proof we present is simple in the sense that it only uses the defining property $\Delta u = 0$ of a harmonic function u, but it is technically sophisticated, because it requires knowledge of the Fourier transformation of tempered distributions. It is due to Chernoff and Waterhouse (1972).

Of course, Corollary 12.16 implies in particular that bounded harmonic functions on $\mathbb{R}^2$ are constant. We shall later (see Example 16.7) need a result in which the boundedness condition is relaxed: We require the function to be sublinearly bounded. It is an easy consequence of Theorem 12.18, but we present a direct, more elementary proof, once again inspired by Nelson (1961).

Theorem 12.19. *Let u be a harmonic function on $\mathbb{R}^2$ satisfying: To each $\varepsilon > 0$ there exist $R > 0$ such that $|u(x)| \leq \varepsilon\|x\|$ for all $x \in \mathbb{R}^2$ for which $\|x\| \geq R$. Then u is a constant.*

Proof. We reuse the notation from the proof of Theorem 12.15 and as there we want to prove that $u(0) = u(1)$. Estimating the right hand side of the identity (12.2) we obtain for any $R > 2$ that

$$
\begin{aligned}
&\pi R^2 |u(0) - u(1)| \\
&\leq \int_{B[0,R] \setminus B[1,R]} |u(x,y)|\, dx dy + \int_{B[1,R] \setminus B[0,R]} |u(x,y)|\, dx dy \\
&\leq \int_{A(0,R)} |u(x,y)|\, dx dy + \int_{A(1,R)} |u(x,y)|\, dx dy \\
&\leq 2 \int_{\{z \in \mathbb{C} \mid R-2 \leq |z| \leq R+1\}} |u(x,y)|\, dx dy.
\end{aligned}
$$

Let $\varepsilon > 0$ be given. Choose R so large that $|u(x,y)| \leq \varepsilon\|(x,y)\|$ whenever $\|(x,y)\| \geq R-2$. For any such R we find using polar coordinates to integrate that

$$\pi R^2 |u(0) - u(1)| \le 2 \int_{\{z \in \mathbb{C} \mid R-2 \le |z| \le R+1\}} |u(x,y)| \, dx \, dy$$

$$\le 2 \int_{\{z \in \mathbb{C} \mid R-2 \le |z| \le R+1\}} \varepsilon \sqrt{x^2 + y^2} \, dx \, dy$$

$$= 4\pi\varepsilon[(R+1)^3 - (R-2)^3] = 4\pi\varepsilon[9R^2 - 9R + 9)],$$

so that

$$|u(0) - u(1)| \le 36\varepsilon[1 - R^{-1} + R^{-2}].$$

Letting $R \to \infty$ we get that $|u(0) - u(1)| \le 36\varepsilon$. We are done, since ε is arbitrary. $\square$

12.4   Jensen's formula

If f is holomorphic and without zeros in an open set containing the ball $B[0, R]$ then $\log|f|$ is harmonic inside the ball, so by the Mean Value Property

$$\log|f(0)| = \frac{1}{2\pi} \int_0^{2\pi} \log|f(Re^{i\phi})| \, d\phi.$$

In the case where f has zeros *Jensen's inequality*

$$\log|f(0)| \le \frac{1}{2\pi} \int_0^{2\pi} \log|f(Re^{i\phi})| \, d\phi$$

holds as a consequence of Jensen's formula (Theorem 12.20 below). Jensen's inequality also follows from the theory of subharmonic functions (next chapter).

Jensen's formula tells what the difference is between the right hand side and the left hand side of Jensen's inequality. It is useful tool for providing information on the zeros of analytic functions on a disc, when restrictions on their sizes are imposed.

Theorem 12.20 (Jensen's Formula). *Let f be meromorphic in an open set containing $B[0, R]$. Let $z_1, z_2, \ldots, z_n$ and $p_1, p_2, \ldots, p_m$ be the zeros and poles respectively of f in $B(0, R)$ (counted with multiplicity). Assume none of them is zero.*

Then $\phi \mapsto \log|f(Re^{i\phi})|$ is integrable over $(0, 2\pi)$, and

$$\log|f(0)| + \log\left|R^{n-m}\frac{p_1\cdots p_m}{z_1\cdots z_n}\right| = \frac{1}{2\pi}\int_0^{2\pi}\log|f(Re^{i\phi})|\,d\phi.$$

Proof. Let $z_1, z_2, \ldots, z_n, \ldots, z_N$ and $p_1, p_2, \ldots, p_m, \ldots, p_M$ be the zeros and poles respectively of f in the closed ball $B[0, R]$, counted with multiplicity. The function

$$F(z) := f(z)\prod_{j=1}^m \frac{R(z-p_j)}{R^2-\overline{p_j}z}\prod_{j=m+1}^M \frac{p_j-z}{p_j}\left\{\prod_{k=1}^n \frac{R(z-z_k)}{R^2-\overline{z_k}z}\prod_{k=n+1}^N \frac{z_k-z}{z_k}\right\}^{-1}$$

is holomorphic and zero free in a neighborhood of $B[0, R]$, so

$$\log|F(0)| = \frac{1}{2\pi}\int_0^{2\pi}\log|F(Re^{i\phi})|\,d\phi.$$

Putting $z = 0$, we find from the expression for F that

$$|F(0)| = |f(0)|\prod_{j=1}^m \frac{|p_j|}{R}\prod_{k=1}^n \frac{R}{|z_k|}$$

so $\log|F(0)|$ is the left hand side of the desired formula.

Writing $p_j = Re^{i\theta_j}$ for $j = m+1, \ldots, M$ and $z_k = Re^{i\phi_k}$ for $k = n+1, \ldots, N$ we find that

$$|F(Re^{i\theta})| = |f(Re^{i\theta})|\prod_{j=m+1}^M |1 - e^{i(\theta-\theta_j)}|\prod_{k=n+1}^N |1 - e^{i(\theta-\phi_k)}|^{-1}$$

so taking logarithms we get

$$\log|F(Re^{i\theta})| = \log|f(Re^{i\theta})| + \sum_{j=m+1}^M \log|1 - e^{i(\theta-\theta_j)}|$$

$$- \sum_{k=n+1}^N \log|1 - e^{i(\theta-\phi_k)}|.$$

To finish the proof it now suffices to show that $\phi \mapsto \log|1 - e^{i\phi}|$ is integrable over $]-\pi, \pi[$ and that

$$\int_{-\pi}^{\pi} \log|1 - e^{i\phi}| \, d\phi = 0.$$

(Now, this is a book on complex function theory, so we shall here evaluate the integral by help of the Cauchy Theorem. But the integral can actually be computed by elementary means Arora *et al.*, 1988, Young, 1986.)

Since $x \mapsto \log|x|$ is locally integrable, and the function $(1 - e^{i\phi})/\phi$ is continuous and never 0 on $[-\pi, \pi]$, the function

$$\phi \mapsto \log|1 - e^{i\phi}| = \log\left|\frac{1 - e^{i\phi}}{\phi}\right| + \log|\phi|$$

is integrable over $]-\pi, \pi[$.

To show that the value of the integral is 0 we note that

$$z \mapsto \frac{\text{Log } z}{z - 1}$$

is analytic in the right half plane, so that its integral along any circle $|z - 1| = r$, where $r \in]0, 1[$, is zero by the Cauchy Theorem, i.e.,

$$0 = \frac{1}{i} \int_{|z-1|=r} \frac{\text{Log } z}{z - 1} \, dz = \int_{-\pi}^{\pi} \text{Log}(1 - re^{i\phi}) \, d\phi.$$

Taking real parts we get

$$\int_{-\pi}^{\pi} \log|1 - re^{i\phi}| \, d\phi = 0.$$

We will apply the Dominated Convergence Theorem to this for a sequence $r_n \uparrow 1$. The desired result is then an immediate consequence, once we dominate the integrands by an integrable function.

By differentiation we see that the function

$$\phi \mapsto \left|\frac{1 - re^{i\phi}}{1 - e^{i\phi}}\right|$$

for fixed r is decreasing for $\phi > 0$, so

$$\frac{2}{|1 - e^{i\phi}|} > \left|\frac{1 - re^{i\phi}}{1 - e^{i\phi}}\right| \geq \frac{1 + r}{2} > \frac{1}{2}$$

and so

$$\log 2 > \log|1 - re^{i\phi}| \geq \log|1 - e^{i\phi}| - \log 2.$$

$\square$

Exercises

12.1. Let u be harmonic in the region Ω. Show that $u(\Omega)$ is open, unless u is constant.

Hint: Connected subsets of the real line are intervals.

12.2. Derive the Maximum Principle for holomorphic functions from the Strong Maximum Principle for harmonic functions.

12.3. Let f be a continuous, bounded and real valued function on the real line. Define for $x \in \mathbb{R}$ and $y > 0$ a function u by

$$u(x + iy) := \begin{cases} \frac{1}{\pi} \int_{-\infty}^{\infty} \frac{yf(t)}{(x-t)^2+y^2}\, dt & \text{for } y > 0, \\ f(x) & \text{for } y = 0. \end{cases}$$

(α) Show that u is bounded and harmonic in the upper half plane.

(β) Show that u is continuous in the closed upper half plane.

12.4. Let u be harmonic in the open region Ω and identically 0 in an open, non-empty subset of Ω. Show that $u = 0$ in all of Ω. (This property of harmonic functions is called the *Weak Unique Continuation Property.*)

Hint: The Unique Continuation for holomorphic functions.

12.5. Define the map $u\colon \{x + iy \in \mathbb{C} \mid y > 0\} \to\,]0, \pi[$ by

$u(z) :=$ the angle under which the interval $[0, 1]$ is seen from z.

Show that u is harmonic.

Hint: Consider the function $\text{Log}((z - 1)/z)$.

12.6. Show that the function

$$u(x, y) := \frac{y}{x^2 + y^2}$$

is harmonic on $\mathbb{R}^2$ except at the origin.

12.7. Is $|z|^n$ harmonic for some n?

12.8. Show that the polynomial $u(x, y) := x^3 - 3xy^2 - x$ is harmonic on $\mathbb{R}^2$. Find an $f \in \text{Hol}(\mathbb{C})$ such that $u = \text{Re}\, f$.

Same problems with $u(x, y) := \sin x \cosh y$.

12.9. Show the following formula for $r \in [0, 1[$ and $\theta \in \mathbb{R}$:

$$r \cos \theta = \frac{1}{2\pi} \int_{-\pi}^{\pi} \frac{(1 - r^2) \cos \phi}{1 - 2r \cos(\phi - \theta) + r^2} \, d\phi$$

Hint: Solve the Dirichlet Problem when the boundary function is $f(x + iy) = x$.

12.10 (Liouville's Theorem for Harmonic Functions). Let u be harmonic on $\mathbb{R}^2$ and assume that u is bounded from below. Show that u is a constant.

Hint: Property (d) of the Poisson kernel. See also Boas and Boas (1988) and Chernoff (1972).

12.11. Show that the Dirichlet Problem on $\{z \in \mathbb{C} \mid 0 < |z| < 1\}$ with the boundary function $f(z) = 0$ for $|z| = 1$, $f(0) = 1$ has no solution.

12.12. Let u be harmonic in the annulus $\{z \in \mathbb{C} \mid R_1 < |z| < R_2\}$. Show that

$$\frac{1}{2\pi} \int_0^{2\pi} u(re^{i\phi}) \, d\phi$$

is a linear function of $\log r$, i.e., there are real constants c and d such that for all $r \in]R_1, R_2[$:

$$\frac{1}{2\pi} \int_0^{2\pi} u(re^{i\phi}) \, d\phi = c \log r + d.$$

Hint: The Logarithmic Conjugation Theorem.

12.13. Let $R \in]0, \infty[$. *Inversion* in the circle $|z| = R$ is the map

$$p_R \colon \mathbb{R}^2 \setminus \{0\} \to \mathbb{R}^2 \setminus \{0\}$$

defined by

$$p_R(z) := \frac{R^2}{|z|^2} z \qquad \text{for } z \in \mathbb{R}^2 \setminus \{0\}.$$

Note that $p_R(z) = z$ when $|z| = R$, that points inside the circle are mapped outside and vice versa, and that

$$p_R \circ p_R = \text{identity} \qquad \text{on all of } \mathbb{R}^2 \setminus \{0\}.$$

(α) Let u be harmonic in $\mathbb{R}^2 \setminus \{0\}$. Show that so is $u \circ p_R$.

(β) Define and prove similar statements for the circle $|z - z_0| = R$ centered at $z_0 \in \mathbb{R}^2$.

12.14. Let u be harmonic in all of $\mathbb{R}^2$, positive on the open upper half space and 0 on the real axis.

(α) Let z_0 be a point in the open lower half plane. Let p denote inversion in the circle $|z - z_0| = R$ (see the previous exercise). Let z be a point in the open upper half plane such that $|z - z_0| < R$. Show that $u(z) \leq u(p(z))$.

   Hint: Apply the Weak Maximum Principle to $u \circ p - u$ in the domain D indicated on Fig. 12.1.

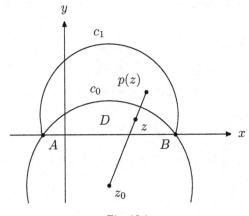

Fig. 12.1

The curve c_1 is the image by p of the interval $[A, B]$.

(β) Let $0 < y_1 < y_2$. Show that

$$u(x_1, y_1) \leq u(x_2, y_2) \qquad \text{for all } x_1, x_2 \in \mathbb{R}.$$

(γ) Show that $u(x, y)$ is independent of x for $y > 0$.

(δ) Show that $u(x, y) = cy$ where c is a constant.

(ε) Let f be an entire function that maps the upper half plane into itself and is real on the real line. Show that f is an affine function, i.e., has the form $f(z) = az + b$ where a and b are constants.

The exercise is taken from Tideman (1954). See also Boas and Boas (1988).

12.15. Prove the following theorem, due to Kellogg. It says that in special situations to get harmonicity, we do not need the Mean Value Property for all sufficiently small circles around the points of the domain, but only for one circle around each point.

> **Kellogg's Theorem.** *Let $u \in C(B[0,1])$ and assume that there for each $z \in B(0,1)$ exists an $r = r(z) \in \,]0, 1 - |z|[\,$ such that*
>
> $$u(z) = \frac{1}{2\pi} \int_{-\pi}^{\pi} u(z + re^{i\phi})\, d\phi.$$
>
> *Then u is harmonic on $B(0,1)$.*

Hint: Show that we may assume $u = 0$ on the unit circle by subtracting a suitable harmonic function. Assume $u(z) > 0$ for some $z \in B(0,1)$; obtain a contradiction by applying the assumption of the theorem to a point in $\{z \in B[0,1] \,|\, u(z) = |u|_\infty\}$ closest to the boundary.

12.16. As a curiosity derive the Fundamental Theorem of Algebra from Jensen's formula.

12.17. Let $a, b > 0$ and put $r(x,y) = 1 - ax^2 - by^2$.

(α) For $n \in \mathbb{N}$, write V_n for the real vector space of real polynomials in two variables of degree at most n. Define a linear operator $T\colon V_n \to V_n$ by setting $Tq = \Delta(rq)$. Show that T is a bijection.

Hint: The polynomial rq vanishes on the boundary of the ellipse $G = \{(x,y) \in \mathbb{R}^2 \,|\, ax^2 + by^2 < 1\}$.

(β) Let p be a polynomial of degree n. Show that there exists a harmonic polynomial q ($\Delta q = 0$) of degree at most n, such that $q(x,y) = p(x,y)$ for all x, y with $ax^2 + by^2 = 1$.

This exercise is adapted from Baker (1999).

12.18. In this exercise we prove Theorem 12.18, so let the assumptions and the notation be as in that theorem.

(α) Note that u is a tempered distribution. Let $\widehat{u}$ denote the Fourier transform of u.

(β) Show that $\|\xi\|^2 \widehat{u}(\xi) = 0$ for $\xi \in \mathbb{R}^n$.
   Hint: $\Delta u = 0$.

(γ) Show that the support of $\widehat{u}$ is contained in $\{0\}$.

(δ) Note $\widehat{u}$ then is a linear combination of derivatives of the delta function.

   Hint: Hörmander (1963, Theorem 1.5.3) or Hörmander (1990, Theorem 2.3.4).

(ε) Conclude that u is a polynomial.

(ζ) Show that the degree can be at most N.

   Hint: The growth assumption.

12.19. Let f be an entire function and let us write it in the form $f(x+iy) = u(x,y) + iv(x,y)$, where u and v are real-valued functions on $\mathbb{R}^2$. In this exercise we extend Liouville's theorem about f. In contrast to the earlier result (Theorem 1.11) we impose growth conditions only on the real part u of f, not on f. So at our point of departure nothing is assumed about the function v.

(α) Assume that $u = 0$. Show that f is a constant.

   Hint: The Cauchy-Riemann equations.

(β) Assume that u is constant. Show that f is a constant.

(γ) Assume that u is a polynomial in the two variables x and y. Show that $f(z)$ is a polynomial in z.

   Hint: Differentiate to get a function with vanishing real part.

(δ) Assume that u satisfies the growth condition of Theorem 12.18. Show that $f(z)$ is a polynomial in z of degree at most N.

12.20. Let u be harmonic and bounded in $\mathbb{C} \setminus \{0\}$. Show that u is constant. *Hint:* Consider the function $z \mapsto u(e^z)$. A related result is in Theorem 12.5.

Chapter 13

Subharmonic Functions

13.1 Technical results on upper semi-continuous functions

Definition 13.1. *Let X be a topological space. A function $u\colon X \to [-\infty, \infty[$ is said to be **upper semi-continuous** (usc) if the following holds:*

> *To every $x_0 \in X$ and $M > u(x_0)$ there exists a neighborhood U of x_0 such that $M > u(x)$ for all $x \in U$.*

A continuous real valued function is clearly usc. Note for later use that an usc function is Borel measurable – indeed, the set $\{z \in \Omega \mid u(z) < \alpha\}$ is open for any $\alpha \in \mathbb{R}$.

Lemma 13.2.

(a) *If u_1 and u_2 are usc then so are $u_1 + u_2$ and $\max\{u_1, u_2\}$.*

(b) *If u is usc and $\lambda \in [0, \infty[$ then λu is usc.*

(c) *If u_α is usc for each α in a non-empty index set A then $\inf_{\alpha \in A} u_\alpha$ is also usc.*

Theorem 13.3. *An upper semi-continuous function on a compact topological space attains its supremum. In particular it is bounded from above.*

Proof. As the standard proof for continuous functions. □

Theorem 13.4. *Let $u\colon \Omega \to [-\infty, \infty[$ where Ω is an open subset of the complex plane.*
Then u is usc if and only if there exists a decreasing sequence $u_1 \geq u_2 \geq \cdots$ of continuous real valued functions on Ω such that $u_n(z) \to u(z)$ as $n \to \infty$ for each $z \in \Omega$.

Proof. The theorem is trivial in case $u(z) = -\infty$ for all z, so we will from now on assume that there exists a $z_0 \in \Omega$ such that $u(z_0) \in \mathbb{R}$.

The "if" direction of the theorem is part of the lemma above. So assume from now on that u is usc.

We will first treat the special case of u being bounded from above, say $u(z) < M$ for all $z \in \Omega$. The trick is here to consider the functions $u_n, n \in \mathbb{N}$ given by

$$u_n(z) := \sup_{y \in \Omega}\{u(y) - n|y - z|\} \qquad \text{for } z \in \Omega.$$

u_n is finite valued, since $M \geq u_n(z) \geq u(z_0) - n|z_0 - z|$.

We next prove that u_n is continuous: Given $z \in \Omega$ and $\varepsilon > 0$ we choose $y \in \Omega$ such that

$$u_n(z) < u(y) - n|y - z| + \varepsilon$$

Then for any $z' \in \Omega$ we have

$$u_n(z) - u_n(z') < \{u(y) - n|y - z| + \varepsilon\} - \{u(y) - n|y - z'|\}$$
$$= n\{|z' - y| - |y - z|\} + \varepsilon \leq n|z' - z| + \varepsilon$$

and since $\varepsilon > 0$ is arbitrary we see that

$$u_n(z) - u_n(z') \leq n|z - z'|.$$

Interchanging z and z' we get

$$|u_n(z) - u_n(z')| \leq n|z - z'|$$

proving the continuity.

Next we note that

$$u(y) - n|y - z| \geq u(y) - (n + 1)|y - z|$$

which implies that $u_n(z) \geq u_{n+1}(z)$, i.e., that $\{u_n\}$ is a decreasing sequence.

Since

$$u_n(z) = \sup_y\{u(y) - n|y - z|\} \geq u(z) - n|z - z| = u(z)$$

it is left to prove that $\lim_{n \to \infty} u_n(z) \leq u(z)$, i.e., for any prescribed $K > u(z)$ that $u_n(z) \leq K$ for sufficiently large n. For that we use that u is usc: There is a ball $B(z, \delta)$ such that $u(y) < K$ for all $y \in B(z, \delta)$. Now,

$$u_n(z) = \sup_y\{u(y) - n|y - z|\}$$
$$= \max\{\sup_{|y-x|<\delta}\{u(y) - n|y - z|\}, \sup_{|y-x|\geq\delta}\{u(y) - n|y - z|\}\}$$
$$\leq \max\{K, M - n\delta\}$$

so if n is so large that $M - n\delta < K$ we have $u_n(z) \leq K$ as desired.

We shall finally deal with the general case in which u is no longer necessarily bounded from above. Choose an increasing homeomorphism $\phi \colon [-\infty, \infty] \to [-\infty, 0]$ (e.g., $\phi(t) = -\exp(-t)$). The function $\phi \circ u$ is usc and bounded from above, so the construction in the special case treated above yields a sequence $\{v_n\}$ of continuous real valued functions with the property that $v_n(z) \downarrow \phi(u(z))$ as $n \to \infty$ for all $z \in \Omega$. It suffices to verify that $-\infty < v_n(z) < 0$ for all $z \in \Omega$, because we may then choose $u_n := \phi^{-1} \circ v_n$.

The left inequality is trivial because v_n is finite valued. To settle the other inequality we use the definition of v_n, i.e.,

$$v_n(z) = \sup_y \{\phi(u(y)) - n|y - z|\}.$$

By hypothesis $u(z) < \infty$, so $a := \phi(u(z)) < 0$, and by upper semi-continuity of $\phi \circ u$ there exists $\delta > 0$ such that

$$\phi(u(y)) < \tfrac{a}{2} < 0 \qquad \text{for all } y \in B(z, \delta).$$

Thus

$$v_n(z) = \max\Big\{ \sup_{|y-z|<\delta} \{\phi(u(y)) - n|y - z|\}, \ \sup_{|y-z|\geq\delta} \{\phi(u(y)) - n|y - z|\}\Big\}$$
$$\leq \max\{\tfrac{a}{2}, 0 - n\delta\} < 0. \qquad \qquad \square$$

13.2   Introductory properties of subharmonic functions

Throughout the remainder of this chapter Ω denotes an open, connected, non-empty subset of the complex plane.

Definition 13.5. *A map* $u \colon \Omega \to [-\infty, \infty[$ *is said to be* **subharmonic in** Ω *if*

(a) *u is upper semi-continuous (usc) and not identically $-\infty$.*

(b) *For each $z \in \Omega$ there is a ball $B(z, R_z) \subseteq \Omega$ such that*

$$u(z) \leq \frac{1}{2\pi} \int_0^{2\pi} u(z + re^{i\theta}) \, d\theta \qquad \text{for all } r \in \,]0, R_z[.$$

Remark. The integral in (b) makes sense: Fix z and $r > 0$. Since u is usc, then the function $\theta \mapsto u(z + re^{i\theta})$ is Borel measurable and bounded from above. Thus u is essentially a negative function. The value of the integral may a priori be $-\infty$; as we shall see in Proposition 13.9, it is actually finite. There we shall also see that the inequality (b) holds for all r such that $B(z,r) \subseteq \Omega$, not just for sufficiently small r.

A harmonic function is subharmonic by the Mean Value Property.

Theorem 13.6 (The Strong Maximum Principle). *Let u be subharmonic in Ω. If there is a $z_0 \in \Omega$ such that $u(z_0) = \sup_{z \in \Omega} u(z)$, then u is a constant.*

In particular, if $\overline{\Omega}$ is compact, u subharmonic on Ω, usc in $\overline{\Omega}$ and ≤ 0 on $\partial\Omega$, then $u \leq 0$ throughout Ω.

The Maximum Principle is one of the crucial properties of subharmonic functions.

Proof. The set $M := \{z \in \Omega \mid u(z) = u(z_0)\}$ is closed in Ω since u is usc. We will show that M is open as well, so let $z \in M$ be arbitrary. Claim:

$$u(z + re^{i\theta}) = u(z) \qquad \text{for all } \theta \in \mathbb{R} \text{ and } r \in \,]0, R_z[.$$

We prove the claim by contradiction. If it is false then there are $\theta_0 \in [0, 2\pi[$, $r_0 \in \,]0, R_z[$ and $\varepsilon > 0$ so that

$$u(z + r_0 e^{i\theta_0}) < u(z) - \varepsilon.$$

Now, u being usc,

$$u(z + r_0 e^{i\theta}) < u(z) - \varepsilon$$

for all θ in interval $I \subseteq \,]0, 2\pi[$.

Letting $J := \,]0, 2\pi[\,\setminus I$ we get

$$
\frac{1}{2\pi} \int_0^{2\pi} u(z + r_0 e^{i\theta})\, d\theta = \frac{1}{2\pi}\left(\int_I + \int_J\right) u(z + r_0 e^{i\theta})\, d\theta
$$

$$
\leq \frac{1}{2\pi} \int_I (u(z) - \varepsilon)\, d\theta + \frac{1}{2\pi} \int_J u(z)\, d\theta
$$

$$
= u(z) - \varepsilon \frac{|I|}{2\pi}
$$

which contradicts (b) and hence proves the claim.

The claim shows that u equals $u(z_0)$ in a ball around z so M is open. By connectedness $M = \Omega$. $\qquad\square$

Theorem 13.7. *Let u be subharmonic in Ω, and let $B[a, R]$ be a closed ball contained in Ω. Then*

$$u(z) \leq \int_0^{2\pi} u(a + Re^{i\theta}) P_R(\theta, z - a)\, d\theta \qquad \text{for all } z \in B(a, R)$$

where P_R denotes the Poisson kernel.

Proof. Since u is usc it is the pointwise limit of a decreasing sequence $\{u_n\}$ of continuous functions. The function v_n defined by

$$v_n(z) := \begin{cases} \displaystyle\int_0^{2\pi} u_n(a + Re^{i\theta}) P_R(\theta, z - a)\, d\theta & \text{for } z \in B(a, R), \\[2ex] u_n(z) & \text{for } |z - a| = R, \end{cases}$$

is by Theorem 12.9 continuous in $B[a, R]$ and harmonic in its interior. $u - v_n$ is usc in $B[a, R]$, subharmonic in its interior and ≤ 0 on its boundary. So by Theorem 13.6 $u - v_n \leq 0$ throughout $B(a, R)$. The result follows now from the Monotone Convergence Theorem. $\qquad\square$

Theorem 13.7 explains the terminology subharmonic: The graph of u is below the harmonic function which has the same boundary values on the circle $|z - a| = R$ as u. So subharmonic functions are related to harmonic functions in the same way as convex functions are related to linear functions in one dimension.

13.3 On the set where $u = -\infty$

In this Section we deduce that the set of points in which a subharmonic function takes the value $-\infty$ is so small that all our integrals are finite.

Proposition 13.8. *A function u which is subharmonic on Ω, is locally integrable in Ω.*

In particular, the set $\{z \in \Omega \mid u(z) = -\infty\}$ is a null set, and u cannot be identically $-\infty$ on any non-empty open subset of Ω.

It can be proved that $\{z \in \Omega \mid u(z) = -\infty\}$ is a set of capacity 0 (see Hayman and Kennedy, 1976, Theorem 5.32).

Proof. We first observe that u is integrable over $B(a, R)$ if $B[a, R]$ is contained in Ω and $u(a) > -\infty$:

Indeed, multiplying the inequality

$$u(a) \leq \frac{1}{2\pi} \int_0^{2\pi} u(a + re^{i\theta}) \, d\theta$$

by πr and integrating from 0 to $R = R_a$ we get

$$-\infty < \tfrac{1}{2}\pi R^2 u(a) \leq \frac{1}{2} \int_0^R \int_0^{2\pi} u(a + re^{i\theta}) \, d\theta \, r \, dr$$

$$= \iint_{B(a,R)} u(x,y) \, dx \, dy.$$

As a consequence, if u is not locally integrable at a then u must be identically $-\infty$ throughout some neighborhood of a. Thus the complement in Ω of the set

$$\{z \in \Omega \mid u \text{ is locally integrable at } z\}$$

is the set

$$\{z \in \Omega \mid u \text{ is identically } -\infty \text{ in some neighborhood of } z\}.$$

Both these sets are clearly open, so by the connectedness of Ω one of them is empty. But u is not identically $-\infty$ by assumption, so the latter set is empty, and so the first one is Ω. □

The integral in the defining inequality (b) of Definition 13.5 may a priori have the value $-\infty$. It is, however, finite:

Proposition 13.9. *Let u be subharmonic on Ω. Let $a \in \Omega$ and $R > 0$ be such that $B[a, R] \subseteq \Omega$, and let I be the function given by*

$$I(r) := \frac{1}{2\pi} \int_0^{2\pi} u(a + re^{i\theta}) \, d\theta \qquad \text{for } r \in [0, R].$$

Then I is increasing, finite and continuous for $r \in \,]0, R]$, and the map $t \mapsto I(e^t)$ is convex in the interval $\,]-\infty, \operatorname{Log} R]$.

   In particular

$$u(a) \leq \frac{1}{2\pi} \int_0^{2\pi} u(a + re^{i\theta}) \, d\theta \qquad \text{for } r \in [0, R]$$

a result that extends (b) from Definition 13.5.

Proof. To establish that I is increasing we use the notation of the proof of Theorem 13.7. We saw that $u \leq v_n$. For $r < R$ we have by the Mean Value Property for harmonic functions that

$$\frac{1}{2\pi} \int_0^{2\pi} u_n(a + Re^{i\theta}) \, d\theta = v_n(a) = \frac{1}{2\pi} \int_0^{2\pi} v_n(a + re^{i\theta}) \, d\theta$$

$$\geq \frac{1}{2\pi} \int_0^{2\pi} u(a + re^{i\theta}) \, d\theta$$

and the Monotone Convergence Theorem takes over.

We next show by contradiction that I is finite valued on $]0, R]$. Assume that $I(r_0) = -\infty$ for some $r_0 \in]0, R]$. Then $I(r) = -\infty$ for all $r \in [0, r_0]$, because I is increasing, i.e.,

$$\frac{1}{2\pi} \int_0^{2\pi} u(a + re^{i\theta}) \, d\theta = -\infty \qquad \text{for all } r \in [0, r_0].$$

This implies (integrate with respect to r from 0 to r_0) that

$$\int_{B(a,r_0)} u(x, y) \, dx dy = -\infty$$

contradicting the local integrability (Proposition 13.8).

A convex function on an open interval is continuous, so left is only the convexity statement. We postpone a proof of that to the next section where we will possess more machinery. $\qquad\square$

13.4 Approximation by smooth functions

The following characterization of smooth subharmonic functions is very useful and connects once more the concepts of subharmonic and harmonic functions.

Theorem 13.10. *Let $u \in C^2(\Omega, \mathbb{R})$. Then u is subharmonic on Ω if and only if $-\Delta u \leq 0$ on Ω.*

Proof. Suppose first that $\Delta u \geq 0$. We shall for $z \in \Omega$ verify that

$$u(z) \leq \frac{1}{2\pi} \int_0^{2\pi} u(z + re^{i\theta}) \, d\theta \qquad \text{for all sufficiently small } r > 0.$$

Let us for simplicity take $z = 0$.

The expression for the Laplacian Δ in polar coordinates (r, θ) is

$$\Delta u = u_{rr} + \tfrac{1}{r} u_r + \tfrac{1}{r^2} u_{\theta\theta}. \tag{13.1}$$

When we integrate (13.1) with respect to θ from 0 to 2π (noting that the last term drops out because $\phi \mapsto u_\theta(re^{i\phi})$ is periodic in ϕ) we get with the notation

$$I(r) := \frac{1}{2\pi} \int_0^{2\pi} u(re^{i\theta}) \, d\theta$$

that

$$\frac{1}{2\pi} \int_0^{2\pi} \Delta u(re^{i\theta}) \, d\theta = I'' + \tfrac{1}{r} I' = \tfrac{1}{r}(rI')'. \tag{13.2}$$

Combining (13.2) with $\Delta u \geq 0$ we see that $r \mapsto rI'(r)$ is an increasing function. Since $I'(r)$ clearly is bounded (for small r), $rI'(r) \to 0$ as $r \to 0$. But then $rI'(r) \geq 0$ so that $I'(r) \geq 0$ and thus I is an increasing function of r. Now, as desired

$$u(0) = I(0) \leq I(r) = \frac{1}{2\pi} \int_0^{2\pi} u(re^{i\theta}) \, d\theta.$$

Let next u be subharmonic. We assume that $-\Delta u(z) > 0$ for some $z \in \Omega$ and shall derive a contradiction. Take again $z = 0$ for the sake of simplicity.

Applying the first part of the proof to $-u$ we get that I is decreasing for small r. But u is subharmonic so I is actually increasing (Proposition 13.9); hence I is constant and the right hand side of (13.2) is zero. But that contradicts the assumption $-\Delta u(0) > 0$. $\square$

A subharmonic function need not be C^2. But it can be approximated by smooth subharmonic functions:

Theorem 13.11. *Let u be subharmonic in Ω. Let ω be an open subset of Ω with the property that $\mathrm{dist}(\omega, \partial\Omega) > 0$.*

Then there is a decreasing sequence $u_1 \geq u_2 \geq \cdots$ of subharmonic functions in $C^\infty(\omega)$ such that $u_n(z) \to u(z)$ as $n \to \infty$ for all $z \in \Omega$.

Proof. The standard mollifying method (convolution) works. It goes as follows:

Let $0 < R < \tfrac{1}{2} \mathrm{dist}(\omega, \partial\Omega)$. Then the function

$$U(z) := \begin{cases} u(z) & \text{when } \mathrm{dist}(z, \omega) < R, \\ 0 & \text{otherwise on } \mathbb{R}^2, \end{cases}$$

is locally integrable in $\mathbb{R}^2$, since u is locally integrable in Ω. Let $\phi \in C_0^\infty(\mathbb{R}^2)$ be a positive radial function such that $\operatorname{supp}\phi \subseteq B[0, R]$ and $\int_{\mathbb{R}^2}\phi = 1$ and define for $n = 1, 2, \ldots$ the functions

$$\phi_n(z) := n^2 \phi(nz) \qquad \text{for } z \in \mathbb{R}^2.$$

Then

$$\phi_n \in C_0^\infty(\mathbb{R}^2), \quad \operatorname{supp}\phi_n \subseteq B\left[0, \tfrac{1}{n}R\right] \quad \text{and} \quad \int_{\mathbb{R}^2}\phi_n = 1.$$

Let U_n be the convolution of U and ϕ_n, i.e., the function given by the expression (the integration is over $\mathbb{R}^2$):

$$U_n(z) := \int U(y)\phi_n(z - y)\, dy \qquad \text{for } z \in \mathbb{R}^2.$$

Clearly $U_n \in C^\infty(\mathbb{R}^2)$, because we may differentiate under the integral sign.

Let us next prove that U_n is subharmonic on ω. Let $z \in \Omega$ and let $0 < r \leq R$. Then (using Proposition 13.9 and that ϕ is radial)

$$
\begin{aligned}
U_n(z) &= \int U(y)\phi_n(z-y)dy = \int U(z+y)\phi_n(y)\, dy \\
&= \int_{B(0,\frac{R}{n})} U(z+y)\phi_n(y)\, dy = \int_{B(0,\frac{R}{n})} u(z+y)\phi_n(y)\, dy \\
&\leq \int_{B(0,\frac{R}{n})} \left\{ \frac{1}{2\pi} \int_0^{2\pi} u(z+y+re^{i\theta})d\theta \right\} \phi_n(y)\, dy \\
&= \frac{1}{2\pi} \int_0^{2\pi} \int_{B(0,\frac{R}{n})} u(z+y+re^{i\theta})\phi_n(y)\, dy d\theta \\
&= \frac{1}{2\pi} \int_0^{2\pi} U_n(z+re^{i\theta})\, d\theta
\end{aligned}
$$

showing that U_n is subharmonic on ω.

We continue by proving that $U_n(z)$ is decreasing in n when $z \in \omega$:

$$
\begin{aligned}
U_n(z) &= \int_{B(0,\frac{R}{n})} u(z+y)\phi_n(y)\, dy = \int_{B(0,\frac{R}{n})} u\left(z+\frac{y}{n}\right)\phi(y)\, dy \\
&= \frac{1}{2} \int_0^R \int_0^{2\pi} u\left(z+\frac{\rho e^{i\theta}}{n}\right) d\theta\, \phi(\rho)\, \rho d\rho.
\end{aligned}
$$

The inner integral is increasing with ρ (Proposition 13.9), so we get

$$U_n(z) \geq \frac{1}{2} \int_0^R \int_0^{2\pi} u\left(z+\frac{\rho e^{i\theta}}{n+1}\right) d\theta\, \phi(\rho)\, \rho d\rho = U_{n+1}(z).$$

The computation just made also reveals that

$$U_n(z) \geq \frac{1}{2} \int_0^R 2\pi u(z)\phi(\rho)\rho\, d\rho = u(z)$$

so now it is only left to show that

$$\lim_{n\to\infty} U_n(z) \leq u(z) \qquad \text{for all } z \in \omega$$

i.e., given $K > u(z)$ there exists an n so that $U_n(z) \leq K$.

By the usc $u < K$ in a neighborhood V of z. When n is so large that $B(z, R/n) \subseteq V$ we have

$$U_n(z) = \int_{B(0, \frac{R}{n})} u(z+y)\phi_n(y)\, dy \leq \int_{B(0, \frac{R}{n})} K\phi_n(y)\, dy = K.$$

□

We shall now keep a promise:

Proof of Proposition 13.9 (continued). For simplicity of exposition we take $a = 0$ as usual. Assume first that $u \in C^\infty$. We shall show that the function

$$h(t) := \frac{1}{2\pi} \int_0^{2\pi} u(e^t e^{i\theta})\, d\theta$$

is convex for $-\infty < t < \operatorname{Log} R$.

During the proof of the first part of Theorem 13.10 we observed that $r \mapsto rI'(r)$ is an increasing function. Hence so is

$$t \mapsto e^t I'(e^t) = \frac{d}{dt}(I(e^t)) = h'(t)$$

implying that h is convex.

If u is no longer assumed C^∞ we approximate by a sequence of C^∞ subharmonic functions $u_n \downarrow u$. Then

$$h_n(t) := \frac{1}{2\pi} \int_0^{2\pi} u_n(e^t e^{i\theta})\, d\theta$$

is a decreasing sequence of convex functions, so its limit, which is h, is convex as well. □

13.5 Constructing subharmonic functions

Proposition 13.12.

(a) *If u_1 and u_2 are subharmonic in Ω then so are u_1+u_2 and $\max\{u_1, u_2\}$.*

(b) *If u is subharmonic in Ω then so is λu for any $\lambda \in [0, \infty[$.*

(c) *If $\{u_n\}$ is a decreasing sequence of subharmonic functions in Ω then $\lim u_n$ is subharmonic in Ω, unless it is identically $-\infty$.*

Proof. Left to the reader. $\qquad\qquad\square$

For the next result we need (a particular case of) Jensen's inequality for integrals.

Theorem 13.13 (Jensen's Inequality for Integrals). *If f is a real valued function in $L^1(0, 2\pi)$ such that $a < f(t) < b$ and if ϕ is convex on $]a, b[$, then*

$$\phi\left\{\int_0^{2\pi} f(\theta)\frac{d\theta}{2\pi}\right\} \le \int_0^{2\pi} (\phi \circ f)(\theta)\frac{d\theta}{2\pi}.$$

The cases $a = -\infty$ and $b = \infty$ are allowed.

Proof. For a proof consult Rudin (1974, Theorem 3.3). $\qquad\qquad\square$

Theorem 13.14.

(a) *If u is harmonic in Ω and ϕ is convex on the range of u, then $\phi \circ u$ is subharmonic in Ω.*

(b) *If u is subharmonic in Ω and ϕ is a convex increasing function on the range of u, then $\phi \circ u$ is subharmonic in Ω. (We put $\phi(-\infty) = \lim_{t \to -\infty} \phi(t)$.)*

Proof. Ad (a): A convex function on the real line is automatically continuous, so $\phi \circ u$ is continuous. From the Mean Value Property

$$u(a) = \int_0^{2\pi} u(a + re^{i\theta})\frac{d\theta}{2\pi}$$

we deduce via Jensen's inequality for integrals that

$$(\phi \circ u)(a) \le \int_0^{2\pi} (\phi \circ u)(a + re^{i\theta})\frac{d\theta}{2\pi}$$

so $\phi \circ u$ is subharmonic.

Ad (b): Similar arguments to the ones from (a) work here. $\qquad\qquad\square$

Theoretical Examples 13.15.

(a) Let $f \in \text{Hol}(\Omega)$. If f is not identically 0 then $\text{Log}|f|$ and $\text{Log}^+|f| := \max\{\text{Log}|f|, 0\}$ are subharmonic in Ω.

(b) If f and g are holomorphic in Ω then $|f|^p|g|^q$ is subharmonic for all $p, q \geq 0$.

(c) If u is harmonic in Ω then $|u|^p$ is subharmonic in Ω for all $p \geq 1$.

Proof. Ad (a): $\text{Log}|f|$ is clearly usc and not identically $-\infty$. If $f(z) = 0$ at a point $z \in \Omega$ then $\text{Log}|f(z)| = -\infty$, so the inequality (b) from Definition 13.5 is obviously satisfied at z. If $f(z) \neq 0$ then $\text{Log}|f| = \text{Re}(\text{Log } f)$ is harmonic around z, hence subharmonic.

$\text{Log}^+|f|$ is the composition of $\text{Log}|f|$ with the convex increasing function $t \mapsto t1_{]0,\infty[}(t)$.

Ad (b): Assume first that neither f nor g is identically 0. Then $\text{Log}|f|$ and $\text{Log}|g|$ are subharmonic, hence so is $p\,\text{Log}|f| + q\,\text{Log}|g|$. Left is just to compose with the exponential map. The modifications in the cases where f and/or g are identically 0 are left to the reader.

Ad (c): Theorem 13.14(a). □

Practical examples are now numerous, e.g, $u(x, y) = |x|$ or more generally $u(x, y) = |x|^p$ for $p \geq 1$.

13.6   Applications

We present three applications: One is a proof of Radó's theorem, the other is a treatment of H^p-spaces, and the third is a proof of the F. and R. Nevanlinna Theorem.

13.6.1   *Radó's theorem*

Our first application of the theory of subharmonic functions is a proof of Radó's theorem.

Theorem 13.16 (Radó's Theorem). *If $f \in C(\Omega)$ is holomorphic in $\{z \in \Omega \mid f(z) \neq 0\}$ then f is holomorphic in all of Ω.*

Proof. We may assume that f is not identically 0. We write $f = u + iv$, where u and v are real valued.

The function $s := \mathrm{Log}|f|$ is subharmonic in Ω (same proof as in (a) of Theoretical Examples 13.15 above). So is $u + \varepsilon s$ for any $\varepsilon > 0$ (same proof).

We will show that f is holomorphic in any ball in Ω; for simplicity of notation we assume that the ball is $B(0,1)$ and that $B[0,1] \subseteq \Omega$. We assume furthermore that $|f(z)| \le 1$ for $z \in B[0,1]$. Note that then $s \le 0$ on $B(0,1)$.

Now, for any $z \in B[0,1]$ we have by Theorem 13.7 that

$$u(z) + \varepsilon s(z) \le \frac{1}{2\pi} \int_0^{2\pi} \{u(e^{i\theta}) + \varepsilon s(e^{i\theta})\} P_1(\theta, z) \, d\theta$$

$$\le \frac{1}{2\pi} \int_0^{2\pi} u(e^{i\theta}) P_1(\theta, z) \, d\theta.$$

At those points $z \in B(0,1)$ where $f(z) \ne 0$ we get by letting ε decrease to 0 that

$$u(z) \le \frac{1}{2\pi} \int_0^{2\pi} u(e^{i\theta}) P_1(\theta, z) \, d\theta.$$

However, u is continuous and the points where $f(z) \ne 0$ are dense in $B(0,1)$ (Theorem 13.10), so the inequality holds for all points $z \in B(0,1)$.

Our arguments work in the same way for $-f$, i.e., with $-u$ instead of u. Thus

$$u(z) = \frac{1}{2\pi} \int_0^{2\pi} u(e^{i\theta}) P_1(\theta, z) \, d\theta$$

which shows (Lemma 7.8) that u is harmonic. In particular $u \in C^\infty(B(0,1))$. Replacing f by if we get $v \in C^\infty(B(0,1))$, so $f \in C^\infty(B(0,1))$. The function f satisfies the Cauchy-Riemann equations (2.2) on the dense set $\{z \in B(0,1) \,|\, f(z) \ne 0\}$, and hence on all of $B(0,1)$. So f is holomorphic. $\square$

13.6.2   Hardy spaces

Definition 13.17. *Let f be holomorphic in the unit disc. f is said to belong to the **Hardy space** $H^p, p \in \,]0, \infty[$, if*

$$\sup_{0 < r < 1} \frac{1}{2\pi} \int_0^{2\pi} |f(re^{i\theta})|^p \, d\theta < \infty.$$

We leave it to the reader to verify that H^p is a linear subspace of $\mathrm{Hol}(B(0,1))$.

Theorem 13.18. *If $f \in H^p$ where $p \geq 1$ then there exists exactly one $f^* \in L^p(S^1)$ such that*

$$f(z) = \frac{1}{2\pi} \int_0^{2\pi} P_1(t, z) f^*(e^{it}) \, dt \qquad \text{for all } |z| < 1.$$

Proof. We prove the theorem for $p = 1$; for $p > 1$ the argument is similar, but simpler.

Let $0 < r_1 < r_2 < \cdots < 1$ be an increasing sequence such that $r_n \to 1$ as $n \to \infty$, and define

$$g_n(e^{it}) := |f(r_n e^{it})|^{\frac{1}{2}} \qquad \text{for } t \in \mathbb{R}.$$

That $f \in H^1$ means that $\{g_n\}$ is a bounded subset of $L^2(S^1)$. Since any closed ball in L^2 is weakly sequentially compact (see Dunford and Schwartz, 1958, Corollary IV.8.4), we may (possibly by taking a subsequence) assume that there exists a function $g \in L^2(S^1)$ such that $g_n \to g$ weakly in L^2. Recalling that $|f|^{\frac{1}{2}}$ is subharmonic (Theoretical Examples 13.15(b)) we get by Theorem 13.7 for $m < n$ that

$$g_m(e^{it}) \leq \int_0^{2\pi} g_n(e^{i\theta}) P_{r_n}(\theta, z) \, d\theta \qquad \text{where } z = r_m e^{it}$$

and letting $n \to \infty$ it follows that

$$g_m(e^{it}) \leq \int_0^{2\pi} g(e^{i\theta}) P_1(\theta, z) \, d\theta.$$

By the Cauchy-Schwarz inequality we get

$$g_m(e^{it})^2 \leq \int_0^{2\pi} \{g(e^{i\theta})(P_1(\theta, z)^{\frac{1}{2}})\}^2 \, d\theta \int_0^{2\pi} P_1(\theta, z) \, d\theta$$

$$= \int_0^{2\pi} g(e^{i\theta})^2 P_1(\theta, z) \, d\theta$$

which we integrate with respect to t to get $|g_m|_2^2 \leq |g|_2^2$.

This inequality combined with the fact that g_m converges weakly to g implies that $g_m \to g$ in L^2. And then $g_m^2 \to g^2$ in L^1. Indeed,

$$\int |g_m^2 - g^2| \, dt = \int |g_m - g||g_m + g| \, dt$$

$$\leq |g_m - g|_2 |g_m + g|_2 \leq 2|g_m - g|_2 |g|_2.$$

In particular the sequence $\{g_n^2 = |f(r_n e^{it})|\}$ is uniformly integrable. Hence so is the sequence $\{t \mapsto f(r_n e^{it})\}$. By the Dunford-Pettis compactness criterion (see Dunford and Schwartz, 1958, Theorem IV.8.9, p. 292) we may (possibly by taking a subsequence) assume that there exists an $f^* \in L^1(S^1)$ such that $f(r_n e^{it}) \to f^*(e^{it})$ weakly in L^1.

Letting $n \to \infty$ in the Poisson formula

$$f(z) = \frac{1}{2\pi} \int_0^{2\pi} f(r_n e^{it}) \frac{r_n^2 - |z|^2}{|r_n e^{it} - z|^2} \, dt$$

we get the existence part of the theorem.

The uniqueness will follow once we check for any L^1-function f^* that

$$\frac{1}{2\pi} \int_0^{2\pi} P_1(t, z) f^*(e^{it}) \, dt = 0 \qquad \text{for all } |z| < 1$$

implies $f^* = 0$. It suffices to prove the statement for f^* real valued. In that case we get by property (c) of the Poisson kernel that

$$
\begin{aligned}
0 &= \frac{1}{2\pi} \int_0^{2\pi} P_1(t, z) f^*(e^{it}) \, dt \\
&= \mathrm{Re}\Big\{ \frac{1}{2\pi} \int_0^{2\pi} f^*(e^{it}) \, dt + 2 \sum_{n=1}^{\infty} \frac{1}{2\pi} \int_0^{2\pi} e^{-int} f^*(e^{it}) \, dt\, z^n \Big\}.
\end{aligned}
$$

The parenthesis is an analytic function with real part 0, so it is a purely imaginary constant c, i.e.,

$$c = \frac{1}{2\pi} \int_0^{2\pi} f^*(e^{it}) \, dt + 2 \sum_{n=1}^{\infty} \frac{1}{2\pi} \int_0^{2\pi} e^{-int} f^*(e^{it}) \, dt\, z^n.$$

Taking $z = 0$ we see that $c = 0$, because f^* is real. Now,

$$\frac{1}{2\pi} \int_0^{2\pi} e^{-int} f^*(e^{it}) \, dt = 0 \qquad \text{for all } n \geq 0.$$

Since f^* is real, complex conjugation shows that this holds for all $n \in \mathbb{Z}$, i.e., that all the Fourier coefficients of f^* are 0. But then $f^* = 0$. $\qquad \square$

The following corollary which is equivalent to the theorem just proved, has been generalized extensively and is useful in prediction theory.

Theorem 13.19 (F. and M. Riesz's Theorem). *Let μ be a complex Borel measure on S^1 for which*

$$\int_{S^1} z^n \, d\mu(z) = 0 \qquad \text{for all } n = 1, 2, \ldots$$

Then μ is absolutely continuous with respect to the linear measure on S^1.

Proof. Expanding $P_1(t, z)$ in powers of z and $\bar{z}$ and using the condition on μ we see that

$$f(z) := \int_0^{2\pi} P_1(t, z) \, d\mu(e^{it}) \qquad \text{for } |z| < 1$$

is a power series in z alone, so that f is holomorphic in $B(0, 1)$. Using Fubini we find that $f \in H^1$.

When we apply Theorem 13.18 and once again expand P_1 in z and $\bar{z}$ we find that

$$\frac{1}{2\pi} \int_0^{2\pi} e^{int} \, d\mu(t) = \frac{1}{2\pi} \int_0^{2\pi} e^{int} f^*(e^{it}) \, dt \qquad \text{for all } n \in \mathbb{Z}.$$

Since the trigonometric polynomials are dense in the periodic continuous functions we get for all $g \in C(S^1)$ that

$$\frac{1}{2\pi} \int_0^{2\pi} g(t) \, d\mu(t) = \frac{1}{2\pi} \int_0^{2\pi} g(t) f^*(e^{it}) \, dt$$

which finishes the proof. □

The f^* of Theorem 13.18 above cannot be an arbitrary function in $L^1(S^1)$ as the following result shows:

Theorem 13.20. *Let f and f^* be as in Theorem 13.18. Assume that f is not identically 0. Then*

$$\text{Log}|f(0)| \leq \frac{1}{2\pi} \int_0^{2\pi} \text{Log}|f^*(e^{i\theta})| \, d\theta$$

and f^* cannot vanish on any set of positive linear measure on S^1.

Proof. In the course of proving Theorem 13.18 we showed that the sequence $\{|f(r_n e^{i\theta})|\}$ converges to $|f^*(e^{i\theta})|$ in L^1. Choosing a subsequence if necessary we may assume that

$$|f(r_n e^{i\theta})| \xrightarrow[n\to\infty]{} |f^*(e^{i\theta})| \qquad \text{for almost all } \theta.$$

Since

$$|f(r_n e^{i\theta})| - \mathrm{Log}|f(r_n e^{i\theta})| \geq 0.$$

Fatou's lemma implies that

$$\limsup_{n\to\infty} \int_0^{2\pi} \mathrm{Log}|f(r_n e^{i\theta})|\, d\theta \leq \int_0^{2\pi} \mathrm{Log}|f^*(e^{i\theta})|\, d\theta.$$

By Jensen's inequality,

$$\mathrm{Log}|f(0)| \leq \frac{1}{2\pi} \int_0^{2\pi} \mathrm{Log}|f(r_n e^{i\theta})|\, d\theta$$

finishing the proof of the first statement.

The last statement is a triviality if $f(0) \neq 0$. If f has a zero of order N at $z = 0$ we apply the above result to $F(z) := f(z)z^{-N}$, noting that $F^*(e^{it}) = f^*(e^{it})e^{-iNt}$. □

If $f \in H^p$ then $f \in \mathrm{Hol}(D) \cap L^p(D)$ for all $p \in [1,\infty[$ (use polar coordinates). The converse is not true:

Example 13.21. The function $f(z) := (1-z)^{-1}$ belongs to $\mathrm{Hol}(D) \cap L^1(D)$, but not to H^1.

Proof. Clearly $f \in \mathrm{Hol}(D)$, and we leave it to the reader to check that $f \in L^p(D)$ for all $p \in [1,2[$.

We assume now that $f \in H^1$ and derive a contradiction from that assumption: During the proof of Theorem 13.18 we saw that $f(r_n e^{it}) \to f^*(e^{it})$ weakly in L^1 as $n \to \infty$. It follows that

$$f^*(e^{it}) = \frac{1}{1 - e^{it}}, \qquad \text{so} \left(t \mapsto \frac{1}{1 - e^{it}} \right) \in L^1(-\pi, \pi).$$

But

$$|1 - e^{it}| = \left| \sum_{n=1}^{\infty} \frac{(it)^n}{n!} \right| = |t| \left| \sum_{n=0}^{\infty} \frac{(it)^n}{(n+1)!} \right| \leq 2|t|$$

for small t, so (again for t small)

$$\left| \frac{1}{1 - e^{it}} \right| \geq \frac{1}{2|t|}$$

and the right hand side is not integrable over any open interval around 0. □

13.6.3   *F. and R. Nevanlinna's theorem*

Theorem 13.22 (F. and R. Nevanlinna's Theorem). *Let f be analytic in the unit disc. Then f is the quotient between two bounded analytic functions, iff*

$$\sup_{0<r<1}\left\{\frac{1}{2\pi}\int_0^{2\pi}\text{Log}^+|f(re^{it})|\,dt\right\}<\infty$$

Proof. Let us first assume that

$$\sup_{0<r<1}\left\{\frac{1}{2\pi}\int_0^{2\pi}w(re^{it})\,dt\right\}<\infty,\qquad\text{where }w:=\text{Log}^+|f|.\qquad(13.3)$$

Note that w is subharmonic (Theoretical Examples 13.15).

The harmonic function u_r on $B(0,r)$ that on the boundary coincides with w, increases with r. To see this let $0<r<R<1$. We shall prove that $u_r-u_R\le 0$ in the ball $B(0,r)$. Since w is subharmonic we get by the remark immediately after Theorem 13.7 that $w\le u_R$ in $B(0,R)$. In particular that $w\le u_R$ on the boundary $\partial B(0,r)$ of $B(0,r)$. But on that boundary $w=u_r$, so $u_r-u_R\le 0$ on $\partial B(0,r)$. From the maximum principle for harmonic functions (Theorem 12.8 or Theorem 13.6) we read that $u_r-u_R\le 0$ throughout $B(0,r)$, as desired. The values at 0 of the functions u_r are by the Mean Value Property (Theorem 7.6) bounded by (13.3), so according to Harnack's theorem (Theorem 7.10) $u_r\to u$ as $r\to 1_-$, where u is harmonic on $B(0,1)$ and nonnegative, because $w\le u$.

Choose a holomorphic function g such that $u=\text{Re}\,g$ (Theorem 12.2(b)). Then $|h|=e^u\ge 1$ and $|f|\le|h|$, so both $1/h$ and f/h are analytic and bounded by 1. We are through, since their quotient is f.

Suppose conversely that $f=a/b$ where a and b are bounded analytic functions. We may assume that $|a|\le 1,|b|\le 1$ and that $b(0)\ne 0$. Then

$$\text{Log}^+|f|\le-\text{Log}|b|$$

and by the subharmonicity of $\text{Log}|b|$ (13.15)

$$-\infty<\text{Log}|b(0)|\le\frac{1}{2\pi}\int_0^{2\pi}\text{Log}|b(re^{i\theta})|\,d\theta.$$

Thus

$$\frac{1}{2\pi}\int_0^{2\pi}\text{Log}^+|f(re^{i\theta})|\,d\theta\le-\frac{1}{2\pi}\int_0^{2\pi}\text{Log}|b(re^{i\theta})|\,d\theta\le-\text{Log}|b(0)|$$

and the right hand side is independent of r.   □

Exercises

13.1. For which values of $a, b, c \in \mathbb{R}$ is the polynomial

$$u(x, y) := ax^2 + 2bxy + cy^2$$

a harmonic function? A subharmonic function?

13.2. Let $\phi \colon \Omega \to \Omega_0$ be a holomorphic map between two domains in the complex plane. Let u be subharmonic on Ω_0.

Show that $u \circ \phi$ is subharmonic on Ω unless it is identically $-\infty$.

Hint: $\Delta(u \circ \phi) = |\phi'|^2 \Delta u$, when ϕ is holomorphic.

13.3. Let K be a compact subset of Ω. Let u be subharmonic on Ω. Let h be a real valued continuous function on K such that h is harmonic on $\mathrm{int}(K)$ and $u \leq h$ on ∂K. Show that $u \leq h$ on all of K.

13.4. Let $f \in \mathrm{Hol}(B(0,1))$ have the power series expansion

$$f(z) = \sum_{n=0}^{\infty} a_n z^n \qquad \text{for } |z| < 1.$$

Show that

$$f \in H^2 \quad \Leftrightarrow \quad \sum_{n=0}^{\infty} |a_n|^2 < \infty.$$

13.5. Let $f \in \mathrm{Hol}(B(0,1))$ and let $p \in [1, \infty]$.

(α) Let $f \in H^p$. Show that there exists a harmonic function u such that $|f|^p \leq u$.

Hint: Theorem 8.20.

(β) Assume that there exists a harmonic function u such that $|f|^p \leq u$. Show that $f \in H^p$.

This exercise gives us a hint as to how to define the Hardy space $H^p(\Omega)$ when Ω is a domain in $\mathbb{C}$: $f \in \mathrm{Hol}(\Omega)$ is in $H^p(\Omega)$ if there exists a harmonic function u on Ω such that $|f|^p \leq u$.

13.6. Let u be a convex function on an open convex subset of $\mathbb{R}^2$. Show that u is subharmonic.

Hint: A convex function on an open set is automatically continuous, so

$$\frac{1}{2\pi}\int_0^{2\pi} u(a + re^{i\theta})\, d\theta$$

makes sense as a Riemann integral.

13.7. Let u be usc, and let f be continuous and increasing on the range of u. Show that $f \circ u$ is usc.

13.8. Show that a continuous function with the Mean Value Property is harmonic.

Hint: Theorem 13.7.

13.9. Prove

Schwarz' Reflection Principle. *Let Ω be an open subset of $\mathbb{R}^2$ such that Ω is symmetric around the x-axis. Let u be harmonic on $\{(x,y) \in \Omega \mid y > 0\}$, continuous on $\{(x,y) \in \Omega \mid y \geq 0\}$ and 0 on $\{(x,y) \in \Omega \mid y = 0\}$.*
Then there exists a harmonic function U on Ω such that $U = u$ on $\{(x,y) \in \Omega \mid y > 0\}$.

Hint: Show that

$$U(x,y) := \begin{cases} u(x,y) & \text{for } y \geq 0, \\ -u(x,-y) & \text{for } y < 0, \end{cases}$$

has the Mean Value Property.

13.10. Let u be subharmonic on an open subset Ω of $\mathbb{R}^2$. Let $B[0,r] \subseteq \Omega$, where $r > 0$. Define v on Ω by

$$v(z) := \begin{cases} \dfrac{1}{2\pi}\displaystyle\int_0^{2\pi} u(re^{i\theta}) P_r(\theta, z)\, d\theta & \text{for } z \in B(0,r), \\[2mm] u(z) & \text{for } z \in \Omega \setminus B(0,r). \end{cases}$$

Show that v is subharmonic on Ω.

13.11. We have seen that $|h|$ is subharmonic if h is harmonic. Show the following more general statement:
If $\alpha \geq 1$ and $h_1, h_2, \ldots, h_n$ are harmonic, then

$$\left(|h_1|^\alpha + |h_2|^\alpha + \cdots + |h_n|^\alpha\right)^{\frac{1}{\alpha}}$$

is subharmonic.

13.12. Use the procedure from the proof of Radó's theorem to derive the theorem about an isolated singularity for a harmonic function (Theorem 7.5).

Chapter 14

Various Applications

14.1 The Phragmén-Lindelöf principle

The Weak Maximum Principle (Corollary 4.16), which says that an analytic function never exceeds its maximal boundary value, does not hold in general for unbounded domains. For example, if Ω is the lower half plane, so that $\partial\Omega$ is the real axis, then $f(z) := \exp(iz)$ is bounded on $\partial\Omega$ but not on Ω. So we need extra restrictions on the function and the domain to get maximum principles for unbounded domains. The so-called Phragmén-Lindelöf Principle is a way of finding such maximum principles. Our version is the following.

Theorem 14.1 (Phragmén-Lindelöf's Principle). *Let Ω be a connected open subset of $\mathbb{C}$ and let ϕ be a function in $C(\overline{\Omega}) \cap \mathrm{Hol}(\Omega)$ which is bounded and not identically 0.*

If $f \in C(\overline{\Omega}) \cap \mathrm{Hol}(\Omega)$ and the constant $M \geq 0$ satisfy that

(i) $|f(z)| \leq M$ *for all $z \in \partial\Omega$, and*

(ii) $|f(z)||\phi(z)|^{\eta} \to 0$ *as $z \to \infty$ in Ω for each fixed $\eta > 0$,*

then $|f(z)| \leq M$ for all $z \in \Omega$.

Proof. The case $M = 0$ is a consequence of the case $M > 0$, so we may assume that $M > 0$. Let $B > 0$ be a bound of $|\phi|$. Consider for $\eta > 0$ the continuous function

$$u_\eta(z) := |f(z)| \left(\frac{|\phi(z)|}{B} \right)^{\eta} \qquad \text{for } z \in \overline{\Omega}$$

which is subharmonic on Ω (Theoretical Examples 13.15(b)). By assumption (ii) we may choose $R > 0$ so large that

$$u_\eta(z) \leq M \qquad \text{for all } z \in \Omega \text{ such that } |z| \geq R.$$

Since $\partial(\Omega \cap B(0, R)) \subseteq \partial\Omega \cup \{z \in \Omega \mid |z| = R\}$ we get by the Maximum Principle for subharmonic functions (Theorem 12.6) that

$$u_\eta(z) \leq \max\{M, M\} = M \qquad \text{for } z \in \Omega \cap B(0, R).$$

This inequality is also true outside of the ball $B(0, R)$ by our choice of R, so $u_\eta(z) \leq M$ for $z \in \Omega$, i.e.,

$$|f(z)| \left(\frac{|\phi(z)|}{B} \right)^\eta \leq M \qquad \text{for } z \in \Omega.$$

Letting $\eta \to 0$ for fixed $z \in \Omega$ we get that $|f(z)| \leq M$ for all those $z \in \Omega$ for which $\phi(z) \neq 0$. However, the zeros of ϕ are isolated in Ω (Theorem 4.11), so the desired inequality holds by the continuity of f everywhere in Ω. $\quad\square$

The connectivity assumption was only used at the very end of the proof. We may delete that assumption if we in return require that ϕ vanishes nowhere.

Corollary 14.2. *Let $\Omega \neq \mathbb{C}$ be an open subset of the complex plane. Let $f \in C(\overline{\Omega}) \cap \mathrm{Hol}(\Omega)$ be bounded.*

If $|f| \leq M$ on $\partial\Omega$, then $|f| \leq M$ on all of Ω.

Proof. We may of course assume that $\Omega \neq \emptyset$, so that also $\partial\Omega \neq \emptyset$. Let $\varepsilon > 0$. By continuity of f there exists a point $z_0 \in \Omega$ such that $|f(z_0)| < M + \varepsilon$ and then there is even a ball $B[z_0, r] \subseteq \Omega$ such that $|f| \leq M + \varepsilon$ on that ball. Applying Theorem 14.1 to the set $\Omega \setminus B[z_0, r]$ and the function $\phi(z) := (z - z_0)^{-1}$ we get that

$$|f(z)| \leq M + \varepsilon \qquad \text{for all } z \in \Omega \setminus B[z_0, r].$$

This inequality also holds for $z \in B[z_0, r]$ by the very construction of that ball, so $|f(z)| \leq M + \varepsilon$ for all $z \in \Omega$. But $\varepsilon > 0$ was arbitrary. $\quad\square$

Corollary 14.3 (The Three Lines Theorem). *Let $\Omega := \{z \in \mathbb{C} \mid 0 < \mathrm{Re}\, z < 1\}$. Let f be bounded and holomorphic on Ω and continuous on the closure $\overline{\Omega}$ of Ω and put*

$$M(x) := \sup_{-\infty < y < \infty} \{|f(x + iy)|\}. \tag{14.1}$$

Then $\log M$ *is a convex function on* $[0,1]$. *In particular*

$$M(x) \leq M(0)^{1-x} M(1)^x \qquad for\ all\ x \in [0,1]. \qquad (14.2)$$

Proof. Let $g(z) := e^{az} f(z)$, where a is a real constant. Applying Corollary 14.2 to g we see for any $x \in]0,1[$ that

$$|g(x+iy)| \leq \max\{M(0), M(1)e^a\}$$

and so that

$$M(x) \leq \max\{M(0)e^{-ax}, M(1)e^{a(1-x)}\}. \qquad (14.3)$$

If $M(0) = 0$ then (14.3) says that $M(x) \leq M(1)e^{a(1-x)}$. Since this holds for all real numbers a, we see that $M(x) = 0$, and so (14.2) is trivially satisfied. The same kind of arguments works if $M(1) = 0$.

If both $M(0)$ and $M(1)$ are different from 0 we choose a so that $M(0) = M(1)e^a$ (this minimizes the right hand side of the inequality (14.3); then (14.3) simplifies as desired to

$$M(x) \leq e^{-ax} M(0) = M(0)^{1-x} M(1)^x.$$

If we apply the same arguments to the strip

$$\{z \in \mathbb{C} \mid s \leq \operatorname{Re} z \leq t\} \qquad \text{where } 0 \leq s < t \leq 1$$

as we did to the entire strip $\{z \mid 0 \leq \operatorname{Re} z \leq 1\}$ we find that

$$M(x)^{t-s} \leq M(s)^{t-x} M(t)^{x-s}$$

i.e., that $\log M$ is convex. $\qquad\square$

14.2 The Riesz-Thorin interpolation theorem

Using the Three Lines Theorem we shall now prove the M. Riesz-Thorin Interpolation Theorem. The Interpolation Theorem holds for general measure spaces, but for simplicity we will only consider $\mathbb{R}^d$ with Lebesgue measure. A proof for the general case can be found in Dunford and Schwartz (1958, VI.10).

We need some preliminaries:

A simple function is a (finite) linear combination of indicator functions for measurable sets of finite measure. We let S denote the complex vector space of simple functions.

If $1 \leq q_0 < q_1 \leq \infty$ and $f \in L^{q_0} \cap L^{q_1}$, then $f \in L^r$ for each $r \in [q_0, q_1]$: Indeed, if $q_1 < \infty$, then

$$|f|^r \leq |f|^{q_0} 1_A + |f|^{q_1} 1_B$$

where

$$A := \{x \in \mathbb{R}^d \mid |f(x)| \leq 1\} \quad \text{and} \quad B := \{x \in \mathbb{R}^d \mid |f(x)| > 1\}.$$

If $q_1 = \infty$, then we may scale f down so that we may assume that $|f| \leq 1$; and then $|f|^r \leq |f|^{q_0}$, so we are through also in this case.

Theorem 14.4 (Riesz-Thorin's Interpolation Theorem). *Let* $1 \leq p_0$, $q_0, p_1, q_1 \leq \infty$, *and let* $T: S \to L^{q_0} \cap L^{q_1}$ *be a linear operator such that the mappings* $T: S \to L^{q_0}$ *and* $T: S \to L^{q_1}$ *have finite norms* M_0 *and* M_1, *when* S *is equipped with the norms from* L^{p_0} *and* L^{p_1} *respectively.*

For $t \in [0, 1]$ *we define* p_t *and* q_t *by the formulas*

$$\frac{1}{p_t} = \frac{1-t}{p_0} + \frac{t}{p_1} \quad \text{and} \quad \frac{1}{q_t} = \frac{1-t}{q_0} + \frac{t}{q_1}.$$

Then T *maps* S *into* L^{q_t}, *and its norm is at most* $M_0^{1-t} M_1^t$ *when* S *is given the norm from* L^{p_t}.

Proof. We let $\Omega := \{z \in \mathbb{C} \mid 0 < \operatorname{Re} z < 1\}$, and for $z \in \overline{\Omega}$ we put

$$\frac{1}{p(z)} = \frac{1-z}{p_0} + \frac{z}{p_1} \quad \text{and} \quad \frac{1}{q(z)} = \frac{1-z}{q_0} + \frac{z}{q_1}.$$

Furthermore we let r_t denote the dual exponent to q_t, i.e.,

$$\frac{1}{r_t} + \frac{1}{q_t} = 1.$$

Clearly

$$\frac{1}{r_t} = \frac{1-t}{r_0} + \frac{t}{r_1}$$

which we extend to $z \in \overline{\Omega}$ by

$$\frac{1}{r(z)} = \frac{1-z}{r_0} + \frac{z}{r_1}.$$

Note that $p(t) = p_t$, $q(t) = q_t$ and $r(t) = r_t$.

Fix $t \in]0, 1[$. Let us for convenience write $p = p_t$, $q = q_t$ and $r = r_t$.

Let f be in S. We must show that $Tf \in L^q$ has norm less than or equal to $M_0^{1-t}M_1^t|f|_p$. It suffices to prove that

$$\left|\int (Tf)(x)g(x)\,dx\right| \le M_0^{1-t}M_1^t|g|_r|f|_p \qquad \text{for all } g \in S.$$

For each $z \in \overline{\Omega}$ we define

$$f_z := \text{sign}(f)|f|^{\frac{p}{p(z)}} \quad \text{and} \quad g_z := \text{sign}(g)|g|^{\frac{r}{r(z)}}.$$

Note that $f_t = f$ and $g_t = g$; and also that

$$|f_z| = |f_{\text{Re}\,z}| = |f|^{\frac{p}{P(\text{Re}\,z)}} \quad \text{and} \quad |g_z| = |g_{\text{Re}\,z}| = |g|^{\frac{r}{r(\text{Re}\,z)}}.$$

If

$$f = \sum_j a_j 1_{A_j}$$

then

$$f_z = \sum_j \text{sign}(a_j)|a_j|^{\frac{p}{p(z)}} 1_{A_j}$$

so by the linearity of T we get

$$T(f_z) = \sum_j \text{sign}(a_j)|a_j|^{\frac{p}{p(z)}} T(1_{A_j}).$$

Since a similar formula holds for g_z we clearly have that the function ϕ, defined on $\overline{\Omega}$ by

$$\phi(z) := \int T(f_z)g_z, \qquad z \in \overline{\Omega}$$

is holomorphic on Ω, and continuous and bounded on $\overline{\Omega}$. By the Three Lines Theorem applied to ϕ and the point $z = t$ we get

$$|\phi(t)| \le N_0^{1-t}N_1^t \qquad \text{where } N_0 = \sup_y|\phi(iy)| \text{ and } N_1 = \sup_y|\phi(1 + iy)|.$$

$$(14.4)$$

Let us estimate $|\phi(1 + iy)|$: By Hölder

$$|\phi(1 + iy)| \le |T(f_{1+iy})|_{q(1)}|g_{1+iy}|_{r(1)} \le M_1|f_{1+iy}|_{p(1)}|g_{1+iy}|_{r(1)}$$
$$\le M_1|f_{\text{Re}(1+iy)}|_{p(1)}|g_{\text{Re}(1+iy)}|_{r(1)} \le M_1|f|_p^{\frac{p}{p(1)}}|g|_r^{\frac{r}{r(1)}}$$

so

$$N_1 \le M_1|f|_p^{\frac{p}{p(1)}}|g|_r^{\frac{r}{r(1)}}.$$

We estimate N_0 similarly. Substituting these estimates into (14.4) and recalling how p and r are defined we get

$$|\phi(t)| \leq M_0^{1-t} M_1^t |f|_p |g|_r$$

proving the theorem.   □

A bounded linear operator mapping of L^p into L^q is said to be of type (p, q). In this language Theorem 14.4 reads: If T is simultaneously of type (p_0, q_0) and of type (p_1, q_1), then it is also of type (p_t, q_t) for any $t \in [0, 1]$ where

$$\frac{1}{p_t} = \frac{1-t}{p_0} + \frac{t}{p_1} \quad \text{and} \quad \frac{1}{q_t} = \frac{1-t}{q_0} + \frac{t}{q_1}$$

and we have an estimate for the the operator norm of T.

A very simple consequence of Theorem 14.4 is

Theorem 14.5 (The Hausdorff-Young inequality). *If $p \in [1, 2]$ and $f \in L^p(\mathbb{R}^n)$ then its Fourier transform Ff belongs to $L^{p'}(\mathbb{R}^n)$ and*

$$|Ff|_{p'} \leq |f|_p \quad \text{where} \quad \frac{1}{p'} + \frac{1}{p} = 1.$$

Proof. The Fourier transformation $f \mapsto Ff$ is simultaneously of type $(1, \infty)$ and of type $(2, 2)$. Apply the Riesz-Thorin Interpolation Theorem.   □

14.3   Schwarz's integral formula

Any harmonic function u on the unit disc is the real part of an analytic function f (Theorem 12.2). The imaginary part v of $f = u + iv$ is not unique, but any two choices of v differ by a real constant. The *conjugate function* (or *harmonic conjugate*) of u is the v which is fixed by the requirement that $v(0) = 0$, or in other words by the requirement that $f(0)$ should be real. As simple examples we mention that the conjugate function of $u(z) := \operatorname{Re}(z^n)$ is $v(z) = \operatorname{Im}(z^n)$, and that the one of $u(z) := \operatorname{Im}(z^n)$ is $v(z) = -\operatorname{Re}(z^n)$, $n = 0, 1, \ldots$; more generally, if u is a trigonometric polynomial with real coefficients,

$$u(re^{i\theta}) = a_0 + \sum_{n=1}^{N} r^n \{a_n \cos(n\theta) + b_n \sin(n\theta)\}$$

then its conjugate function is

$$v(re^{i\theta}) = \sum_{n=1}^{N} r^n \{-b_n \cos(n\theta) + a_n \sin(n\theta)\}.$$

The holomorphic functions that have u as real part are characterized by *Schwarz' formula*:

Theorem 14.6 (Schwarz' Formula). *Let $f = u + iv$ be holomorphic on the open ball $B(0, R)$ and let u be continuous on the closed ball $B[0, R]$. Then*

$$f(z) = iv(0) + \frac{1}{2\pi} \int_0^{2\pi} u(Re^{i\theta}) \frac{Re^{i\theta} + z}{Re^{i\theta} - z} d\theta \qquad \text{for all } |z| < R.$$

Proof. By Poisson's formula (Theorem 12.6)

$$\begin{aligned}
\text{Re}(f(z)) = u(z) &= \frac{1}{2\pi} \int_0^{2\pi} u(Re^{i\theta}) \frac{R^2 - |z|^2}{|Re^{i\theta} - z|^2} d\theta \\
&= \frac{1}{2\pi} \int_0^{2\pi} u(Re^{i\theta}) \, \text{Re}\left(\frac{Re^{i\theta} + z}{Re^{i\theta} - z}\right) d\theta \\
&= \text{Re}\left(\frac{1}{2\pi} \int_0^{2\pi} u(Re^{i\theta}) \frac{Re^{i\theta} + z}{Re^{i\theta} - z} d\theta\right)
\end{aligned}$$

so the real part of the analytic function

$$z \mapsto f(z) - \frac{1}{2\pi} \int_0^{2\pi} u(Re^{i\theta}) \frac{Re^{i\theta} + z}{Re^{i\theta} - z} d\theta$$

is 0. Hence the function is a constant. Taking $z = 0$ we see that this constant is $iv(0)$. □

Schwarz's formula tells us how to recover a holomorphic function, up to an arbitrary, imaginary constant, from the boundary values on the unit ball of its real part. We shall below derive a related result with the ball replaced by the closed upper half-plane. However, since the half-plane in contrast to the ball is unbounded, we must impose a decay condition at infinity to give sense to the integrals. The arbitrary constant from the case of a ball is not present here.

Theorem 14.7 (Schwarz' Formula for the Upper Half-plane). *Let $\mathbb{H} := \{z \in \mathbb{C} \mid \text{Im } z > 0\}$. Let $\overline{\mathbb{H}}$ denote the closure of $\mathbb{H}$ in the Riemann sphere $\mathbb{C}_\infty$, i.e.,*

$$\overline{\mathbb{H}} = \{z \in \mathbb{C} \mid \text{Im } z \geq 0\} \cup \{\infty\}.$$

Let $f \in \text{Hol}(\mathbb{H})$ and write $f = u + iv$ with u and v real-valued. Assume that $u \in C(\overline{\mathbb{H}})$ and that there exist constants $C > 0$ and $a > 0$ such that $|z|^a |f(z)| \leq C$ for all $z \in \mathbb{H}$. Then

$$f(z) = \frac{1}{\pi i} \int_{-\infty}^{+\infty} \frac{u(t)}{t - z} \, dt \qquad \text{for all } z \in \mathbb{H}.$$

Proof. Note for use below that $f(\infty) = 0$ due to the decay restriction on f.

We have earlier derived Schwarz' formula for a disc (Theorem 14.6). We shall use that to get Schwarz' formula for the upper half-plane. We will do so by help of the Cayley transformation $C \colon \mathbb{C}_\infty \to \mathbb{C}_\infty$ (see Example 8.13) which is given by

$$C(z) := \frac{z - i}{z + i} \qquad \text{for } z \in \mathbb{C}_\infty.$$

Its inverse is

$$C^{-1}(w) := \frac{1}{i} \frac{w + 1}{w - 1} \qquad \text{for } w \in \mathbb{C}_\infty.$$

The Cayley transformation C maps $\mathbb{H}$ conformally onto $B(0, 1)$ and it maps $\mathbb{R} \cup \{\infty\}$ bijectively onto the unit circle with $C(\infty) = 1$.

The function $f \circ C^{-1} \in \text{Hol}(B(0, 1)) \cap C(B[0, 1])$, so from Schwarz' formula for a disc we know that

$$(f \circ C^{-1})(w) = i(v \circ C^{-1})(0) + \frac{1}{2\pi} \int_0^{2\pi} (u \circ C^{-1})(e^{i\theta}) \frac{e^{i\theta} + w}{e^{i\theta} - w} \, d\theta.$$

If we here write $z = C^{-1}(w)$ we get for all $z \in \mathbb{H}$ that

$$f(z) = i(v \circ C^{-1})(0) + \frac{1}{2\pi} \int_0^{2\pi} (u \circ C^{-1})(e^{i\theta}) \frac{e^{i\theta} + C(z)}{e^{i\theta} - C(z)} \, d\theta. \qquad (14.5)$$

To identify the constant $i(v \circ C^{-1})(0)$ on the right hand side we let $z \to \infty$ in (14.5). The left hand side converges to $f(\infty) = 0$. The right hand side is more obstinate technically, because the denominator $e^{i\theta} - C(z)$ in the limit $z \to \infty$ becomes $e^{i\theta} - 1$, so that the integral has singularities for $e^{i\theta} = 1$, i.e., at $\theta = 0$ and $\theta = 2\pi$. As we shall see the behaviour of u at infinity compensates for the singularities. To make the limit transition simple we take $z_n := ni$ for $n = 1, 2, \ldots$

Let us estimate the crucial part of the denominator:

$$|e^{i\theta} - C(z_n)| = \left|e^{i\theta} - \frac{z_n - i}{z_n + i}\right| = \left|e^{i\theta} - \frac{n-1}{n+1}\right|$$

$$= \left|\cos\theta + i\sin\theta - \frac{n-1}{n+1}\right| \geq |\sin\theta| \geq \tfrac{1}{2}|e^{i\theta} - 1|.$$

We take u into account by the following estimate:

$$|u(C^{-1}(e^{i\theta}))| = \left|u\left(\frac{1}{i}\frac{e^{i\theta}+1}{e^{i\theta}-1}\right)\right| \leq C\left|\frac{e^{i\theta}+1}{e^{i\theta}-1}\right|^{-a} = C\frac{|e^{i\theta}-1|^a}{|e^{i\theta}+1|^a}.$$

Since $a > 0$ we get by Lebesgue's dominated convergence that the function

$$\theta \mapsto (u \circ C^{-1})(e^{i\theta})\frac{e^{i\theta}+1}{e^{i\theta}-1}$$

is integrable over $]0, 2\pi[$ and that

$$i(v \circ C^{-1})(0) = -\frac{1}{2\pi}\int_{]0,2\pi[}(u \circ C^{-1})(e^{i\theta})\frac{e^{i\theta}+1}{e^{i\theta}-1}\,d\theta.$$

We shall change variables so that it becomes an integral over $]-\infty, \infty[$ like in the statement of the theorem. Consider the diffeomorphism $\psi\colon \mathbb{R} \to \,]0, 2\pi[$ given by the composition of $C|_{\mathbb{R}}\colon \mathbb{R} \to \{w \in \mathbb{C}\,|\,|w| = 1\} \setminus \{1\}$ with the map $\exp i\theta \mapsto \theta$. Thus

$$e^{i\psi(t)} = C(t) = \frac{t-i}{t+i} \qquad \text{for all } t \in \mathbb{R}. \tag{14.6}$$

Differentiating this with respect to t we find that

$$\psi'(t) = \frac{2}{1+t^2} \qquad \text{for all } t \in \mathbb{R}.$$

When we apply the change of variables formula

$$\int_{]0,2\pi[} F(\theta)\,d\theta = \int_{\mathbb{R}} F(\psi(t))|\psi'(t)|\,dt = \int_{\mathbb{R}} F(\psi(t))\frac{2dt}{1+t^2}$$

to the second term on the right of (14.5)

$$\frac{1}{2\pi}\int_0^{2\pi}(u \circ C^{-1})(e^{i\theta})\frac{e^{i\theta}+C(z)}{e^{i\theta}-C(z)}\,d\theta$$

$$= \frac{1}{2\pi}\int_{]0,2\pi[}u(C^{-1}(e^{i\theta}))\frac{e^{i\theta}+C(z)}{e^{i\theta}-C(z)}\,d\theta,$$

we get that

$$\frac{1}{2\pi} \int_{]0,2\pi[} u(C^{-1}(e^{i\theta})) \frac{e^{i\theta} + C(z)}{e^{i\theta} - C(z)} \, d\theta$$

$$= \frac{1}{\pi} \int_{\mathbb{R}} u(C^{-1}(e^{i\psi(t)})) \frac{e^{i\psi(t)} + C(z)}{e^{i\psi(t)} - C(z)} \frac{dt}{1 + t^2}.$$

Using the formula (14.6) we continue the computations and get

$$\frac{1}{2\pi} \int_{]0,2\pi[} u(C^{-1}(e^{i\theta})) \frac{e^{i\theta} + C(z)}{e^{i\theta} - C(z)} \, d\theta$$

$$= \frac{1}{\pi} \int_{\mathbb{R}} u(t) \frac{\frac{t-i}{t+i} + \frac{z-i}{z+i}}{\frac{t-i}{t+i} - \frac{z-i}{z+i}} \frac{dt}{1+t^2} = \frac{1}{\pi} \int_{\mathbb{R}} u(t) \frac{tz+1}{i(t-z)} \frac{dt}{1+t^2}.$$

For the first term on the right of (14.5), corresponding to $z = \infty$, we get similarly that

$$-\frac{1}{2\pi} \int_{]0,2\pi[} (u \circ C^{-1})(e^{i\theta}) \frac{e^{i\theta} + 1}{e^{i\theta} - 1} \, d\theta = -\frac{1}{\pi} \int_{\mathbb{R}} u(t) \frac{t}{-i} \frac{dt}{1+t^2}.$$

Adding the two terms we obtain that

$$f(z) = -\frac{1}{\pi} \int_{\mathbb{R}} u(t) \frac{t}{-i} \frac{dt}{1+t^2} + \frac{1}{\pi} \int_{\mathbb{R}} u(t) \frac{tz+1}{i(t-z)} \frac{dt}{1+t^2}$$

$$= \frac{1}{\pi i} \int_{\mathbb{R}} u(t) \frac{1+t^2}{t-z} \frac{dt}{1+t^2} = \frac{1}{\pi i} \int_{\mathbb{R}} \frac{u(t)}{t-z} \, dt,$$

which finishes the proof.   □

14.4   M. Riesz' theorem

For any complex valued function ϕ on the unit disc we let

$$|\phi|_p := \sup_{0<r<1} \left\{ \frac{1}{2\pi} \int_0^{2\pi} |\phi(re^{i\theta})|^p \, d\theta \right\}^{\frac{1}{p}} \qquad \text{for } p \in [1, \infty[.$$

With this notation a holomorphic function f on the unit disc is in H^p iff $|f|_p < \infty$. If f is holomorphic or harmonic then $|f|^p$ is subharmonic (Theoretical Examples 13.15) and so the supremum is the limit as $r \to 1_-$

(Proposition 13.9). If f furthermore is continuous on the closed unit disc then clearly

$$|f|_p^p = \frac{1}{2\pi} \int_0^{2\pi} |f(e^{i\theta})|^p \, d\theta \qquad \text{for } p \in [1, \infty[$$

i.e., the norm equals the L^p-norm of the restriction of f to the unit circle.

Let $f = u + iv$ be holomorphic on the unit disc and such that $f(0)$ is real, i.e., v is the conjugate function of u. The problem that we shall now consider is whether $|u|_p < \infty$ implies $|v|_p < \infty$, or said differently: Must $f \in H^p$ if $|u|_p < \infty$?

Let us analyze the special case $p = 2$, where we can compute all the norms involved. We consider an $f \in \text{Hol}(B(0,1))$,

$$f(z) = \sum_{n=0}^{\infty} a_n z^n$$

such that $f(0) = a_0 \in \mathbb{R}$. Then

$$u(z) = \text{Re}(f(z)) = \sum_{n=0}^{\infty} \text{Re}(a_n z^n).$$

From the theory for power series we know that the series $\sum_n a_n r^n e^{in\theta}$ for each fixed $r \in [0, 1[$ converges uniformly with respect to θ to $f(re^{i\theta})$; its real part will then of course converge uniformly to $u(re^{i\theta})$. Since

$$\frac{1}{2\pi} \int_0^{2\pi} e^{in\theta} e^{-im\theta} \, d\theta = \delta_{n,m}$$

we get

$$\frac{1}{2\pi} \int_0^{2\pi} |f(re^{i\theta})|^2 \, d\theta = \lim_{N \to \infty} \frac{1}{2\pi} \int_0^{2\pi} \left(\sum_{n=0}^{N} r^n e^{in\theta} \right) \left(\sum_{m=0}^{N} r^m \overline{a_m} e^{-im\theta} \right) d\theta$$

$$= \lim_{N \to \infty} r^{n+m} a_n \overline{a_m} \frac{1}{2\pi} \int_0^{2\pi} e^{in\theta} e^{-im\theta} \, d\theta$$

$$= \lim_{N \to \infty} \sum_{n=0}^{N} r^{2n} |a_n|^2 = \sum_{n=0}^{\infty} r^{2n} |a_n|^2$$

so

$$|f|_2^2 = \sum_{n=0}^{\infty} |a_n|^2.$$

In a similar way we can compute (we leave the details to the reader) that

$$|u|_2^2 = a_0^2 + \frac{1}{2} \sum_{n=1}^{\infty} |a_n|^2.$$

In particular

$$|f|_2^2 = 2|u|_2^2 - a_0^2 \le 2|u|_2^2$$

so $f \in H^2$ if $|u|_2 < \infty$, answering our question above affirmatively.

Marcel Riesz showed in 1927 that this phenomenon persists for p in the range $]1, \infty[$:

Theorem 14.8 (M. Riesz' Theorem). *For each $p \in]1, \infty[$ there exists a positive constant C_p such that*

$$|f|_p \le C_p |\operatorname{Re} f|_p \qquad \text{for all } f \in \operatorname{Hol}(B(0,1)) \text{ such that } f(0) \in \mathbb{R}.$$

Remark. We have not specified any bound on the constant C_p, except in the case $p = 2$. It can be shown that $C_p = (p/(p-1))^{1/p}$ works, Hayman and Kennedy (1976).

Proof. We let for convenience $\mathcal{R}$ denote the set of those $f \in \operatorname{Hol}(B(0,1))$ for which $f(0)$ is real.

We divide the proof into 3 steps.

Step 1: Here we prove the following claim.

Claim. *Let p be fixed and assume that there is a constant $C > 0$ such that $|f|_p \le C |\operatorname{Re} f|_p$ for all $f \in \mathcal{R}$. Then*

$$|f|_{2p} \le (4C + 1)|\operatorname{Re} f|_{2p} \qquad \text{for all } f \in \mathcal{R}.$$

Proof of the claim. Consider first the case of $f(0) = 0$. Possibly replacing f by $z \mapsto f(rz)$ for $r \in]0, 1[$ we may assume that f is holomorphic in a neighborhood of the closed unit disc, so that

$$
\begin{aligned}
|f|_{2p}^{2p} &= \frac{1}{2\pi} \int_0^{2\pi} |f(e^{i\theta})|^{2p}\, d\theta = \frac{1}{2\pi} \int_0^{2\pi} |f(e^{i\theta})^2|^p\, d\theta \\
&= \frac{1}{2\pi} \int_0^{2\pi} |u^2 - v^2 + 2iuv|^p\, d\theta = \frac{2^p}{2\pi} \int_0^{2\pi} \left| uv + \frac{u^2 - v^2}{2i} \right|^p d\theta \\
&= 2^p \left| uv + \frac{u^2 - v^2}{2i} \right|_p^p
\end{aligned}
$$

which by our assumption this can be estimated by

$$2^p C^p |uv|_p^p = (2C)^p \frac{1}{2\pi} \int_0^{2\pi} |u|^p |v|^p \, d\theta$$

and further by the Cauchy-Schwarz inequality

$$\leq (2C)^p \left\{ \frac{1}{2\pi} \int_0^{2\pi} |u|^{2p} \, d\theta \right\}^{\frac{1}{2}} \left\{ \frac{1}{2\pi} \int_0^{2\pi} |v|^{2p} \, d\theta \right\}^{\frac{1}{2}}$$

$$= (2C)^p |u|_{2p}^p |v|_{2p}^p \leq (2C)^p |u|_{2p}^p |f|_{2p}^p.$$

Dividing through by $|f|_{2p}^p$ we get that $|f|_{2p} \leq 2C|u|_{2p}$. For a general $f \in \mathcal{R}$ we thus have established the inequality

$$|f - f(0)|_{2p} \leq 2C|u - u(0)|_{2p}.$$

Now, $f(0) = u(0)$, so

$$|f|_{2p} \leq 2C|u|_{2p} + (1 + 2C)|u(0)|.$$

To deal with the last term we note that $|u|^{2p}$ is subharmonic (Theoretical Examples 13.15(c)), so that

$$|u(0)|^{2p} \leq \frac{1}{2\pi} \int_0^{2\pi} |u(re^{i\theta})|^{2p} \, d\theta \leq |u|_{2p}^{2p}$$

and finally $|f|_{2p} \leq (1 + 4C)|u|_{2p}$, proving the claim. $\qquad\square$

Step 1: We know that the inequality of M. Riesz' theorem holds for $p = 2$ (with the constant $1/2$). From Step 1 we see that it then also holds for $p = 4, 8, 16, \ldots$ In the present step we use the Riesz-Thorin Interpolation Theorem to show that the inequality holds for any p in the interval $[2, \infty[$.

The map that interests us, i.e., $u \mapsto f = u + iv$, where v is the conjugate function of u, has according to Schwarz' formula (Theorem 14.5) the explicit expression

$$f(z) = \frac{1}{2\pi} \int_0^{2\pi} u(e^{i\theta}) \frac{e^{i\theta} + z}{e^{i\theta} - z} \, d\theta \qquad \text{for } |z| < 1 \qquad (14.7)$$

if u is continuous on the closed unit disc. However, u and f are in general only defined on the open disc, so we must consider the functions on the circle $S_r := \{z \in \mathbb{C} \mid |z| = r\}$ and afterwards let $r \to 1_-$. More precisely, we define for each $r \in \,]0,1[$ a linear map $T_r: L^1(S^1) \to C(S^1)$ by

$$(T_r\phi)(e^{i\theta}) := \frac{1}{2\pi} \int_0^{2\pi} \phi(e^{i\theta}) \frac{e^{it} + re^{i\theta}}{e^{it} - re^{i\theta}}\, dt$$

and shall for each $p \in [2, \infty[$ study its restriction

$$T_{r,p} := T_r|_{L^p}: L^p(S^1) \to L^p(S^1).$$

That restriction is a continuous linear operator from $L^p(S^1)$ into $L^p(S^1)$.

Let now $\phi \in C(S^1)$ be real valued and let f denote the analytic function from (14.7) with u replaced by ϕ. Then

$$|T_{r,p}\phi|_{L^p} = \left\{ \frac{1}{2\pi} \int_0^{2\pi} |f(re^{i\theta})|^p\, d\theta \right\}^{\frac{1}{p}} \le |f|_p.$$

If, furthermore, p is of the form $p = 2^N$ for some $N = 1, 2, \ldots$ we may by Step 1 estimate as follows:

$$|f|_p \le A_p |\operatorname{Re} f|_p = A_p |\phi|_{L^p}$$

where A_p is a constant.

If ϕ is not real valued we split it into its real and imaginary parts to get

$$|T_{r,p}\phi|_{L^p} \le 2A_p |\phi|_{L^p}$$

so that

$$|T_{r,p}|_{L^p, L^p} \le 2A_p \qquad \text{for } p = 2^N, N = 1, 2, \ldots$$

By the Riesz-Thorin Interpolation Theorem there exists to each $p \in [2, \infty[$ a constant C_p, not depending on r, such that

$$|T_{r,p}\phi|_{L^p} \le C_p |\phi|_{L^p} \qquad \text{for all } \phi \in L^p(S^1).$$

If f is holomorphic on a neighborhood of the closed unit disc and $f(0)$ is real then

$$|f|_p = \lim_{r \to 1_-} |f|_{S_r}|_{L^p} = \lim_{r \to 1_-} |T_{r,p}(\operatorname{Re} f)|_{L^p} \le C_p |\operatorname{Re} f|_p.$$

If f is holomorphic on $B(0,1)$, but not necessarily on a neighborhood of the closed unit disc, then we apply the result just derived to the analytic function $f_\lambda(z) := f(\lambda z)$ and let $\lambda \to 1_-$.

With Step 1 we have proved M. Riesz' theorem for p in the range $[2, \infty[$.

Step 2: Here we derive M. Riesz' theorem in the range $]1, 2[$ by the standard technique of duality, knowing it to be true in the range $[2, \infty[$.

For exponents $q < 2$ we consider the conjugate exponent $p = q/(q-1) > 2$. Let $f = u + iv \in H^q$ where $v(0) = 0$. We may assume that f is continuous on the closed disc $|z| \leq 1$ (if not consider $f(rz)$ for $r < 1$), so that the norms reduce to the relevant L^s-norms of the functions on the unit circle.

Since trigonometric polynomials are dense in $L^p(S^1)$ we have

$$|v|_q = \sup_g \left| \int vg \right|$$

where $\int$ denotes integration over S^1 with respect to $\frac{d\theta}{2\pi}$ and where the supremum ranges over all real trigometric polynomials g with L^p-norm ≤ 1. Let

$$g(\theta) = a_0 + \sum_{n=1}^{N} \{a_n \cos(n\theta) + b_n \sin(n\theta)\}$$

be such a polynomial and consider the harmonic function given by

$$g(re^{i\theta}) = a_0 + \sum_{n=1}^{N} r^n \{a_n \cos(n\theta) + b_n \sin(n\theta)\}.$$

Its conjugate function h is

$$h(re^{i\theta}) = \sum_{n=1}^{N} r^n \{-b_n \cos(n\theta) + a_n \sin(n\theta)\}.$$

Since $g + ih$ is analytic so is the product $f(g + ih)$. Its imaginary part $vg + uh$ is harmonic and vanishes at 0, so by the Mean Value Property for harmonic functions (Theorem 12.6)

$$\int vg = -\int uh.$$

Combining Hölder's inequality $\left| \int uh \right| \leq |u|_q |h|_p$ with the following result, which we have already proved

$$|h|_p \leq |g + ih|_p \leq C_p |g|_p$$

we get

$$\left| \int vg \right| \leq C_p |u|_q |g|_p \leq C_p |u|_q$$

which implies that $|v|_q \leq C_p |u|_q$. Hence

$$|f|_q \leq |u|_q + |v|_q \leq (C_p + 1)|u|_q$$

which is the desired statement.   □

Exercises

14.1. Let

$$\Omega := \{z \in \mathbb{C} \setminus \{0\} \mid |\operatorname{Arg} z| < \tfrac{\pi}{2a}\}$$

where $a \geq \frac{1}{2}$ is a constant. Let $f \in C(\overline{\Omega}) \cap \operatorname{Hol}(\Omega)$ satisfy that

$$|f(z)| \leq C \exp(A|z|^b) \qquad \text{for all } z \in \Omega$$

where A and C are positive constants and $b \in [0, a[$.

Show that $|f| \leq M$ throughout Ω, if $|f| \leq M$ on $\partial\Omega$.

Hint: Consider the function $\phi(z) := \exp(-z^c)$ for $c \in]b, a[$.

14.2. Let Ω be the strip

$$\Omega := \{z \in \mathbb{C} \mid 0 < \operatorname{Re} z < L\}$$

where $L \in]0, \infty[$. Assume that $f \in C(\overline{\Omega}) \cap \operatorname{Hol}(\Omega)$ satisfies that

$$|f(z)| \leq C \exp(Ae^{a|z|}) \qquad \text{for } z \in \Omega$$

where A and C are positive constants and $a \in [0, \tfrac{\pi}{2L}[$.

Show that $|f| \leq M$ throughout Ω, if $|f| \leq M$ on $\partial\Omega$.

Hint: Consider for $b \in]a, \tfrac{\pi}{2L}[$ the function

$$\phi(z) := \exp(-e^{-ibz} - e^{ibz}).$$

Chapter 15

Jordan Regions

The present chapter is devoted to the plane topology surrounding Jordan's curve theorem, and is a natural extension to our study of continuous logarithms and winding number in Section 3.3 and Section 3.4. While it is topically a departure from the earlier material, we shall need stronger topological tools before we can delve further into the theory of complex functions. For further material in this direction we refer the reader to Moise (1977) and Newman (1954).

15.1 Jordan's curve theorem

We recall that by a curve $\gamma\colon I \to \mathbb{C}$ we mean a continuous function defined on an interval I. (By convention, points are not considered intervals.) If $I = [a, b]$ and $\gamma(a) = \gamma(b)$, we say that γ is a closed curve.

We call a curve $\gamma\colon I \to \mathbb{C}$ piecewise linear if for any compact subinterval $J = [a, b] \subseteq I$, there exist finitely many points $a = t_0 < t_1 < \cdots < t_N = b$, such that $\gamma|_{[t_n, t_{n+1}]}$ is a line segment for $n = 0, 1, 2, \ldots, N - 1$.

The goal of this section is to establish Jordan's curve theorem, cf. Theorem 15.8 below. The line of argument we follow is closely inspired by Maehara (1984). See also García Arenas and Luz Puertas (1998).

Definition 15.1. *Let $a < b$ be real numbers and $\gamma\colon I \to \mathbb{C}$ a curve defined on an interval I with interior $]a, b[$, such that γ restricted to $]a, b[$ is injective. We say that γ is a*

(a) *closed Jordan curve if $I = [a, b]$ and $\gamma(a) = \gamma(b) \notin \gamma(]a, b[)$.*

(b) *Jordan arc if $I = [a, b]$ and γ is injective on I.*

(c) *Jordan half-interval if γ is the restriction to $[a, b[$ of a Jordan arc $\gamma\colon [a, b] \to \mathbb{C}$.*

Remark 15.2. Let I be an interval and $\gamma\colon I \to \mathbb{C}$ a curve. Equip γ^* with the relative topology inherited from $\mathbb{C}$. It is easy to see that if γ is a Jordan arc, then γ^* is homeomorphic to $[0,1]$. If γ is closed Jordan curve, then γ^* is homeomorphic to S^1, the unit circle. If γ is a Jordan half-interval, then γ^* is homeomorphic to $[0,1[$.

Remark 15.3. Let $a < b$ and $\gamma\colon [a,b[\to \mathbb{C}$ be a simple curve. Then γ is a Jordan half-interval if and only if the limit $\xi := \lim_{t\to b_-} \gamma(t)$ exists and $\xi \notin \gamma^*$.

Notation. For two distinct points z_1 and z_2 in $\mathbb{C}$, we write $[z_1, z_2]$ both for the line segment connecting z_1 and z_2 and for the curve $[z_1, z_2]\colon [0,1] \to \mathbb{C}$, given by $[z_1, z_2](t) = (1 - t)z_1 + tz_2$. Note that we do not write $[z_1, z_2]^*$ for the line segment, in order to conform with the usual interval notation. Similarly, we write $[z_1, z_2[,\ ]z_1, z_2]$ and $]z_1, z_2[$ for the obvious curves defined on $[0,1[,\ ]0,1]$ and $]0,1[$, respectively.

Suppose now γ is either a Jordan arc, a Jordan half-interval or a closed Jordan curve and $z_1, z_2 \in \gamma^*$. Denote by a, b the endpoints of the interval I on which γ is defined. Let $t_1, t_2 \in I$ be such that $\gamma(t_1) = z_1$ and $\gamma(t_2) = z_2$. If $t_1 < t_2$, we write $\gamma_{[z_1,z_2]}\colon [t_1, t_2] \to \mathbb{C}$, for the curve $\gamma|_{[t_1,t_2]}$.

We will frequently below have to compose two or more consecutive curves $\gamma_1, \gamma_2, \ldots, \gamma_n$, defined on closed intervals $[a_1, b_1], [a_2, b_2], \ldots, [a_n, b_n]$, respectively. To do this, we extend the construction from Exercise 3.12(γ). Let $a_1' < b_1' = a_2' < b_2' \cdots < b_{n-1}' = a_n' < b_n'$ be defined inductively by setting $a_1 = a_1'$, $a_j' = b_{j-1}'$ and $b_j' = a_j' + b_j - a_j$, for $j = 1, 2, \ldots, n$. Finally, let $a = a_1' = a_1$ and $b = b_n'$ and define by the expression $\gamma = \gamma_1 + \gamma_2 + \cdots + \gamma_n$, the curve $\gamma\colon [a,b] \to \mathbb{C}$ that for $t \in [a_j', b_j']$ is given by $\gamma(t) = \gamma_j(a_j + t - a_j')$. The choice of the interval $[a, b]$ and the particular parametrization is not of importance, as long as the orientation of the curve is preserved. We chose this particular formulation to conform with earlier notation.

For a countable collection of curves $\gamma_j\colon [a_j, b_j] \to \mathbb{C}$, we set instead $b = \lim_{j\to\infty} b_j' \le \infty$ and define the sum $\gamma_1 + \gamma_2 + \cdots + \gamma_n + \cdots$ to be the obvious curve $\gamma\colon [a, b[\to \mathbb{C}$.

Proposition 15.4. *There does not exist any set $U \subseteq \mathbb{C}$ with the following two properties:*

(i) *U is nonempty, open and bounded.*

(ii) *There exists a Jordan arc $\gamma\colon [0,1] \to \mathbb{C}$, such that $\partial U \subseteq \gamma^*$.*

Proof. Suppose a set U satisfying the two listed properties does exist. Since γ is a homeomorphism onto its image, the inverse map $\theta = \gamma^{-1} \colon \gamma^* \to [0,1]$ is continuous. We now invoke Tietze's extension theorem (see Dugundji, 1966, Theorem 5.1), to get a continuous map $\bar{\theta} \colon \mathbb{C} \to [0,1]$, with $\bar{\theta}(z) = \theta(z)$, for $z \in \gamma^*$. We define a map $f \colon \mathbb{C} \to \mathbb{C}$ by setting

$$f(z) = \begin{cases} \gamma(\bar{\theta}(z)), & z \in \overline{U}, \\ z, & z \in \mathbb{C} \setminus U. \end{cases}$$

Observe that f is well defined and continuous, since $\overline{U} \cap (\mathbb{C} \setminus U) = \partial U \subseteq \gamma^*$ and $\gamma(\bar{\theta}(z)) = \gamma(\theta(z)) = z$, for $z \in \gamma^*$.

Fix a point w in the nonempty set U. Let $R > 0$ be so large that $\gamma^* \cup \overline{U} \subseteq B(w, R)$, which is possible since U is bounded. Define a continuous retraction map $r \colon B[w, R] \setminus \{w\} \to \partial B(w, R)$, by setting

$$r(z) = R \frac{z - w}{|z - w|} + w,$$

which is the identity on $\partial B(w, R)$. We write $a \colon \partial B(w, R) \to \partial B(w, R)$ for the continuous antipodal map $a(z) = -(z - w) + w = 2w - z$. Using these three maps, we may construct a continuous function $F \colon B[w, R] \to B[w, R]$ by setting $F(z) = a \circ r \circ f(z)$ for $z \in B[w, R]$. By Brouwer's fixed point theorem (Theorem 3.11(c)), there exists $z_0 \in B[w, R]$ with $F(z_0) = z_0$. Since F takes values on the boundary circle $\partial B(w, R)$, we must have $|z_0 - w| = R$. Then $z_0 \in \mathbb{C} \setminus U$ and therefore $f(z_0) = z_0$. Since also $r(z_0) = z_0$, we must have $F(z_0) = a(z_0) = 2w - z_0 \neq z_0$, which is a contradiction. $\qquad\square$

Corollary 15.5. *Let* $\gamma \colon [0,1] \to \mathbb{C}$ *be a Jordan arc. Then* $\mathbb{C} \setminus \gamma^*$ *is connected.*

Proof. Abbreviate $\Omega = \mathbb{C} \setminus \gamma^*$. Let U and U' be two disjoint open subsets of $\mathbb{C}$ with $\Omega = U \cup U'$. We assume towards a contradiction that both U and U' are nonempty. We may assume that U' either is identical to or contains the unbounded connected component of Ω. Then U must be an open and bounded set. We have $\overline{U} \cap U' = \emptyset$, which implies that $\partial U \subseteq \partial \Omega \subseteq \mathbb{C} \setminus \Omega = \gamma^*$. We have thus verified the two properties listed in Proposition 15.4, which then yields a contradiction. $\qquad\square$

Corollary 15.6. *Let* $\gamma \colon [0,1] \to \mathbb{C}$ *be a closed Jordan curve. Suppose* $\Omega = \mathbb{C} \setminus \gamma^*$ *has at least two connected components. Then all connected components of* $\mathbb{C} \setminus \gamma^*$ *has boundary equal to* γ^*.

Proof. Let U be a bounded connected component of $\mathbb{C} \setminus \gamma^*$, and write U' for the union of all other connected components. Since $\overline{U} \cap U' = \emptyset$, we must have $\partial U \subseteq \partial \Omega \subseteq \gamma^*$. Proposition 15.4 implies that in fact $\partial U = \gamma^*$, since ∂U otherwise would be a subset of the range of a Jordan arc.

To deal with the unbounded connected component U of $\mathbb{C} \setminus \gamma^*$, we fix a point w in one of the bounded components of $\mathbb{C} \setminus \gamma^*$. At least one bounded component exists. Then $U' = U \cup \{\infty\}$ is a connected component of $\mathbb{C}_\infty \setminus \gamma^*$. The Möbius transformation $m(z) = (z - w)^{-1}$ is a homeomorphism of the Riemann sphere onto itself, taking γ into $\sigma = m \circ \gamma$ and the connected components of $\mathbb{C}_\infty \setminus \gamma^*$ into connected components of $\mathbb{C}_\infty \setminus \sigma^*$. In particular, it takes U' into a bounded connected component of $\mathbb{C}_\infty \setminus \sigma^*$ and hence, also of $\mathbb{C} \setminus \sigma^*$. Since σ is also a closed Jordan curve, we conclude that $\partial m(U') = \sigma^*$. Going back with the inverse Möbius transform, we conclude that $\partial U' = \gamma^*$ (in the Riemann sphere), but this implies that $\partial U = \gamma^*$. $\square$

Lemma 15.7. *Let $a, b > 0$ be real numbers and write $E = [-a, a] + i[-b, b] \subseteq \mathbb{C}$, $\partial_\mathrm{n} E = \{z \in \partial E \mid \mathrm{Im}\, z > 0\}$ and $\partial_\mathrm{s} E = \{z \in \partial E \mid \mathrm{Im}\, z < 0\}$. Suppose $\gamma_j \colon [-1, 1] \to E$, $j = 1, 2$, are two curves with*

$$\gamma_1(-1) = -a, \quad \gamma_1(1) = a, \quad \gamma_2(-1) \in \partial_\mathrm{s} E, \quad and \quad \gamma_2(1) \in \partial_\mathrm{n} E.$$

Then $\gamma_1^* \cap \gamma_2^* \neq \emptyset$.

Proof. We may assume without loss of generality that $\mathrm{Im}\, \gamma_2(-1) = -b$ and $\mathrm{Im}\, \gamma_2(1) = b$. If this is not true, we may pass to a larger rectangle $\widetilde{E} = [-a - 1, a + 1] + i[-b, b]$ and extend the curves by adding appropriate Jordan arcs. More precisely: We replace γ_1 by $\tilde{\gamma}_1 = [-a - 1, -a] + \gamma_1 + [a, a + 1]$ and γ_2 by $\tilde{\gamma}_2 = \sigma_1 + \sigma_\mathrm{r}$, where σ_1 and σ_r are defined as follows. If $\mathrm{Im}\, \gamma_2(-1) = -b$, then we put $\sigma_1 = \gamma_2|_{[-1,0]}$. If $\mathrm{Im}\, \gamma_2(-1) \in \,]-b, 0[$, then $\mathrm{Re}\, \gamma_2(-1) \in \{-a, a\}$ and, writing $s_\mathrm{s} = \mathrm{sign}(\gamma_2(-1)) \in \{\pm 1\}$, we put $\sigma_1 = [s_\mathrm{s} a - ib, \gamma_2(-1)] + \gamma_2|_{[-1,0]}$. If $\mathrm{Im}\, \gamma_2(1) = b$, then we put $\sigma_\mathrm{r} = \gamma_2|_{[0,1]}$. If $\mathrm{Im}\, \gamma_2(1) \in \,]0, b[$, then $\mathrm{Re}\, \gamma_2(1) \in \{-a, a\}$ and, writing $s_\mathrm{n} = \mathrm{sign}(\gamma_2(1)) \in \{\pm 1\}$, we put $\sigma_\mathrm{r} = \gamma_2|_{[0,1]} + [s_\mathrm{n} a + ib, \gamma_2(1)]$. (Both curves $\tilde{\gamma}_1$ and $\tilde{\gamma}_2$ should be reparametrized to be defined on $[-1, 1]$.) Note that $\gamma_1^* \cap \gamma_2^* \neq \emptyset$ if and only if $\tilde{\gamma}_1^* \cap \tilde{\gamma}_2^* \neq \emptyset$. We have now reduced to the case $\mathrm{Im}\, \gamma_2(\pm 1) = \pm b$.

Suppose, towards a contradiction, that $\gamma_1^* \cap \gamma_2^* = \emptyset$. Then

$$\forall s, t \in [-1, 1]: \quad M(s, t) = |\gamma_1(s) - \gamma_2(t)|_\infty > 0,$$

where $|z|_\infty = \max\{|\mathrm{Re}\, z|, |\mathrm{Im}\, z|\}$ is the maximum norm on $\mathbb{C}$, as a real vector space. Let $R = [-1, 1] \times [-1, 1] \subseteq \mathbb{R}^2$ be a closed square in $\mathbb{R}^2$. Define a continuous function $F\colon R \to R$ by setting

$$F(s,t) = \begin{pmatrix} \frac{\mathrm{Re}(\gamma_2(t) - \gamma_1(s))}{M(s,t)} \\ \frac{\mathrm{Im}(\gamma_1(s) - \gamma_2(t))}{M(s,t)} \end{pmatrix}.$$

By Brouwer's fixed point theorem, cf. Theorem 3.11(c), the function F has a fixed point $(s_0, t_0) \in R$. Since F takes values in the boundary of R, we must have $(s_0, t_0) \in \partial R$. That is, (s_0, t_0) is an element of (at least) one of the 4 edges of R.

We distinguish between vertical and horizontal edges. Suppose $|s_0| = 1$. That is, the fixed point sits on a vertical edge. Then $\gamma_1(s_0) = s_0 a$. This implies that

$$1 = s_0^2 = s_0 \frac{\mathrm{Re}(\gamma_2(t_0) - \gamma_1(s_0))}{M(s_0, t_0)} = \frac{s_0 \, \mathrm{Re}(\gamma_2(t_0)) - a}{M(s_0, t_0)},$$

which is a contradiction since the right-hand side is negative (or zero). If (s_0, t_0) is an element of one of the two horizontal edges, then $|t_0| = 1$. In this case $t_0 \, \mathrm{Im} \, \gamma_2(t_0) = b$. We therefore have

$$1 = t_0^2 = t_0 \frac{\mathrm{Im}(\gamma_1(s_0) - \gamma_2(t_0))}{M(s_0, t_0)} = \frac{t_0 \, \mathrm{Im}(\gamma_1(s_0)) - b}{M(s_0, t_0)},$$

which is a contradiction since the right-hand side is negative (or zero). □

Theorem 15.8 (Jordan's Curve Theorem). *Let $\gamma\colon [a,b] \to \mathbb{C}$ be a closed Jordan curve. Then the open set $\mathbb{C} \setminus \gamma^*$ has exactly two connected components, one bounded and one unbounded, which we abbreviate G_0 and G_∞, respectively. Furthermore,*

$$\partial G_0 = \partial G_\infty = \gamma^*.$$

Remark 15.9. In fact, the index of points z in the bounded component is either $+1$ or -1. This extra piece of information will be established in the following section, cf. Theorem 15.16.

Proof. Due to Corollary 15.6, it suffices to show that $\mathbb{C} \setminus \gamma^*$ has exactly one bounded connected component. Let $A, B \in \gamma^*$ be two points on the closed Jordan curve maximizing the distance $|A - B|$. These two points exist (and are distinct) since γ^* is compact. We may assume that $A = -1$ and $B = 1$, after a (continuous) scaling/translation/rotation of the complex plane. Then

$$\gamma^* \subseteq E := \{z \in \mathbb{C} \mid |\mathrm{Re}\, z| \le 1, |\mathrm{Im}\, z| \le 2\}$$

and γ intersects the boundary ∂E only at the two points -1 and 1.

We choose a new parametrization of γ such that $\gamma \colon [-1, 1] \to \mathbb{C}$, $\gamma(\pm 1) = -1$ and $\gamma(0) = 1$. The two Jordan arcs $\gamma|_{[-1,0]}$ and $\gamma|_{[0,1]}$ both start and end at the same points -1 and $+1$, which are the only two points where the two Jordan arcs intersect.

By Lemma 15.7, the line segment $[-2i, 2i]$ intersects both these Jordan arcs at least once. Let ni be the intersection point in $\gamma^* \cap [-2i, 2i]$ with the largest imaginary part n. We label the Jordan arc containing ni as γ_n and the other Jordan arc we call γ_s. The reader may think of the label "s" as *south* and the label "n" as *north*.

Write $n'i$ for the point in $\gamma_\mathrm{n}^* \cap [-2i, 2i]$ with smallest imaginary part, and note that $n' \leq n$ and that they may coincide.

The Jordan arc $[-2i, n'i] + \gamma_{[n'i, ni]} + [ni, 2i]$, must intersect γ_s. Since γ_s does not intersect the last two legs of the journey from $-2i$ to $2i$, it must intersect the first $[-2i, n'i]$. Let $s'i$ and si be the points in $\gamma_\mathrm{s}^* \cap [-2i, n'i]$ with the largest, respectively smallest, imaginary part

Finally, let $m = (n' + s')/2$, such that $mi \notin \gamma^*$, and write U for the connected component of $\mathbb{C} \setminus \gamma^*$ containing mi. To sum up

$$-2 < s \leq s' < m < n' \leq n < 2.$$

The choice of the points $ni, n'i, si, s'i$ and mi is illustrated in Fig. 15.1.

We claim that U is a bounded set. If not, then U must be the unbounded connected component of $\mathbb{C} \setminus \gamma^*$. Then there exists a curve $\sigma \colon [0, 1] \to U$ with $\gamma(0) = mi$ and $\gamma(1) \notin E$, where we used that $\gamma^* \subseteq E$. By the Intermediate Value Theorem, we have $\sigma^* \cap \partial E \neq \emptyset$. Let $0 < t < 1$ be the smallest t with $w = \sigma(t) \in \partial E$.

Write $\partial_\mathrm{n} E = \{z \in \partial E \mid \mathrm{Im}\, z > 0\}$ and $\partial_\mathrm{s} E = \{z \in \partial E \mid \mathrm{Im}\, z < 0\}$. Then we must have either $w \in \partial_\mathrm{n} E$ or $w \in \partial_\mathrm{s} E$, since $\pm 1 \in \gamma^*$.

Assume that $w \in \partial_\mathrm{n} E$. Let $\gamma_2 = [-2i, mi] + \sigma_{[mi, w]}$. The first leg of the curve does not intersect γ_n^*, and the second does not intersect γ^*. Hence, γ_2^* does not intersect γ_n^*, which contradicts Lemma 15.7. If $w \in \partial_\mathrm{s} E$, we choose $\gamma_2 = [2i, mi] + \sigma_{[mi, w]}$. Then γ_2^* does not intersect γ_s^*, which again contradicts Lemma 15.7. This proves that U is a bounded connected component.

It remains to prove that U is the only bounded connected component. Let V be another bounded connected component. Being bounded, we must have $V \subseteq E$. By Corollary 15.6, $\pm 1 \in \partial V$.

Let $0 < \varepsilon < 1$ be such that $B(\pm 1, \varepsilon)$ does not contain any points on γ_n^* lying between ni and $n'i$ nor any points on γ_s^* lying between si and $s'i$. This is possibly since the two arc segments are compact and do not contain ± 1.

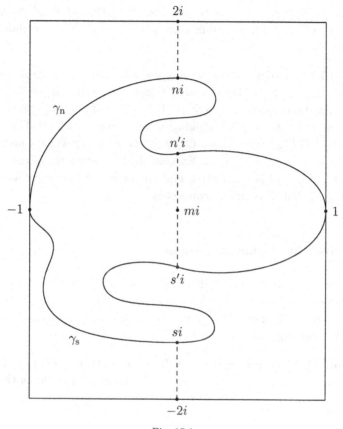

Fig. 15.1

There exist $C \in B(-1,\varepsilon) \cap V$ and $D \in B(1,\varepsilon) \cap V$. Let $\sigma \colon [0,1] \to V$ be a curve with $\sigma(0) = C$ and $\sigma(1) = D$.

We construct two curves $\gamma_1 = [-1, C] + \sigma + [D, 1]$ and $\gamma_2 = [-2i, si] + \gamma_{s,[si,s'i]} + [s'i, n'i] + \gamma_{n,[n'i,ni]} + [ni, 2i]$. By Lemma 15.7, we must have $\gamma_1^* \cap \gamma_2^* \neq \emptyset$. But this is impossible, since γ_2^* only runs through U, the unbounded connected component of $\mathbb{C} \setminus \gamma^*$ (along the imaginary axis) and the curves γ_s and γ_n between ni and $n'i$ and between si and $s'i$, respectively. But γ_1^*, outside of $B(\pm 1, \varepsilon)$ stays inside V. This concludes the proof. $\quad\square$

Definition 15.10 (Jordan Region). *An open bounded set $G \subseteq \mathbb{C}$ is called a Jordan region if there exists a closed Jordan curve γ with $\gamma^* = \partial G$.*

Note that a Jordan region G coincides with the bounded connected component of $\mathbb{C} \setminus \gamma^*$, cf. Theorem 15.8, where γ is a closed Jordan curve with $\gamma^* = \partial G$.

Remark 15.11. Jordan regions are not just connected, they are also simply connected. Indeed, let G be a Jordan region and write $G_\infty = \mathbb{C} \setminus \overline{G}$ for the unbounded connected component of $\mathbb{C} \setminus \partial G$. Let $z \in G_\infty$. Since ∂G is compact, there exists $w \in \partial G$ minimizing the distance $|w - z|$. For this w, we have $[z, w[\subseteq G_\infty$. Since $[z, w]$, G_∞ and ∂G are all (path) connected, so is $G_\infty \cup [z, w] \cup \partial G = \mathbb{C} \setminus G$, cf. Exercise 15.1. Adding the point ∞, we observe that $\mathbb{C}_\infty \setminus G$ is connected, and therefore we may conclude from Corollary 9.12 that G is simply connected.

15.2   Cross cuts in Jordan regions

In this section we show that Jordan regions are cut in half by Jordan arcs running from boundary to boundary. As an application, we argue that for a Jordan region G, the index of a point in G with respect to a closed Jordan curve, parametrizing the boundary of G, is either $+1$ or -1.

Definition 15.12 (Cross Cut in a Jordan Region). *Let $G \subseteq \mathbb{C}$ be a Jordan region and $\sigma \colon [a, b] \to \mathbb{C}$ a curve. We call σ a cross cut in G if*

(a) σ *is a Jordan arc.*

(b) $\sigma(a), \sigma(b) \in \partial G$.

(c) $\sigma(t) \in G$ *for all $t \in \,]a, b[$.*

Proposition 15.13. *Let G be a Jordan region and $\sigma \colon [a, b] \to \mathbb{C}$ a cross cut in G. Then the open set $G \setminus \sigma^*$ consists of exactly two connected components, both of which are Jordan regions.*

Proof. First we argue that $G \setminus \sigma^*$ consists of exactly two connected components. Since G is bounded, connected and simply connected, cf. Remark 15.11, we may choose a biholomorphic map $f \colon G \to B(0, 1)$. Writing $g(z) = z/(1 - |z|)$, we construct a homeomorphism $g \circ f \colon G \to \mathbb{C}$. Since $|f(\sigma(t))| \to 1$ for $t \to b_-$ and for $t \to a_+$, we find that $\gamma = g \circ f \circ \sigma \colon \,]a, b[\to \mathbb{C}$ is a simple curve with $\gamma(t) \to \infty$ for $t \to b_-$ and for $t \to a_+$. Setting $\gamma(a) = \gamma(b) = \infty$, we obtain a simple closed curve in the Riemann sphere. Pick $w \in \mathbb{C} \setminus \gamma^*$ (noting that such a point exists!) and put $m(z) = (w - z)^{-1}$.

Then $m \circ \gamma$ defines a closed Jordan curve, which splits $\mathbb{C}_\infty$ into exactly two connected components $\widetilde{G}_1$ and $\widetilde{G}_2$, where we add the point ∞ to the unbounded of the two connected components we get from Theorem 15.8. Going back with the homeomorphisms, we have now split $G \setminus \sigma^*$ into two connected components $G_j = f^{-1} \circ g^{-1} \circ m^{-1}(\widetilde{G}_j)$, $j = 1, 2$.

It remains to argue that G_1 and G_2 are Jordan regions.

Let $\gamma_0 : [0, 1] \to \mathbb{C}$ denote a closed Jordan curve with $\gamma_0^* = \partial G$. We may assume without loss of generality that $a = 0, b = 1/2$, $\gamma_0(0) = \sigma(0)$ and $\gamma_0(1/2) = \sigma(1/2)$.

For notational convenience, we define three Jordan arcs $\sigma_j : [0, 1/2] \to \mathbb{C}$, $j = 0, 1, 2$, by setting

$$\sigma_0(t) = \sigma(t), \quad \sigma_1(t) = \gamma_0(1/2 - t) \quad \text{and} \quad \sigma_2(t) = \gamma_0(1/2 + t),$$

for $t \in [0, 1/2]$. Using these three arcs, we construct two new closed Jordan curves

$$\gamma_1 = \sigma_0 + \sigma_2 \quad \text{and} \quad \gamma_2 = \sigma_0 + \sigma_1.$$

Note that $\gamma_0 = -\sigma_1 + \sigma_2$, where $-\sigma_1$ means σ_1 with reversed orientation.

Denote by H_j, $j = 1, 2$, the bounded connected components of $\mathbb{C} \setminus \gamma_j^*$. Since the boundaries of H_1 and H_2 are different, we must have $H_1 \neq H_2$. Since $\sigma^* \subseteq \partial H_j$, $j = 1, 2$, we must also have $G \cap H_j \neq \emptyset$, for $j = 1, 2$. Indeed, if $w \in \sigma_*$ then there exists $\{z_n\}_{n \in \mathbb{N}} \subseteq H_j$ such that $z_n \to w$. Since $w \in G$, we must have $z_n \in G$ eventually. Since G, H_1 and H_2 are connected and bounded we must have $H_1, H_2 \subseteq G$ and consequently, H_1 and H_2 must coincide with G_1 and G_2. This shows that G_1 and G_2 are Jordan regions. $\qquad\square$

Lemma 15.14. *Let G be a Jordan region and γ_1, γ_2 two closed Jordan curves with $\gamma_1^* = \gamma_2^* = \partial G$. Then, for any $z \in G$, the indices $\mathrm{Ind}_{\gamma_1}(z)$ and $\mathrm{Ind}_{\gamma_2}(z)$ are equal up to a sign.*

Proof. By Theorem 3.18(a), we may assume that both closed Jordan curves γ_1, γ_2 are defined on $[0, 1]$.

Put $A = \gamma_1(0)$ and let $s \in [0, 1]$ be such that $\gamma_2(s) = A$. Extend γ_2 from $[0, 1]$ to $\mathbb{R}$ by periodicity. If $s \neq 0$, we define for $t, \theta \in [0, 1]$ a homotopy $\Gamma(t, \theta) = \gamma_2(t + \theta s)$ between γ_2 and $\Gamma(\cdot, 1)$, which is a closed Jordan curve with the same image as γ_2 and satisfying that $\Gamma(0, 1) = A$. Hence, by Theorem 3.18(b), we may assume that $\gamma_2(0) = A$ as well.

Let $\kappa_j : [0, 1[\to \partial G$ be bijections defined by $\kappa_j(t) = \gamma_j(t)$. The composition $\phi = \kappa_2^{-1} \circ \kappa_1 : [0, 1[\to [0, 1[$ is also a bijection satisfying $\phi(0) = 0$.

Consequently, ϕ must be continuous and strictly increasing. Furthermore, setting $\phi(1) = 1$ defines a bijection $\phi\colon [0,1] \to [0,1]$. Since $\gamma_1 = \gamma_2 \circ \phi$, we may conclude the lemma from Theorem 3.18(a). $\qquad\square$

The lemma above will of course be superfluous when we have established, in Theorem 15.16 below, that the index of a point $z \in G$ (with respect to a closed Jordan curve parametrizing ∂G) is either $+1$ or -1.

Lemma 15.15. *Let G_0 be a Jordan region and $\sigma\colon [a,b] \to \mathbb{C}$ a cross cut in G_0. Denote by G_1 and G_2 the two connected components of $G_0 \setminus \sigma^*$, and by $\gamma_0, \gamma_1, \gamma_2$ closed Jordan curves with $\gamma_j^* = \partial G_j$, $j = 0, 1, 2$. Then, for any $j \in \{1, 2\}$ and $z \in G_j$, the index $\mathrm{Ind}_{\gamma_j}(z)$ is up to a sign equal to $\mathrm{Ind}_\gamma(z)$.*

Proof. By Theorem 3.18(a), we may assume that the three closed Jordan curves $\gamma_0, \gamma_1, \gamma_2$ are all defined on $[0, 1]$.

Put $A = \sigma(a)$ and let $r \in [0, 1]$ be such that $\gamma_0(r) = A$. We may assume that $r = 0$. Indeed, as in the proof of Lemma 15.14, we may achieve this with a homotopy. Let now $B = \sigma(b)$. and choose $s \in [0, 1]$ such that $B = \gamma_0(s)$. Note that $s > 0$. Define a continuous strictly increasing function $\phi\colon [0, 1] \to [0, 1]$ by setting

$$\phi(t) = \begin{cases} 2st, & t \in [0, 1/2], \\ 2(1-s)t + 2s - 1, & t \in [1/2, 1]. \end{cases}$$

Then $\phi(0) = 0$, $\phi(1/2) = s$ and $\phi(1) = 1$. Replacing γ_0 by $\gamma_0 \circ \phi$, and invoking Theorem 3.18(a) again, we may assume that $\gamma_0(0) = A$ and $\gamma_0(1/2) = B$.

Appealing to Lemma 15.14, we observe that we may replace γ_1 and γ_2 by the two closed Jordan curves

$$\gamma_1(t) = \begin{cases} \gamma_0(t), & t \in [0, 1/2], \\ \sigma(1-t), & t \in [1/2, 1], \end{cases} \quad \text{and} \quad \gamma_2(t) = \begin{cases} \sigma_0(t), & t \in [0, 1/2], \\ \gamma_0(t), & t \in [1/2, 1], \end{cases}$$

which have the same images and hence, up to a sign, the same indices as the original curves (after possibly switching the labels 1 and 2).

Let $z \in \mathbb{C} \setminus (\gamma_0^* \cup \sigma^*)$ and denote by $L_0\colon [0, 1] \to \mathbb{C}$ and $L_\sigma\colon [0, 1/2] \to \mathbb{C}$ continuous logarithms of $\gamma_0(t) - z$ and $\sigma(t) - z$, respectively.

Since $z \notin \gamma_j^*$, $j = 1, 2$, we can define continuous logarithms $L_j\colon [0, 1] \to \mathbb{C}$ for $\gamma_j(t) - z$ by

$$L_1(t) = \begin{cases} L_0(t), & t \in [0, 1/2], \\ L_\sigma(1-t) - L_\sigma(1/2) + L_0(1/2), & t \in [1/2, 1], \end{cases}$$

and

$$L_2(t) = \begin{cases} L_\sigma(t), & t \in [0, 1/2], \\ L_0(t) - L_0(1/2) + L_\sigma(1/2), & t \in [1/2, 1]. \end{cases}$$

We can now compute

$$2\pi i \, \mathrm{Ind}_{\gamma_1}(z) = L_1(1) - L_1(0)$$
$$= L_\sigma(0) - L_\sigma(1/2) + L_0(1/2) - L_0(0)$$

and

$$2\pi i \, \mathrm{Ind}_{\gamma_2}(z) = L_2(1) - L_2(0)$$
$$= L_0(1) - L_0(1/2) + L_\sigma(1/2) - L_\sigma(0).$$

Hence

$$\mathrm{Ind}_{\gamma_1}(z) + \mathrm{Ind}_{\gamma_2}(z) = \mathrm{Ind}_\gamma(z). \tag{15.1}$$

Now let $z \in G_1$. Then z is in the unbounded component of $\mathbb{C} \setminus \gamma_2$ and consequently, it follows from Theorem 3.18(c) that $\mathrm{Ind}_{\gamma_2}(z) = 0$. Equation (15.1) now implies the lemma. A similarly argument applies if $z \in G_2$. $\square$

Theorem 15.16. *Let G be a Jordan region and $\gamma: [a, b] \to \mathbb{C}$ a closed Jordan curve parametrizing the boundary ∂G. Then $\mathrm{Ind}_\gamma(z) = \pm 1$ for all $z \in G$.*

Proof. Let $z \in G$. For $r > 0$, we introduce 4 lines in $\mathbb{C}$

$$L_1 = z + r + i\mathbb{R}, \quad L_2 = z + ir + \mathbb{R},$$
$$L_3 = z - r + i\mathbb{R} \quad \text{and} \quad L_4 = z - ir + \mathbb{R}.$$

These four lines contain the 4 edges of an open square S of side length $2r$ centered at z. We pick $r > 0$ small enough such that S is a subset of G. Let $\sigma_1(s) = z + r + is$, $\sigma_2(s) = z + ir + s$, $\sigma_3(s) = z - r - is$ and $\sigma_4(s) = z - ir - s$ be (counterclockwise) parametrizations of the four lines.

Put $G_0 = G$, $a_0 = a$, $b_0 = b$ and $\gamma_0 = \gamma$. We proceed inductively to construct new Jordan regions G_1, G_2, G_3, G_4 with boundary curves $\gamma_1, \gamma_2, \gamma_3, \gamma_4$ and the property that $S \subseteq G_j$, $G_j \subseteq G_{j-1}$ and $\mathrm{Ind}_{\gamma_j}(z) = \pm \mathrm{Ind}_{\gamma_{j-1}}(z)$ for $j = 1, 2, 3, 4$.

Define $a_j = \max\{s < 0 \mid \sigma_j(s) \in \partial G_{j-1}\} \leq -r$ and $b_j = \min\{s > 0 \mid \sigma_j(s) \in \partial G\} \geq r$. Then $\sigma_j: [a_j, b_j] \to \mathbb{C}$ is a cross cut in G_{j-1}. By Proposition 15.13, this splits G_{j_1} into two Jordan regions. The one containing z we choose as G_j. By a connectedness argument, $S \subseteq G_{j-1} \setminus \sigma_j^*$ implies that

$S \subseteq G_j$ as well. That the indices of z with respect to closed Jordan boundary curves of G_{j-1} and G_j are equal up to a sign follows from Lemma 15.15.

With this construction, and since G_4 is bounded, we must have $G_4 = S$! The theorem now follows, since the result is easily verified for $G = S$. $\square$

15.3   Simple and accessible boundary points

In this section we investigate properties of Jordan regions that will be crucial in connection with the extension theorem in Chapter 17. The primary result of this section, due to Schönflies, is about simplicity of boundary points in Jordan regions. See Theorem 15.19 below. Recall from Definition 15.10 the notion of a Jordan region.

The following key proposition we learned from §346 in Carathéodory (1954).

Proposition 15.17. *Let G be a Jordan region and $A \in \partial G$. There exists a function $m \colon]0, \infty[\to]0, \infty[$, with $m(r) \leq r$ and non-decreasing, such that: for any $r > 0$ and any two points $z_1, z_2 \in G \cap B(A, m(r))$, there exists a curve $\gamma \colon [0, 1] \to G \cap B[A, r]$, such that $\gamma(0) = z_1$ and $\gamma(1) = z_2$.*

Proof. Suppose the boundary of G is parametrized by the closed Jordan curve $\gamma \colon [-1, 1] \to \mathbb{C}$ with $\gamma(0) = A$.

Put $M = \max_{|t| \leq 1} |\gamma(t) - A| < +\infty$. Define $\delta \colon]0, M[\mapsto [-1, 1]$ by

$$\delta(r) = \min\{|t| \mid t \in [-1, 1], \ |\gamma(t) - A| = r\},$$

which is the first time in modulus that the boundary curve intersects the circle of radius r around A. Clearly δ is non-decreasing.

The function m in the proposition is now defined for $r \in]0, M]$ as

$$m(r) = \min_{\delta(r) \leq |t| \leq 1} |\gamma(t) - A|. \tag{15.2}$$

Since δ is non-decreasing, m is necessarily non-decreasing as well. Furthermore, $m(r) \leq r$. For $r \geq M$ we simply define $m(r) = M$ and the proposition obviously holds true for $r \geq M$ since Jordan regions are path connected. The choice of $m(r)$ is illustrated in Fig. 15.2.

Now fix an $r \in]0, M[$. The set

$$\{\theta \in \mathbb{R} \mid A + re^{i\theta} \in G\}$$

is an open 2π-periodic subset of $\mathbb{R}$. Hence, it is an at most countable union of open disjoint intervals $I_k = \,]a_k, b_k[$, where we only count intervals modulo 2π. Note that $A + re^{ia_k}, A + re^{ib_k} \in \partial G$.

For each k, we can now define cross cuts $\sigma_k \colon [a_k, b_k] \to \mathbb{C}$ in G by setting $\sigma_k(t) = A + re^{it}$ for $t \in [a_k, b_k]$. By Proposition 15.13, each cross cut σ_k divides the Jordan region G into two connected components G_k and G'_k. We choose to label them such that $A \in \partial G_k$ and $A \notin \partial G'_k$. In the following, we write $J_k = \sigma_k(\,]a_k, b_k[) \subseteq G$ for the k'th arc without its endpoints. In Fig. 15.2 an example with three arcs J_1, J_2 and J_3 is depicted. The sets G_2 and G'_2 are indicated as well.

Note that

$$\forall k \colon \quad G \cap B(A, m(r)) \subseteq G_k. \tag{15.3}$$

Indeed, by the choice of G_k there exists at least one $z_1 \in G_k \cap B(A, m(r))$. Assume that there is a $z_2 \in G'_k \cap B(A, m(r))$ as well. The line segment $[z_1, z_2] \subseteq B(A, m(r))$ must then intersect $\partial G'_k \subseteq J_k \cup \gamma(\{t \in [-1, 1] \mid |t| \geq \delta(r)\})$. But by the construction (15.2) of $m(r)$ this is impossible and the claim is established.

Pick $z_1, z_2 \in G \cap B(A, m(r))$. Let $\tilde{\gamma} \colon [0, 1] \to G$ be a piecewise linear curve with $\tilde{\gamma}(0) = z_1$ and $\tilde{\gamma}(1) = z_2$. (See Exercise 15.2.) Since each of the finitely many line segments can have at most 2 intersections with $\partial B(A, r)$, we find that the curve $\tilde{\gamma}$ intersects at most finitely many arcs J_k. We may assume, after reordering, that it is the first N arcs $J_1, \ldots, J_N$. We keep only those arcs J_k, where $\tilde{\gamma}^* \cap G'_k$ is not empty.

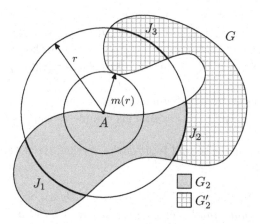

Fig. 15.2

Let $t_1^{(1)}$ be the first time $\tilde{\gamma}$ hits J_1 and $t_2^{(1)}$ the last time it hits J_1. Since the curve has to eventually reach $z_2 \in G_1$, cf. (15.3), the curve must cut back out through J_1 at least once.

Replace the part of the curve $\tilde{\gamma}$ from $\tilde{\gamma}(t_1^{(1)})$ to $\tilde{\gamma}(t_2^{(1)})$ by the arc segment inside $J_1 \subseteq G$. This gives a new curve $\tilde{\gamma}^{(1)}$, which never meets G_1' Repeat the process with J_2 creating a new curve $\tilde{\gamma}^{(2)}$ never meeting G_1' and G_2'. Doing this finitely many times we arrive at a curve $\tilde{\gamma}^{(N)}$ never entering G_j', $j = 1, \ldots, N$. Since by construction $\tilde{\gamma}^{(N)*} \subseteq \tilde{\gamma}^* \cup J_1 \cup \cdots \cup J_N$, we in particular conclude that $\tilde{\gamma}^{(N)*} \subseteq G \cap B[A, r]$, and $\tilde{\gamma}^{(N)}$ runs from z_1 to z_2. This concludes the construction. $\qquad\square$

Definition 15.18 (Simple Boundary Point). *Let $G \subseteq \mathbb{C}$ be open. A boundary point $A \in \partial G$ is called simple if: for any convergent sequence $\{z_n\}_{n \in \mathbb{N}} \subseteq G$ with $\lim_{n \to \infty} z_n = A$, there exist $N \in \mathbb{N}$, a curve $\gamma \colon [0, 1[\to G$ with $\lim_{t \to 1_-} \gamma(t) = A$, and a strictly increasing sequence $\{t_n\}_{n=N}^{\infty}$ with $\lim_{n \to \infty} t_n = 1$ and $\gamma(t_n) = z_n$ for all $n \geq N$.*

Theorem 15.19 (Schönflies, 1st form). *Let G be a Jordan region. All boundary points of G are simple.*

Proof. We will make use of the function $m \colon]0, \infty[\to]0, \infty[$ constructed in Proposition 15.17. Let $A \in \partial G$ and $\{z_n\}_{n \in \mathbb{N}} \subseteq G$ a sequence with $\lim_{n \to \infty} z_n = A$. We will verify Definition 15.18 with $N = 1$.

For $j \in \mathbb{N}$, choose $n_j \in \mathbb{N}$ such that $|A - z_n| < m(1/j)$ for all $n \geq n_j$. This is possible since $z_n \to A$. We may furthermore choose the n_j's such that $n_{j+1} > n_j$ for all $j \in \mathbb{N}$.

Let $j \in \mathbb{N}$ and put $t_n = 1 - 1/n$. For $n = 1, \ldots, n_1 - 1$, we choose (arbitrary) curves $\gamma_n \colon [t_n, t_n + 1] \to G$ with $\gamma_n(t_n) = z_n$ and $\gamma_n(t_{n+1}) = z_n + 1$. This we can do since G is connected. For $j \in \mathbb{N}$, we invoke Proposition 15.17 and obtain for each $n \in \{n_j, \ldots, n_{j+1} - 1\}$ a curve $\gamma_n \colon [t_n, t_{n+1}] \to G$, with $\gamma_n(t_n) = z_n$, $\gamma_n(t_{n+1}) = z_{n+1}$ and $\gamma_n^* \subseteq G \cap B[A, 1/j]$.

We can add these curves to get a new curve $\gamma \colon [0, 1[\to G$, with $\gamma(t_n) = z_n$ for all $n \in \mathbb{N}$ and $\lim_{t \to 1_-} \gamma(t) = A$. This completes the proof. $\qquad\square$

We now arrive at the second topic of this section, the notion of *accessible boundary points*.

Definition 15.20. *Let $G \subseteq \mathbb{C}$ be an open set. By an accessible boundary point for G, we understand a pair (A, γ), where $A \in \partial G$ and $\gamma \colon [0, 1[\to G$ is a curve with $\lim_{t \to 1_-} \gamma(t) = A$.*

We introduce an equivalence relation $\sim$ on the set of all accessible boundary points.

Definition 15.21. *We say that* $(A_1, \gamma_1) \sim (A_2, \gamma_2)$ *if and only if* $A_1 = A_2$ *and there exist a curve* $\gamma \colon [0, 1[\to G$ *such that* $\lim_{t \to 1_-} \gamma(t) = A$ *and for any* $r > 0$ *the two sets* $\gamma^* \cap \gamma_j^* \cap B(A, r)$ *are nonempty.*

We write $\partial_A G$ for the set of equivalence classes of accessible boundary points. Observe that the map $\partial_A G \ni [(A, \gamma)] \to A \in \partial G$ is well-defined. A boundary point A is itself called accessible if it is in the range of this map. We say that $A \in \partial G$ defines a unique boundary point if it is accessible and the preimage of $\{A\}$ is a singleton in $\partial_A G$.

We remark that the accessible boundary points are dense in the boundary, cf. Exercise 15.6.

Proposition 15.22. *Let G be an open set and $A \in \partial G$ a simple boundary point. Then A defines a unique accessible boundary point.*

Proof. That simple boundary points are accessible follows directly from Definition 15.18. It remains to argue that A defines only one accessible boundary point.

Assume $\gamma_1, \gamma_2 \colon [0, 1[\to G$ are two curves with $\lim_{t \to 1_-} \gamma_j(t) = A$. If the two curves have a sequence of times of intersection converging to 1, then we are done. We may therefore suppose that there exists $T \in [0, 1[$ such that $\gamma_1(t) \neq \gamma_2(s)$ for all $s, t \in [T, 1[$.

Let $m \colon]0, \infty[\to]0, \infty[$ be the function from Proposition 15.17. Pick two strictly increasing sequences $\{t_n^{(j)}\}_{n \in \mathbb{N}} \subseteq [T, 1[$, $j = 1, 2$, such that $\lim_{n \to \infty} t_n^{(j)} = 1$ and $|\gamma_j(t_n^{(j)}) - A| < m(1/n)$. Put $z_n^{(j)} = \gamma_j(t_n^{(j)})$, $j = 1, 2$.

For n even, we invoke Proposition 15.17 and obtain a curve $\tilde{\gamma}_n \colon [0, 1] \to G$, with $\tilde{\gamma}_n(0) = z_n^{(1)}$ and $\tilde{\gamma}_n(1) = z_n^{(2)}$. For n odd, we obtain a curve $\tilde{\gamma}_n \colon [0, 1] \to G$ with $\tilde{\gamma}_n(0) = z_n^{(2)}$ and $\tilde{\gamma}_n(1) = z_n^{(1)}$. By construction, and in both cases,

$$\sup_{t \in [0,1]} |\tilde{\gamma}_n(t) - A| \leq \tfrac{1}{n}.$$

Finally, we create a new curve $\sigma \colon [0, 1[\to G$ by passing through the points

$$z_1^{(1)} \xrightarrow{\sigma_1} z_1^{(2)} \xrightarrow{\sigma_2} z_2^{(2)} \xrightarrow{\sigma_3} z_2^{(1)} \xrightarrow{\sigma_4} z_3^{(1)} \xrightarrow{\sigma_5} z_3^{(2)} \xrightarrow{\quad} \cdots .$$

To be precise, we construct the curve segments $\sigma_k \colon [S_k, S_{k+1}] \to G$, with $S_k = 1 - k^{-1}$, passing through the above points. Let

$$\sigma_{2k+1}(t) = \tilde{\gamma}_{k+1}\left(\frac{t - S_{2k+1}}{S_{2k+2} - S_{2k+1}}\right)$$

and

$$\sigma_{2k}(t) = \begin{cases} \gamma_1\left(\dfrac{S_{2k+1} - t}{S_{2k+1} - S_{2k}}t_k^{(1)} + \dfrac{t - S_{2k}}{S_{2k+1} - S_{2k}}t_{k+1}^{(1)}\right), & k \text{ even}, \\[4mm] \gamma_2\left(\dfrac{S_{2k+1} - t}{S_{2k+1} - S_{2k}}t_k^{(2)} + \dfrac{t - S_{2k}}{S_{2k+1} - S_{2k}}t_{k+1}^{(2)}\right), & k \text{ odd}, \end{cases}$$

and define the total curve $\sigma \colon [0, 1[\to G$ to be

$$\sigma = \sigma_1 + \sigma_2 + \sigma_3 + \sigma_4 + \cdots.$$

By construction, we observe that $\lim_{t \to 1_-} \sigma(t) = A$ and σ has intersections with γ_1 and γ_2 arbitrarily close to A. This proves the proposition. $\qquad\square$

As an immediate consequence of Theorem 15.19 and Proposition 15.22, we conclude a second form af Schönflies' theorem.

Theorem 15.23 (Schönflies, 2nd form). *Let G be a Jordan region. Each boundary point $A \in \partial G$ defines a unique accessible boundary point.*

Schönflies actually proved a much stronger statement, namely that for any Jordan region G, there exists a homeomorphism $f \colon \overline{G} \to B[0, 1]$ with $f(\partial G) = \partial B(0, 1)$. For a proof, the reader may consult Thomassen (1992), where one can also locate an alternative proof of Jordan's curve theorem. In Chapter 17, we shall in fact prove that Schönflies' function f may be chosen to be holomorphic in G – this follows from Corollary 17.18 (using Theorem 15.23 above).

Theorem 15.24. *Let $\sigma \colon [0, 1] \to \mathbb{C}$ be a Jordan arc and put $\Omega = \mathbb{C} \setminus \sigma^*$. Then $\partial \Omega = \sigma^*$, boundary points $A \in \sigma(]0, 1[)$ are accessible and define exactly two distinct accessible boundary points.*

Proof. Since Ω is open, we have $\partial \Omega \subseteq \mathbb{C} \setminus \Omega = \sigma^*$. Let $A \in \sigma^*$. To see that A is a boundary point, we argue as follows. If we can show that for any $n \in \mathbb{N}$, we have $\partial B(A, 1/n) \not\subseteq \sigma^*$, then we can pick a sequence $\{z_n\}_{n \in \mathbb{N}} \subseteq \Omega$, such that $|A - z_n| = 1/n$. This would prove that $\partial \Omega = \sigma^*$. Suppose towards a contradiction that $\partial B(A, 1/n) \subseteq \sigma^*$. Since the circle is a closed connected subset of σ^*, the set $\{t \in [0, 1] \mid |\sigma(t) - A| = 1/n\}$ is a closed connected subset of $[0, 1]$. See also Remark 15.2. This implies that $\partial B(A, 1/n)$ is homeomorphic to a closed interval, which is a contradiction. Recall Example 3.8 and Proposition 3.10.

Let $s \in \,]0, 1[$ and write $A = \sigma(s)$. To see that the boundary point A is accessible and defines exactly two accessible boundary points, we argue as follows. Let $\delta > 0$ be small enough such that $\partial B(A, \delta) \cap \sigma([0, s]) \neq \emptyset$ and $\partial B(A, \delta) \cap \sigma([s, 1]) \neq \emptyset$. Let $0 \leq r < t < s \leq 1$ be such that $\sigma(r), \sigma(t) \in \partial B(A, \delta)$ and $\sigma(]r, t[) \subseteq B(A, \delta)$. Then $\sigma \colon [r, t] \to \mathbb{C}$ defines a cross cut, cf. Definition 15.12, in the disc $B(A, \delta)$. By Proposition 15.13, this splits $B(A, \delta) \setminus \sigma(]r, t[)$ into two Jordan regions G_1 and G_2.

Since $z \in \sigma([r, t]) = \partial G_1 \cap \partial G_2$, we conclude from Theorem 15.23 that z defines exactly one accessible boundary point for each of the Jordan regions G_1 and G_2. Let γ_1 and γ_2 be two curves defined on $[0, 1[$ with $\gamma_j^* \subseteq G_j$ and $\lim_{T \to 1_-} \gamma_j(T) = z$. Put $M = \min\{|z - \sigma(S)| \mid S \in [0, r] \cup [t, 1]\} > 0$. There exists $0 \leq T' < 1$ such that $|A - \gamma_j(T)| < M$ for $T \in [T', 1[$ and $j = 1, 2$. This implies that $\gamma_j([T', 1[) \subseteq \Omega$, $j = 1, 2$, and hence z is accessible as a boundary point for Ω. We also observe that z defines at least two accessible boundary points, since a third curve $\gamma \colon [0, 1[\to \Omega$ with $\lim_{T \to 1_-} \gamma(T) = A$, must eventually run inside $B(A, \delta)$. Hence, it will eventually no longer be able to intersect both curves γ_1 and γ_2.

Finally to see that z defines exactly two accessible boundary points, we assume that $\gamma_3 \colon [0, 1[\to \Omega$ is a third curve with $\lim_{T \to 1_-} \gamma_3(T) = A$. There exists $0 \leq T' < 1$, such that $\gamma_3(T) \in B(A, \delta)$ for $T \in [T', 1[$. We may therefore assume without loss of generality that $\gamma_3([T', 1[) \subseteq G_1$. Since z defines single boundary point for G_1, there exists a curve $\gamma \colon [0, 1[\to G_1$ with $\lim_{T \to 1_-} \gamma(T) = A$ and such that $\gamma^* \cap \gamma_j([0, 1[) \cap B(A, R) \neq \emptyset$ for all $R > 0$ and $j = 1, 3$. There exists $T' \in [0, 1[$, such that $|z - \gamma(T)| < M$, for all $T \in [T', 1[$. The curve γ restricted to $[T', 1[$ (and reparametrized to $[0, 1[$) now ensures that $(A, \gamma_1) \sim (A, \gamma_3)$. $\qquad\square$

We shall later see (Corollary 17.23) that the endpoints of a Jordan arc are in fact simple boundary points of the complement of the arc. Hence the endpoints are accessible and define unique accessible boundary points.

In the literature, accessible boundary points and the equivalence relation on pairs (A, γ), are often formulated in terms of Jordan half-intervals instead of curves. We end this section by showing that for any accessible boundary point $[(A, \gamma)]$, one may choose a representative (A, γ) with γ a Jordan half-interval. As for the equivalence relation, we refer the reader to Exercise 15.10.

We begin with a measure of distance from one set to another. Given a nonempty set $A \subseteq \mathbb{C}$, we wish to measure how far another nonempty set

$B \subseteq \mathbb{C}$ is from the set A. The function we use is

$$\rho_A(B) = \sup_{b \in B} \inf_{a \in A} |a - b| \in [0, \infty]. \tag{15.4}$$

Due to the asymmetry, we call $\rho_A(B)$ the distance from B to A. Clearly, for nonempty sets A, B and C, we have

$$\forall B' \subseteq B, \ B' \neq \emptyset : \quad \rho_A(B') \leq \rho_A(B) \tag{15.5}$$

and

$$\rho_A(C) \leq \rho_A(B) + \rho_B(C). \tag{15.6}$$

Lemma 15.25. *Let $G \subseteq \mathbb{C}$ be an open set. We have*

(a) *Let $\varepsilon > 0$ and $\gamma \colon [a, b] \to G$ a continuous curve with $\gamma(a) \neq \gamma(b)$. There exists a piecewise linear curve $\tilde{\gamma} \colon [a, b] \to G$ such that $\tilde{\gamma}(a) = \gamma(a)$, $\tilde{\gamma}(b) = \gamma(b)$ and $\rho_{\gamma^*}(\tilde{\gamma}^*) \leq \varepsilon$.*

(b) *Let $\gamma \colon [a, b[\to G$ be a continuous curve with the property that $\xi = \lim_{t \to b_-} \gamma(t)$ exists and $\xi \notin \gamma^*$. There exists a piecewise linear curve $\tilde{\gamma} \colon [a, b[\to G$, with $\tilde{\gamma}(a) = \gamma(a)$, $\lim_{t \to b_-} \tilde{\gamma}(t) = \xi$ and $\xi \notin \tilde{\gamma}^*$.*

(c) *If γ is as in (b) and $\{t_n\}_{n \in \mathbb{N}} \subseteq [a, b[$ is a strictly increasing sequence with $\lim_{n \to \infty} t_n = b$. Then $\tilde{\gamma}$ may be chosen such that $\tilde{\gamma}(t_n) = \gamma(t_n)$ for all $n \in \mathbb{N}$.*

Proof. We may assume that $a = 0$ and $b = 1$.

We begin with (a). Since γ^* is compact and $\mathbb{C} \setminus G$ is closed, the distance $r = \min\{d(\gamma^*, \mathbb{C} \setminus G), \varepsilon\}$ is strictly positive. Then $G_r := \bigcup_{t \in [0,1]} B(\gamma(t), r) \subseteq G$ is open and path connected. In the path connected region G_r, the two points $\gamma(0)$ and $\gamma(1)$ can be connected with a piecewise linear curve $\hat{\gamma} \colon [0, 1] \to G_r$. That is, $\hat{\gamma}(0) = \gamma(0)$, $\hat{\gamma}(1) = \gamma(1)$ and $\rho_{\gamma^*}(\hat{\gamma}^*) \leq \rho_{\gamma^*}(G_r) \leq r \leq \varepsilon$, where we used (15.5) to establish the last assertion.

We prove (b) and (c) at the same time. If an increasing sequence $\{t_n\}_{n \in \mathbb{N}}$ has been supplied, we may assume that $t_1 = 0$. If not, we put $t_n = 1 - 1/n$ for $t \in \mathbb{N}$.

Divide $[0, 1[$ into a union of countable many closed intervals $I_n = [t_n, t_{n+1}]$. Write $\gamma_n = \gamma|_{I_n}$ and let $\varepsilon_n = \inf_{t \in I_n} |\xi - \gamma(t)|/2 > 0$. For each n, we replace γ_n by a piecewise linear curve $\tilde{\gamma}_n$, constructed using (a) with $\varepsilon = \min\{1/n, \varepsilon_n\}$, such that $\tilde{\gamma}_n(t_n) = \gamma(t_n)$, $\tilde{\gamma}(t_{n+1}) = \gamma(t_{n+1})$ and $\rho_{\gamma_n^*}(\tilde{\gamma}_n^*) \leq \min\{1/n, \varepsilon_n\}$. Note that

$$|\tilde{\gamma}_n(s) - \xi| \geq |\xi - \gamma_n(t)| - |\tilde{\gamma}_n(s) - \gamma_n(t)| \geq 2\varepsilon_n - |\tilde{\gamma}_n(s) - \gamma_n(t)|,$$

for all $s, t \in I_n$. Hence, $|\tilde{\gamma}_n(s) - \xi| \geq 2\varepsilon_n - \inf_{t \in I_n} |\tilde{\gamma}_n(s) - \gamma_n(t)| \geq 2\varepsilon_n - \rho_{\gamma_n^*}(\tilde{\gamma}_n^*) \geq \varepsilon_n$. This estimate implies that $\xi \notin \tilde{\gamma}_n^*$. Adding these countable many piecewise linear curves, yields a new piecewise linear curve $\tilde{\gamma} \colon [0, 1[\to G$ with the properties $\tilde{\gamma}(t_n) = \gamma(t_n)$ for all $n \in \mathbb{N}$, and $\lim_{t \to 1_-} \tilde{\gamma}(t) = \xi \notin \tilde{\gamma}^*$. To observe convergence to ξ, we exploit (15.6) and note that

$$\sup_{t \in I_n} |\xi - \tilde{\gamma}(t)| = \rho_{\{\xi\}}(\tilde{\gamma}_n^*) \leq \rho_{\{\xi\}}(\gamma_n^*) + \tilde{\gamma}_{\gamma_n^*}(\tilde{\gamma}_n^*) \leq \sup_{t \in I_n} |\xi - \gamma(t)| + \frac{1}{n}.$$

The right-hand sides converges to zero as $n \to \infty$, which implies that $\lim_{t \to 1_-} \tilde{\gamma}(t) = \xi$. □

Lemma 15.26. *Let $a < b$ be two real numbers.*

(a) *Suppose $\gamma \colon [a, b] \to \mathbb{C}$ is a piecewise linear curve. There exists a piecewise linear Jordan arc $\tilde{\gamma} \colon [a, b] \to \mathbb{C}$, with $\tilde{\gamma}(a) = \gamma(a)$, $\tilde{\gamma}(b) = \gamma(b)$ and $\tilde{\gamma}^* \subseteq \gamma^*$.*

(b) *Let $\gamma \colon [a, b[\to \mathbb{C}$ be a piecewise linear curve. Suppose $\xi := \lim_{t \to b_-} \gamma(t)$ exists in $\mathbb{C}$ and $\xi \notin \gamma^*$. Then there exists a piecewise linear Jordan half-interval $\tilde{\gamma} \colon [a, b[\to \mathbb{C}$, with $\tilde{\gamma}^* \subseteq \gamma^*$, $\tilde{\gamma}(a) = \gamma(a)$ and $\lim_{t \to b_-} \tilde{\gamma}(t) = \xi$.*

Proof. We prove only the second statement. The first claim is simpler and follows by the same argument.

Since the curve is piecewise linear, there exists a sequence $a = t_0 < t_1 < \cdots < t_n < \cdots < b$ such that $t_n \to b$ and $\gamma_n = \gamma|_{[t_n, t_{n+1}]}$ is a line segment connecting $\gamma(t_n)$ and $\gamma(t_{n+1})$, for $n = 0, 1, 2, \ldots$.

We begin by constructing a candidate for $\tilde{\gamma}$. First of all, we may assume that $\gamma(a) \notin \gamma(]a, b[)$. This follows from the assumption that $\xi \notin \gamma^*$.

Put $s_0 = a$. Observe that

$$s_1 = \max\{r \in [t_1, b[\mid \gamma(r) \in \gamma([a, t_1])\},$$

is well defined and that $b > s_1 \geq t_1 > s_0 = a$. The closed line segment ℓ_1 connecting $\gamma(a)$ and $\gamma(s_1)$ lies inside (and may coincide with) γ_1^*. Furthermore, $\gamma(]s_1, b[) \cap \ell_1 = \emptyset$, by construction of s_1.

Denote by $n_1 \in \mathbb{N}$ the natural number for which $s_1 \in [t_{n_1}, t_{n_1+1}[$. We observe that either $\gamma(s_1)$ is an end point of one (perhaps both) of the line segments γ_1^* or $\gamma_{n_1}^*$, or the two aforementioned line segments intersect transversally at $\gamma(s_1)$.

Renaming s_1 and t_0 as a, t_{n_1+j} as t_j, $j \in \mathbb{N}$, and $\gamma|_{[s_1,b[}$ as γ, we may repeat the above argument, and produce a monotone sequence of numbers $a = s_0 < s_1 < \cdots < b$, defining a piecewise linear Jordan half interval $\tilde{\gamma}: [a,b[\to \mathbb{C}$ by

$$\tilde{\gamma}(t) = \frac{s_n - t}{s_n - s_{n-1}}\gamma(s_{n-1}) + \frac{t - s_{n-1}}{s_n - s_{n-1}}\gamma(s_n), \qquad \text{for } t \in [s_{n-1}, s_n].$$

Note that by construction $s_j \geq t_j$, for all j, and consequently, $\lim_{j\to\infty} s_j = b$. $\square$

Proposition 15.27. *Let $G \subseteq \mathbb{C}$ be an open set and $[(A,\gamma)] \in \partial_A G$ an accessible boundary point. Then there exists a Jordan half-interval $\sigma: [0,1[\to G$ with $\lim_{t\to 1_-} \sigma(t) = A$ and $(A,\sigma) \sim (A,\gamma)$.*

Proof. Let $\gamma_1: [0,1[\to G$ be a curve with $\lim_{t\to 1_-} = A$. Applying Lemma 15.25(c) with $t_n = 1 - 1/n$ yields a piecewise linear curve $\gamma_2: [0,1[\to G$ with $\lim_{t\to 1_-} \gamma_2(t) = A$ and $(A,\gamma_1) \sim (A,\gamma_2)$.

Finally, appealing to Lemma 15.26(b), we obtain a Jordan half-interval $\gamma_3: [0,1[\to G$ with $\lim_{t\to 1_-} \gamma_3(t) = A$ and $(A,\gamma_2) \sim (A,\gamma_3)$.

The result now follows, since $\sim$ is an equivalence relation. $\square$

Exercises

15.1. Let X be a topological space. Suppose $A, B \subseteq X$ are connected subsets with $A \cap B \neq \emptyset$. Show that $A \cup B$ is connected. Recall that a subset of a topological space is connected if and only if: it cannot be written as a union of two nonempty open and disjoint subsets.

15.2. Let $U \subseteq \mathbb{R}^n$ be an open set. Show that U is connected if and only if any two points in U can be connected by a piecewise linear curve.

15.3. Let G be an open connected set with $\partial G = \gamma^*$ for some closed Jordan curve γ. Show that either $G = G_0$ or $G = G_\infty$, where G_0 and G_∞ are the bounded and unbounded connected components of $\mathbb{C} \setminus \gamma^*$, respectively.

Hint: Show the following: Let $U \subseteq V \subseteq \mathbb{C}$ be non-empty open and connected sets with the same boundary $\partial U = \partial V$. Then $U = V$.

15.4. Let $\gamma_1, \gamma_2: [0,1] \to \mathbb{C}$ be two Jordan arcs with $\gamma_1^* \cap \gamma_2^*$ either empty or a singleton. Show that there are no bounded nonempty open sets $U \subseteq \mathbb{C}$ with $\partial U \subseteq \gamma_1^* \cup \gamma_2^*$.

Hint: Massage the proof of Proposition 15.4.

15.5. Let G be a Jordan region with boundary curve $\gamma\colon [0,1] \to \mathbb{C}$. Abbreviate $A = \gamma(0) \in \partial G$. Suppose $\sigma\colon [0,1] \to \mathbb{C}$ is a closed Jordan curve with $\sigma(0) = A$ and $\sigma(]0,1[) \subseteq G$. Denote by G_σ and G'_σ the bounded and unbounded connected components of $\mathbb{C} \backslash \sigma^*$, respectively. We aim to show that $G \setminus \sigma^*$ has exactly two connected components, G_σ and $G \cap G'_\sigma$.

(α) Show that $G_\sigma \subseteq G \backslash \sigma^*$.

(β) Let U be a connected component of $G \cap G'_\sigma$. Show that $\partial U \subseteq \gamma^* \cup \sigma^*$ and that ∂U contains either γ^* or σ^* as a subset.
Hint: Employ Exercise 15.4.

(γ) Argue that at most one connected component U of $G \cap G'_\sigma$ satisfies that $\gamma^* \subseteq \partial U$. Similarly for σ^*.
Hint: Use Proposition 15.13.

(δ) Argue there is at least one connected component of $G \cap G'_\sigma$ with boundary containing points from both $\gamma(]0,1[)$ and from $\sigma(]0,1[)$.
Hint: Let B be a point on σ^* maximizing the distance $|A - B|$. Look for the desired boundary points on the line passing through A and B.

(ε) Conclude that $G \cap G'_\sigma$ is connected and that, consequently, $G \backslash \sigma^*$ has exactly two components.

15.6. Let $G \subseteq \mathbb{C}$ be an open nonempty set. Show that the accessible boundary points are dense in the boundary of G.
Hint: For any $A \in \partial G$ and $r > 0$, show that at least one point in $\partial G \cap B(A,r)$ can be accessed from G using a line segment.

15.7. Show that the relation $\sim$ from Definition 15.21 does in fact define an equivalence relation on the accessible boundary points.

15.8. Let γ be a closed Jordan curve and G_∞ the unbounded connected component of $\mathbb{C} \setminus \gamma^*$. Show that all boundary points of G_∞ are simple.

15.9. Define the Hausdorff distance between nonempty sets A and B to be

$$\rho(A, B) = \max\{\rho_A(B), \rho_B(A)\}.$$

(α) Show that ρ, restricted to closed nonempty subsets of $\mathbb{C}$, is a metric.

(β) Let A, B, B' be nonempty closed sets with $B' \subseteq B$. Show that $\rho(A, B') \geq \rho(A, B)$.

15.10. Let $\Omega \subseteq \mathbb{C}$ be an open set and $A \in \partial\Omega$ an accessible boundary point. Suppose $\gamma_1, \gamma_2 \colon [0, 1[\to G$ are two Jordan half-intervals with $\lim_{t \to 1_-} \gamma_j(t) = A$, for $j = 1, 2$, and such that $(A, \gamma_1) \sim (A, \gamma_2)$. Show that there exists a third Jordan half-interval $\gamma_2 \colon [0, 1[\to G$, such that for all $r > 0$, we have $\gamma_j^* \cap \gamma_3^* \cap B(A, r) \neq \emptyset$, for $j = 1, 2$.

Chapter 16

The Dirichlet problem and Green's functions

In this chapter we continue with the study of the Dirichlet Problem, initiated in Section 12.2. As an application we establish existence of Green's functions in quite general planar regions. The chapter can be read in continuation of Chapter 13 on subharmonic functions, although we use a (baby) version of Jordan's curve theorem (Theorem 15.8). We deal exclusively with planar domains, partly because of our focus on complex function theory, and partly because much stronger results are available for planar regions.

16.1   The Dirichlet problem

In this section we establish an existence theorem for the Dirichlet Problem in a wide class of planar regions, a result due to Lichtenstein (1916–17). In the following section we use the existence theorem to construct Green's functions in quite general planar domains. We remark that chronologically, the two results were arrived at in reverse order with the existence theorem for Green's function being due to Osgood (1900).

Proposition 16.1. *Let $\mathcal{O} \subseteq \mathbb{C}$ be open and bounded, and suppose $\varphi \colon \partial\mathcal{O} \to \mathbb{R}$ is continuous. Let $\mathcal{S}_\varphi$ denote the collection of continuous subharmonic functions $u \colon \mathcal{O} \to \mathbb{R}$ satisfying*

$$\forall w \in \partial\mathcal{O} : \quad \limsup_{\substack{z \to w \\ z \in \mathcal{O}}} u(z) \leq \varphi(w).$$

Then, the function defined for $z \in \mathcal{O}$ by

$$u_\varphi(z) = \sup_{u \in \mathcal{S}_\varphi} u(z)$$

is (finite) and harmonic in $\mathcal{O}$.

Proof. Since the constant function $\inf_{w \in \partial \mathcal{O}} \varphi(w) > -\infty$ is an element of $\mathcal{S}_\varphi$, we conclude that $\mathcal{S}_\varphi \neq \emptyset$. By the Strong Maximum Principle for subharmonic functions (Theorem 13.6), we observe that any $u \in \mathcal{S}_\varphi$ is dominated from above by the constant function $\sup_{w \in \partial \mathcal{O}} \varphi(w) < \infty$. Hence, u_φ is well-defined and everywhere finite.

Let $\mathcal{A} \subseteq \mathcal{O}$ be a countable dense set. By a diagonal sequence argument, we may extract a sequence $\{u_n\}_{n \in \mathbb{N}} \subseteq \mathcal{S}_\varphi$ such that $u_n(a) \to u_\varphi(a)$ for all $a \in \mathcal{A}$.

Setting $v_n = \max\{u_1, \ldots, u_n\} \in \mathcal{S}_\varphi$, we obtain a monotone sequence of subharmonic functions also satisfying that $v_n(a) \to u_\varphi(a)$ for all $a \in \mathcal{A}$.

Let $b \in \mathcal{O}$ and $r > 0$ be such that $B[b, r] \subseteq \mathcal{O}$. Using Theorem 12.9, we replace v_n, for each n, with a new function $\tilde{v}_n \in \mathcal{S}_\varphi$ by setting $\tilde{v}_n(z) = v_n(z)$ for $z \in \mathcal{O} \setminus B(b, r)$ and

$$\tilde{v}_n(z) = \int_0^{2\pi} v_n(b + re^{i\theta}) P_r(\theta, z - b)\, d\theta,$$

for $z \in B(b, r)$. Then $\tilde{v}_n$ is harmonic in $B(b, r)$ and a continuous subharmonic function in $\mathcal{O}$ with the same boundary behavior as v_n. By the Strong Maximum Principle (again), we conclude that $\tilde{v}_n \geq v_n$. The sequence $\{\tilde{v}_n\}_{n \in \mathbb{N}}$ is again monotone and satisfies that $\tilde{v}_n(a) \to u_\varphi(a)$ for all $a \in \mathcal{A}$. By Harnack' monotone convergence theorem (Theorem 12.13), we know that the limiting function $v_\varphi := \lim_{n \to \infty} \tilde{v}_n \colon B(b, r) \to \mathbb{R}$ is harmonic. Hence, $v_\varphi \leq u_\varphi$ and u_φ coincides with the harmonic function v_φ on a dense subset of $B(b, r)$. Suppose there exists $c \in B(b, r)$ where $v_\varphi(c) < u_\varphi(c)$. Then there exists $u \in \mathcal{S}_\varphi$ with $v_\varphi(c) < u(c)$. Since both v_φ and u are continuous, we may conclude that $v_\varphi(a) < u(a)$ for some $a \in \mathcal{A} \cap B(b, r)$. Hence, $v_\varphi(a) < u_\varphi(a)$, which is a contradiction. Therefore, we can conclude that the u_φ is well-defined and harmonic in $\mathcal{O}$. $\square$

The function u_φ constructed in Proposition 16.1 is called a *Perron function* and is our candidate for a solution to the Dirichlet Problem with boundary data φ. It is left to the reader as Exercise 16.1 to show that the Perron function u_φ solves the Dirichlet Problem with boundary data φ if and only if a solution exists. In Theorem 16.3 below we will prove existence of solutions to the Dirichlet Problem.

Lemma 16.2. *Let $U \subseteq \mathbb{R}$ be an open set of finite Lebesgue measure. For $z = x + iy \in \mathbb{H}$, put*

$$h_U(z) = \frac{1}{\pi} \int_U \frac{y}{(x-t)^2 + y^2}\, dt.$$

Then $h_U \colon \mathbb{H} \to [0,1]$ is a harmonic function, it extends by continuity to 1 on U and satisfies that $h_U(z) \leq |U|/(\pi y)$, for all $z = x + iy \in \mathbb{H}$.

Proof. That h_U is harmonic in the upper half-plane follows from Fubini's theorem. (Note that $y/((x-t)^2 + y^2) = -\operatorname{Im}(z-t)^{-1}$.) Below we make use of the elementary computation $\int_{\mathbb{R}}(1 + s^2)^{-1}\, ds = \pi$, which in particular implies that $0 \leq h_U \leq 1$.

The decay estimate on h_U follows from dropping $(x-t)^2$ from the denominator in the integrand. It remains to argue that for any $w \in U$, we have $\lim_{z \to w, \operatorname{Im} z > 0} h_U(z) = 1$. Let $\delta > 0$ be such that $(w - \delta, w + \delta) \subseteq U$. Then, for $z = x + iy$ with $|w - x| < \delta/2$ and $0 < y < \delta^2/2$, we have

$$h_U(z) \geq \frac{1}{\pi} \int_{w-\delta}^{w+\delta} \frac{y}{(x-t)^2 + y^2}\, dt = \frac{1}{\pi} \int_{(w-x-\delta)/y}^{(w-x+\delta)/y} \frac{1}{1 + s^2}\, ds$$

$$\geq \frac{1}{\pi} \int_{-\delta/(2y)}^{\delta/(2y)} \frac{1}{1 + s^2}\, ds \geq \frac{1}{\pi} \int_{-\delta^{-1}}^{\delta^{-1}} \frac{1}{1 + s^2}\, ds.$$

Since the right-hand side converges to 1, when δ goes to zero, we may conclude the proof. $\qquad\square$

Theorem 16.3 (Lichtenstein's Theorem). *Let $\mathcal{O} \subseteq \mathbb{C}$ be a nonempty open set, such that no connected component of $\mathbb{C}_\infty \setminus \mathcal{O}$ is a singleton. Then for any continuous function $\varphi \colon \partial\mathcal{O} \to \mathbb{R}$, there exists exactly one continuous function $u \colon \overline{\mathcal{O}} \to \mathbb{R}$, harmonic in $\mathcal{O}$ and equal to φ on $\partial\mathcal{O}$.*

Proof. We begin by establishing the theorem assuming that $\mathcal{O}$ is bounded. The strategy is to show that the associated Perron function $u_\varphi \colon \mathcal{O} \to \mathbb{R}$, extended by φ on the boundary, is in fact a continuous function.

Let $w \in \partial\mathcal{O}$ and $\varepsilon > 0$. We wish to produce a function $u \in \mathcal{S}_\varphi$ with $\limsup_{z \to w, z \in \mathcal{O}} u(z) \geq \varphi(w) - \varepsilon$.

Let $r > 0$ be small enough such that $|\varphi(w') - \varphi(w)| \leq \varepsilon$ for $w' \in \partial\mathcal{O} \cap B(w, r)$. Here we used continuity of φ.

Denote by U the connected component of $\mathbb{C} \setminus \mathcal{O}$ containing w. By assumption, U is not a singleton. Furthermore, U is a closed subset of the Riemann sphere, being the connected component of a closed set.

Let $w' \in U$ with $w' \neq w$ and write $m(z) = (z-w)/(z-w')$ for a Möbius transformation taking w to 0 and w' to ∞. Then $m(U)$ is also a connected closed subset if $\mathbb{C}_\infty$, and by Corollary 9.12 we know that $\Omega = \mathbb{C} \setminus m(U)$ is an open (nonempty) simply connected subset of $\mathbb{C}$. Hence, there exists a holomorphic logarithm $\log\colon \Omega \to \mathbb{C}$. That is, $e^{\log(z)} = z$ for all $z \in \Omega$.

Let $r' > 0$ be such that $B[0,r'] \subseteq m(B(w,r))$ and define $f(z) = -i(\log(z) - \operatorname{Log}(r'))$ for $z \in \Omega$. Then $f(\Omega \cap B(0,r')) \subseteq \mathbb{H}$ and $U := f(\Omega \cap \partial B(0,r')) \subseteq \mathbb{R}$ is open as a subset of $\mathbb{R}$. (We may assume that $U \neq \emptyset$ by possibly choosing r' smaller.) If arg $=$ Im log is the associated argument function, we have $U = \arg(\Omega \cap \partial B(0,r'))$. This shows that the Lebesgue measure of U is finite and in fact bounded by 2π.

Let $h = h_U$ denote the function from Lemma 16.2. The composition $g(z) = -h(f(z))$ is a harmonic function in $\Omega \cap B(0,r')$ with values in $[-1,0]$, satisfying for any $\xi \in \Omega$ with $|\xi| = r'$ that $\lim_{z\to\xi, z\in\Omega\cap B(0,r')} g(z) = -1$.

Since m is a Möbius transformation, there exists $r'' > 0$ such that $m^{-1}(B(0,r')) = B(w,r'')$.

We may now define a harmonic function $v(z) = g(m(z))$ in $\mathcal{O}' := \mathcal{O} \cap B(w,r'')$, which extends by continuously to -1 on $\mathcal{O} \cap \partial B(w,r'')$. It satisfies that $-1 \leq v \leq 0$ and $\lim_{z\to w, z\in\mathcal{O}} v(z) = 0$. Extending v by -1 in $\mathcal{O} \setminus \mathcal{O}'$ therefore defines a continuous subharmonic function in $\mathcal{O}$.

Let $M = \sup_{z\in\partial\mathcal{O}} |\varphi(\xi) - \varphi(w)| < \infty$. Setting $u_-(z) = \varphi(w) - \varepsilon + Mv(z)$ now defines an element of $\mathcal{S}_\varphi$. Since $\lim_{z\to w, z\in\mathcal{O}} u_-(z) = \varphi(w) - \varepsilon$ and $u_\varphi(z) \geq u_-(z)$ for all $z \in \mathcal{O}$, we conclude that

$$\liminf_{\substack{z\to w \\ z\in\mathcal{O}}} u_\varphi(z) \geq \varphi(w) - \varepsilon. \tag{16.1}$$

To obtain an upper bound we set $u_+(z) = \varphi(w) + \varepsilon - Mv(z)$ for $z \in \mathcal{O}$. Let $u \in \mathcal{S}_\varphi$. Then $u - u_+$ is subharmonic and satisfies for any $w' \in \partial\mathcal{O}$ that $\limsup_{z\to w'}(u(z) - u_+(z)) \leq 0$. By the Strong Maximum Principle we may conclude that $u(z) \leq u_+(z)$ for all $z \in \mathcal{O}$. Since $u \in \mathcal{S}_\varphi$ was arbitrary, we arrive at the estimate $u_\varphi(z) \leq u_+(z)$ for all $z \in \mathcal{O}$. Hence $\limsup_{z\to w, z\in\mathcal{O}} u_\varphi(z) \leq \varphi(w) + \varepsilon$. This together with (16.1) concludes the proof for bounded regions, since ε was arbitrary.

It remains to extend the result to unbounded regions. If $\mathcal{O}$ is not dense in $\mathbb{C}$, then we may use a Möbius transform to reduce the problem to a bounded region (composing a harmonic function with a holomorphic function yields a new harmonic function). If $\mathcal{O}$ is dense in $\mathbb{C}$ we must be more careful.

Since $\mathcal{O}$ is unbounded, we have $\infty \in \partial\mathcal{O}$. Denote by C the connected component of $\mathbb{C}_\infty$ containing ∞. Let $\xi \in C \setminus \{\infty\}$. Such a ξ exists by assumption! Put $G = \mathbb{C} \setminus C$, which is an open and simply connected proper subset of $\mathbb{C}$ with $\mathcal{O} \subseteq G$. Denote by $G \ni z \to R(z)$ a (holomorphic) square root of $\xi - z$. Then $\xi - R(z)^2 = z$ for $z \in G$. Put $\widetilde{\mathcal{O}} = R(\mathcal{O})$, which is an open subset of $\mathbb{C}$ not dense in $\mathbb{C}$. Note that

$$\partial\widetilde{\mathcal{O}} \subseteq \{w \in \mathbb{C}_\infty \mid \xi - w^2 \in \partial\mathcal{O}\},$$

and $\partial\widetilde{\mathcal{O}}$ contains the point ∞. Hence, we may define a continuous function

$$\forall w \in \partial\widetilde{\mathcal{O}}: \quad \tilde{\varphi}(w) = \varphi(\xi - w^2)$$

on the boundary of $\widetilde{\mathcal{O}}$. Since $\widetilde{\mathcal{O}}$ is not dense in $\mathbb{C}$, the Dirichlet Problem in $\widetilde{\mathcal{O}}$ with boundary data $\tilde{\varphi}$ has a solution $\tilde{u}$. We now propose $u = \tilde{u} \circ R$, which is harmonic in $\mathcal{O}$, as a solution to the Dirichlet Problem in $\mathcal{O}$. We only need to check that u extended by φ on $\partial\mathcal{O}$ is continuous in $\overline{\mathcal{O}}$. Let $z_0 \in \partial\mathcal{O}$ and $\{z_n\}_{n\in\mathbb{N}} \subseteq \mathcal{O}$ a sequence with $z_n \to z_0$. We must verify that $u(z_n) \to \varphi(z_0)$. Note first that $\tilde{u}$ is a bounded function, being continuous on a compact set (in the Riemann sphere). Let $\{z_{n_k}\}_{k\in\mathbb{N}}$ be a subsequence, such that $\tilde{u}(R(z_{n_k}))$ is convergent. After possibly passing to a further subsequence, we may assume that $R(z_{n_k})$ is convergent (in $\mathbb{C}_\infty$) as well. Let $w = \lim_{k\to\infty} R(z_{n_k}) \in \partial\widetilde{\mathcal{O}}$ and note that $\xi - w^2 = z_0$. Then

$$\lim_{k\to\infty} u(z_{n_k}) = \tilde{u}(w) = \tilde{\varphi}(w) = \varphi(\xi - w^2) = \varphi(z_0).$$

Since the right-hand side does not depend on the subsequence, we conclude that $u(z_n)$ does indeed converge to $\varphi(z_0)$, which concludes the proof. $\square$

The reader is asked to solve the Dirichlet Problem in unbounded regions in Exercise 16.2. While the assumptions of Theorem 16.3 may appear optimal, they are in fact not; as the reader is asked to explore in Exercise 16.3.

Remark 16.4. Let $\mathcal{O}$ be a bounded region in which the Dirichlet Problem is solvable. Suppose $f \colon \mathcal{O} \to \mathbb{R}$ is Hölder continuous and $\varphi \colon \partial\mathcal{O} \to \mathbb{R}$ is continuous. Then one may employ Theorem 16.3 to show that Poisson's equation

$$\begin{cases} \Delta u = f, \\ u|_{\partial\mathcal{O}} = \varphi, \end{cases} \tag{16.2}$$

has a unique continuous solution $u \colon \overline{\mathcal{O}} \to \mathbb{R}$, which is twice continuously differentiable in $\mathcal{O}$. We refer the reader to Gilbarg and Trudinger (2001).

16.2   Green's functions

For an open set $\mathcal{O} \subseteq \mathbb{C}$, we write – in this section – $\overline{\mathcal{O}}$ and $\partial\mathcal{O}$ for the closure and boundary, respectively, of $\mathcal{O}$ in the Riemann sphere $\mathbb{C}_\infty$. So, if $\mathcal{O}$ is unbounded we have $\infty \in \partial\mathcal{O} \subseteq \overline{\mathcal{O}}$.

Definition 16.5. *Let $\mathcal{O} \subseteq \mathbb{C}$ be open and $a \in \mathcal{O}$. A continuous function $\mathcal{G}(\,\cdot\,;a)\colon \overline{\mathcal{O}} \setminus \{a\} \to \mathbb{R}$ is said to be a* **Green's function** *if*

(a) $\mathcal{G}(\,\cdot\,;a)$ *is harmonic in* $\mathcal{O} \setminus \{a\}$.

(b) $\mathcal{G}(z;a) = 0$ *for all* $z \in \partial\mathcal{O}$.

(c) *There exists* $r > 0$ *such that* $B[a,r] \subseteq \mathcal{O}$ *and*

$$\sup_{z\in B[a,r]\setminus\{a\}} |\mathcal{G}(z;a) + \mathrm{Log}(|z - a|)| < \infty.$$

Remark 16.6. We make two remarks regarding Green's functions $\mathcal{G}(\,\cdot\,;a)$:

(A) Since a is a removable singularity for the harmonic function $z \to \mathcal{G}(z;a) + \mathrm{Log}(|z - a|)$, cf. Theorem 12.5, we observe that the last bullet point in Definition 16.5 may be replaced by the (seemingly) stronger demand that there exists a harmonic function $h\colon \mathcal{O} \to \mathbb{R}$ with $\mathcal{G}(z;a) = -\mathrm{Log}(|z - a|) + h(z)$.

(B) It is a consequence of Theorem 12.4 that any function $\mathcal{G}$ harmonic in $\mathcal{O} \setminus \{a\}$ is of the form $\mathcal{G}(z) = -c\,\mathrm{Log}(|z - a|) + h(z)$, where $c \in \mathbb{R}$ and h a harmonic function in $\mathcal{O}$. The requirement $c = 1$ fixes the logarithmic singularity of the Green's function up to a multiplicative constant.

The perhaps most important application of Green's functions is its use in constructing solutions of Poisson's equation (16.2). We will however not pursue this application, and instead refer the reader to Folland (1995) and Gilbarg and Trudinger (2001). See also the review by Gårding (1979).

Suppose that $\mathcal{O}$ is unbounded. By our convention that the boundary is considered in $\mathbb{C}_\infty$ a Green's function has to decay at ∞. This leads in particular to the following example.

Example 16.7. We observe that no Green's functions in $\mathcal{O} = \mathbb{C}$ exist. Indeed, if $a \in \mathbb{C}$ and $\mathcal{G}_\mathbb{C}(\,\cdot\,;a)$ is a Green's function in $\mathbb{C}$, then $h(z) = \mathcal{G}_\mathbb{C}(z;a) + \mathrm{Log}(|z - a|)$ extends to an entire harmonic function. Since $\mathcal{G}_\mathbb{C}(\cdot;a)$ vanishes at ∞, it is bounded outside a neighborhood of a. Hence, there exists $C_1, C_2 > 0$ such that $|h(z)| \leq C_1 \operatorname{Log}(|z|) + C_2$ for all $z \in \mathbb{C}$. By Liouville's theorem, in the guise of Theorem 12.19 for harmonic functions, $h = 0$. Hence, $\mathcal{G}_{\mathbb{C}}(z; a) = -\operatorname{Log}(|z - a|)$, which is a contradiction since this function does not vanish at infinity.

Proposition 16.8. *Let $\mathcal{O} \subseteq \mathbb{C}$ be open and $a \in \mathcal{O}$. There exists at most one Green's function $\mathcal{G}(\,\cdot\,; a)$.*

Proof. Let $\mathcal{G}_1(\,\cdot\,; a)$ and $\mathcal{G}_2(\,\cdot\,, a)$ be two Green's functions. Then $f(z) = \mathcal{G}_2(z; a) - \mathcal{G}_1(z; a) = h_2(z) - h_1(z)$ extends from $\overline{\mathcal{O}} \setminus \{a\}$ to a continuous function in $\overline{\mathcal{O}}$, harmonic in $\mathcal{O}$ and vanishing on the boundary. By Proposition 12.12, we must have $f = 0$. $\qquad\square$

If a Green's function exist for an open set $\mathcal{O} \subseteq \mathbb{C}$ and $a \in \mathcal{O}$, the above proposition suggests that we write $\mathcal{G}_{\mathcal{O}}(\,\cdot\,; a)$ for the unique Green's function.

Example 16.9. We have in fact already seen the prototypical Green's function in Section 12.1. Here $\mathcal{O} = B(0, 1)$ is the unit disc, $a = 0$ and

$$\mathcal{G}_{B(0,1)}(z; 0) = -\operatorname{Log}(|z|).$$

To obtain the Green's function with respect to another point $a \in B(0, 1)$, we use the automorphisms $A(a; \,\cdot\,)$ of the disc from Example 8.14. Then

$$\mathcal{G}_{B(0,1)}(z; a) = \mathcal{G}_{B(0,1)}(A(a; z); 0).$$

The proposed Green's function is clearly harmonic in $B(0, 1) \setminus \{a\}$ and zero on the boundary (since $A(a; \,\cdot\,)$ maps the unit circle into itself). Since $\mathcal{G}_{B(0,1)}(z; a) = -\operatorname{Log}(|z - a|) + \operatorname{Log}(|1 - \bar{a}z|)$ with the second term on the right-hand side being bounded in $B(0, 1)$, we conclude that $\mathcal{G}_{B(0,1)}(z; a)$ is the Green's function.

Example 16.10. As another example, we take $\mathcal{O} = \mathbb{H}$ and $a \in \mathbb{H}$. The Möbius transformation $M(z) = (z - a)/(z + \bar{a})$ takes $\mathcal{O}$ onto $B(0, 1)$ and $M(a) = 0$. Hence,

$$\begin{aligned}
\mathcal{G}_{\mathbb{H}}(z; a) &= \mathcal{G}_{B(0,1)}(M(z); 0) = -\operatorname{Log}(|M(z)|) \\
&= -\operatorname{Log}(|z - a|) + \operatorname{Log}(|z - \bar{a}|)
\end{aligned}$$

is the Green's function.

In the two examples above, we in fact made use of the following lemma. We remark that the topology on $\mathbb{C}_\infty$ is metrizable and denote by d_∞ a metric on $\mathbb{C}_\infty$ with the correct metric topology.

Lemma 16.11. *Let $\mathcal{U}, \mathcal{V} \subseteq \mathbb{C}$ be open sets, $a \in \mathcal{U}$ and $g: \mathcal{U} \to \mathcal{V}$ a biholomorphic function. Suppose the Green's function $\mathcal{G}_{\mathcal{V}}(\,\cdot\,; g(a))$ exists. Then $\mathcal{G}_{\mathcal{U}}(z; a) = \mathcal{G}_{\mathcal{V}}(g(z); g(a))$, $z \in \mathcal{U} \setminus \{a\}$, extends by continuity from $\mathcal{U} \setminus \{a\}$ to $\overline{\mathcal{U}} \setminus \{a\}$ and the extension is a Green's function in $\mathcal{U}$.*

Proof. Write $\mathcal{G}_{\mathcal{V}}(w; g(a)) = -\operatorname{Log}(|w - g(a)|) + h(w)$, where h is harmonic in $\mathcal{V}$.

Clearly, $\mathcal{G}_{\mathcal{U}}(\,\cdot\,; a)$ as defined is harmonic in $\mathcal{U} \setminus \{a\}$. Recall that since g is univalent, we have $g'(z) \neq 0$ for all $z \in \mathcal{U}$. Then, for z in a neighborhood of a, we have the formula

$$\mathcal{G}_{\mathcal{V}}(g(z); g(a)) = -\operatorname{Log}(|g(z) - g(a)|) + h(g(z))$$
$$= -\operatorname{Log}(|z - a|) - \operatorname{Log}(|g'(a)|) - \operatorname{Log}(|1 + O(z - a)|) + h(g(z)),$$

showing that $\mathcal{G}_{\mathcal{U}}(\,\cdot\,; a)$ has the correct normalization at a.

It remains to argue that $\mathcal{G}_{\mathcal{U}}(\,\cdot\,; a)$ extends continuously to zero on $\partial \mathcal{U}$. For this we need the following: For any sequence $\{z_n\}_{n \in \mathbb{N}} \subseteq \mathcal{U}$ with $d_\infty(z, \partial U) \to 0$, we have $d_\infty(g(z_n), \partial \mathcal{V}) \to 0$. Indeed, suppose otherwise; then there exists $\varepsilon > 0$ and a subsequence $\{z_{n_k}\}_{k \in \mathbb{N}}$ such that $g(z_{n_k}) \in V_\varepsilon = \{w \in \mathcal{V} \mid d_\infty(w, \partial \mathcal{V}) \geq \varepsilon\} \subseteq \mathcal{V}$ for all $k \in \mathbb{N}$. Since V_ε is compact, we may assume that $w_k := g(z_{n_k}) \to w \in \mathcal{V}$ for $k \to \infty$. But then we would have $z_{n_k} = g^{-1}(w_k) \to g^{-1}(w) \in \mathcal{U}$, which is absurd.

Let $\xi \in \partial \mathcal{U}$ and $\{z_n\}_{n \in \mathbb{N}}$ be such that $z_n \to \xi$. We must show that $\mathcal{G}_{\mathcal{U}}(z_n; a) \to 0$.

Let $\varepsilon > 0$ and choose $\kappa > 0$ such that $B(a, 2\kappa) \subseteq \mathcal{V}$. Since $\overline{\mathcal{V}} \setminus B(a, \kappa) \subseteq \mathbb{C}_\infty$ is compact, $\mathcal{G}_{\mathcal{V}}(\,\cdot\,; g(a))$ is uniformly continuous in $\overline{\mathcal{V}} \setminus B(a, \kappa)$. Hence, there exists $\delta > 0$ such that for $w \in \partial \mathcal{V}$ and $z \in \mathcal{V}$ with $d_\infty(z, w) \leq \delta$, we have $|\mathcal{G}_{\mathcal{V}}(z; g(a))| \leq \varepsilon$.

Since $d_\infty(z_n, \partial \mathcal{U}) \to 0$ we find an $N \in \mathbb{N}$, such that for $n \geq N$ we have $d_\infty(g(z_n), \partial \mathcal{V}) \leq \varepsilon$ for $n \geq N$. Hence $|\mathcal{G}_{\mathcal{U}}(z_n, a)| = |\mathcal{G}_{\mathcal{V}}(g(z_n); g(a))| \leq \varepsilon$ for $n \geq N$. This concludes the proof. $\qquad \square$

Before we establish existence of Green's functions, we prove a small topological result, which we shall need to deal with unbounded regions.

Lemma 16.12. *Let $C \subseteq \mathbb{C}_\infty$ be a closed connected subset of the Riemann sphere with $\infty \in C$ and $C \cap \mathbb{C} \neq \emptyset$. Let $w \in C \cap \mathbb{C}$. There exists a holomorphic square root $R: \mathcal{O} := \mathbb{C} \backslash C \to \mathbb{C}$ of $z \to z - w$ and $\widetilde{C} = \{z \in \mathbb{C}_\infty \mid w + z^2 \in C\}$ is closed connected subset of $\mathbb{C}_\infty$.*

Proof. That the square root exists follows from Corollary 9.12 and Theorem 5.9.

Let $\widetilde{U}_1$ and $\widetilde{U}_2$ be two (relatively) open subsets of $\widetilde{C}$ with $\widetilde{U}_1 \cup \widetilde{U}_2 = \widetilde{C}$ and $\widetilde{U}_1 \cap \widetilde{U}_2 = \emptyset$. Since $0 \in \widetilde{C}$, we may assume that $0 \in \widetilde{U}_1$. Let $U_j = \{w + z^2 \mid z \in \widetilde{U}_j\}$. Then U_1 and U_2 are (relatively) open subsets of C, $U_1 \cup U_2 = C$, $w \in U_1$ and $w \notin U_2$.

Suppose towards a contradiction that $\widetilde{U}_2 \neq \emptyset$. Then $U_2 \neq \emptyset$. Since U_2 is open, $U_2 \neq C$ and C is connected, we may conclude that U_2 is not closed. Hence, there exists $\xi \in C \backslash U_2 = U_1 \backslash U_2$ and a sequence $\{\xi_n\}_{n \in \mathbb{N}} \subseteq U_2$ with $\xi_n \to \xi$. Then we may find $z_n \in \widetilde{U}_2$ and $z_0 \in \widetilde{U}_1$ such that $\xi_n = w + z_n^2$ and $\xi = w + z_0^2$. Since $\widetilde{C}$ is compact, we may assume that the sequence z_n is convergent. A priori the limit must be either z_0 or $-z_0$, but since $\widetilde{U}_2$ is closed and $z_0 \in \widetilde{U}_1$, we must have $z_n \to -z_0 \in \widetilde{U}_2$. But this implies that $w + (-z_0)^2 = \xi \in U_2$ which is absurd. $\square$

Theorem 16.13 (Osgood's Theorem). *Let $\mathcal{O} \subseteq \mathbb{C}$ and $a \in \mathcal{O}$. Suppose no connected component of $\mathbb{C}_\infty \backslash \mathcal{O}$ consists of a single point. Then there exists a, necessarily unique, Green's function $\mathcal{G}_\mathcal{O}(\,\cdot\,; a)$.*

Proof. Suppose first that $\mathcal{O}$ is a bounded set. Let $a \in \mathcal{O}$. Put $\varphi_a(w) = \mathrm{Log}(|z - a|)$ as a continuous map from $\partial \mathcal{O}$ into $\mathbb{C}_\infty$. Then, by Theorem 16.3, we have a solution h_a of the Dirichlet Problem $\Delta h_a = 0$ and $h_a(w) = \mathrm{Log}(|w - a|)$ for $w \in \partial \mathcal{O}$. With this choice, the function $\mathcal{G}_\mathcal{O}(z; a) = -\mathrm{Log}(|z - a|) + h_a(z)$ is the Green's function.

As for the general case, we proceed as follows. Denote by C the connected component of $\mathbb{C}_\infty \backslash \mathcal{O}$ containing ∞ and let $w \in C \backslash \{\infty\}$. Let $R: \mathbb{C} \backslash C \to \mathbb{C}$ be a holomorphic square root of $z \to z - w$ (Lemma 16.12). Then $R(\mathcal{O})$ is open and $R(\mathcal{O}) \cap (-R(\mathcal{O})) = \emptyset$. Let $\xi \in -R(\mathcal{O})$ and out $m(z) = (\xi - z)^{-1}$. Then $m \circ R$ is a biholomorphic function mapping $\mathcal{O}$ onto the bounded set $\widetilde{\mathcal{O}} = m \circ R(\mathcal{O})$.

The theorem follows from Lemma 16.11 as soon as we have verified that no connected component of $\mathbb{C} \backslash \widetilde{\mathcal{O}}$ is a singleton. Since m is a homeomorphism of the Riemann sphere onto itself, it suffices to show that no connected component of $\mathbb{C}_\infty \backslash R(\mathcal{O})$ is a singleton. Let $z_0 \in \mathbb{C}_\infty \backslash R(\mathcal{O})$. We may assume that $z_0 \notin -R(\mathcal{O})$, and consequently that $\xi = w + z_0^2 \in \mathbb{C}_\infty \backslash \mathcal{O}$.

Let C' denote ξ's connected component, which is not a singleton. If $C' \neq C$, then $z_0 \in R(C')$ which is (defined and) connected. We are left with the case where $C' = C$. But then $z_0 \in \{z \in \mathbb{C}_\infty \,|\, w + z^2 \in C\}$, which by Lemma 16.12 is connected and we are done. $\square$

If $z, a \in \mathcal{O}$ are in different connected components, then $\mathcal{G}_\mathcal{O}(z; a) = 0$. This follows from Proposition 12.12, since $\mathcal{G}_\mathcal{O}(\,\cdot\,; a)$ is harmonic in z's connected component and vanishes on its boundary. The following proposition deals with the situation when z and a are in the same connected component.

Proposition 16.14. *Let $\mathcal{O} \subseteq \mathbb{C}$ be open and connected. Let $a \in \mathbb{C}$. Suppose the associated Green's function $\mathcal{G}_\mathcal{O}(\,\cdot\,; a)$ exists. Then $\mathcal{G}_\mathcal{O}(z; a) > 0$ for all $z \in \mathcal{O} \setminus \{a\}$.*

Proof. Suppose towards a contradiction that there exists a $z_0 \in \mathcal{O} \setminus \{a\}$ with $\mathcal{G}_\mathcal{O}(z_0; a) \leq 0$. Then the function $f(z) = -\mathcal{G}_\mathcal{O}(z; a)$ is harmonic in the connected set $\mathcal{O} \setminus \{a\}$, which has boundary $\partial\mathcal{O} \cup \{a\}$. The function f attains nonnegative values by assumption.

When $z \to a$, $f(z)$ diverges to $-\infty$. Since f is zero on $\partial\mathcal{O}$, we may conclude that f must attain its maximum in $\mathcal{O} \setminus \{a\}$ and therefore is forced to be constant by the Strong Maximum Principle. This, however, collides with the normalization of Green's functions and concludes the proof. $\square$

Theorem 16.15. *Let $\mathcal{O} \subseteq \mathbb{C}$ be an open set, and suppose that for all $a \in \mathbb{C}$ the Green's function $\mathcal{G}_\mathcal{O}(\,\cdot\,; a)$ exists. Then $\mathcal{G}_\mathcal{O}(a; b) = \mathcal{G}_\mathcal{O}(b; a)$ for all $a, b \in \mathcal{O}$ with $a \neq b$.*

Proof. The proof revolves around a construction of polygonal sets that approximate $\mathcal{O}$ from the inside. This construction was in fact central to Osgood's original existence proof for Green's functions, cf. Osgood (1900). For each $N \in \mathbb{N}$ we tile $\mathbb{C}$ with squares $Q_{i,j}^N = \{x + iy \,|\, i \leq 2^N x \leq i + 1,\ j \leq 2^N y \leq j + 1\}$. Let

$$\mathcal{Q}_N = \{(i,j) \in \mathbb{Z}^2 \,|\, Q_{i,j}^N \subseteq \mathcal{O},\ \max\{|i|, |j|\} \leq N 2^N\}$$

and

$$K_N = \bigcup_{(i,j) \in \mathcal{Q}_N} Q_{i,j}^N.$$

Then K_N is a compact subset of $\mathcal{O}$. Denote by $\mathcal{O}_N$ the interior of K_N. Then $\mathcal{O}_N$ is a union over the interiors of squares $Q_{i,j}^N$, open edges shared by two squares in $\mathcal{Q}_N$, as well as the vertices shared by 4 squares in $\mathcal{Q}_N$.

The boundary $\partial \mathcal{O}_N$ is a union of closed edges $L_1, \ldots, L_M$ of exactly one square in $\mathcal{Q}_N$.

For the $\mathcal{O}_N$'s, Green's functions exists for all $a \in \mathcal{O}_N$ by Theorem 16.13. We claim that $\mathcal{G}_{\mathcal{O}_N}(a; b) = \mathcal{G}_{\mathcal{O}_N}(b; a)$ for all $a, b \in \mathcal{O}_N$ with $a \neq b$. In order to verify this claim, we may assume that $\mathcal{O}_N$ is connected. The set $\mathcal{O}_N$ is a polygon with finitely many polygons excised. We write the boundary of $\mathcal{O}_N$ as a union of piecewise linear closed Jordan curves $\gamma_0, \ldots, \gamma_L$ with the following properties: Writing $G_0, G_1, \ldots, G_L$ for the Jordan regions associated with $\gamma_0, \gamma_1 \ldots, \gamma_L$ and $U_1 \ldots, U_L$ for the unbounded connected components of $\mathbb{C} \setminus \gamma_1^*, \ldots, \mathbb{C} \setminus \gamma_L^*$, we have

(i) $G_i \cap G_j = \emptyset$ for $i, j \in \{1, \ldots, L\}$ with $i \neq j$.

(ii) $G_j \subseteq G_0$ for $j = 1, \ldots, L$.

(iii) $\mathcal{O}_N = G_0 \cap U_1 \cap \cdots \cap U_L$.

We orient γ_0 counterclockwise and $\gamma_1, \ldots \gamma_L$ clockwise.

Let $a, b \in \mathcal{O}_N$ with $a \neq b$. Pick $r > 0$ such that $B[c, r] \subseteq \mathcal{O}_N$, for $c = a, b$, and $B[a, r] \cap B[b, r] = \emptyset$. Then the boundary of $\mathcal{O}_N' = \mathcal{O}_N \setminus (B[a, r] \cup B[b, r])$ consists of the same closed Jordan curves together with the two circles of radius r bounding $B[a, r]$ and $B[b, r]$, which we parametrize clockwise with piecewise linear curves γ_{L+1} and γ_{L+2}.

By Green's theorem, identifying $\mathbb{C}$ with $\mathbb{R}^2$, we have

$$0 = \int_{\mathcal{O}_N'} \left(\mathcal{G}_{\mathcal{O}_N}(z; b) \Delta_z \mathcal{G}_{\mathcal{O}_N}(z; a) - \mathcal{G}_{\mathcal{O}_N}(z; a) \Delta_z \mathcal{G}_{\mathcal{O}_N}(z; b) \right) dx dy$$

$$= \sum_{j=0}^{L+2} \int_{\gamma_j} \left(\mathcal{G}_{\mathcal{O}_N}(z; b) \nabla_z \mathcal{G}_{\mathcal{O}_N}(z; a) - \mathcal{G}_{\mathcal{O}_N}(z; a) \nabla_z \mathcal{G}_{\mathcal{O}_N}(z; b) \right) \cdot dz$$

$$= \int_{\partial B(a, r)} \left(\mathcal{G}_{\mathcal{O}_N}(z; b) \nabla_z \mathcal{G}_{\mathcal{O}_N}(z; a) - \mathcal{G}_{\mathcal{O}_N}(z; a) \nabla_z \mathcal{G}_{\mathcal{O}_N}(z; b) \right) \cdot dz$$

$$+ \int_{\partial B(b, r)} \left(\mathcal{G}_{\mathcal{O}_N}(z; b) \nabla_z \mathcal{G}_{\mathcal{O}_N}(z; a) - \mathcal{G}_{\mathcal{O}_N}(z; a) \nabla_z \mathcal{G}_{\mathcal{O}_N}(z; b) \right) \cdot dz.$$

$$(16.3)$$

Note that one only needs Green's theorem for rectangles, where it follows from Fubini and the Fundamental Theorem of Calculus, as well as for a square with a disc excised. (In the latter case, the integrand is smooth on an open neighborhood of the closure of the integration region.)

Observe that $\nabla_z \mathcal{G}_{\mathcal{O}_N}(z; a) = -\frac{z-a}{|z-a|^2} + \nabla_z h_a(z)$ with the second term uniformly bounded in the disc $B[a, r]$. We may now take the limit $r \to 0$

and conclude that

$$\lim_{r \to 0} \int_{\partial B(a,r)} \mathcal{G}_{\mathcal{O}_N}(z;a) \nabla_z \mathcal{G}_{\mathcal{O}_N}(z;b) \cdot dz = 0$$

and

$$\lim_{r \to 0} \Bigg(\int_{\partial B(a,r)} \mathcal{G}_{\mathcal{O}_N}(z;b) \nabla_z \mathcal{G}_{\mathcal{O}_N}(z;a) \cdot dz$$

$$+ \frac{1}{r^2} \int_{\partial B(a,r)} \mathcal{G}_{\mathcal{O}_N}(z;b)(z-a) \cdot dz \Bigg) = 0$$

Compute

$$\lim_{r \to 0} \frac{1}{r^2} \int_{\partial B(a,r)} \mathcal{G}_{\mathcal{O}_N}(z;b)(z-a) \cdot dz = \lim_{r \to 0} \int_0^{2\pi} \mathcal{G}_{\mathcal{O}_N}(a + re^{-i\theta};b) \, d\theta$$

$$= 2\pi \mathcal{G}_{\mathcal{O}_N}(a;b).$$

We used dominated convergence in the last step. Reading the above three computations with the role a and b interchanged, we conclude the symmetry of $\mathcal{G}_{\mathcal{O}_N}$ from (16.3).

Let $\mathcal{O}$ be as in the statement of the theorem and let $a, b \in \mathcal{O}$ with $a \neq b$. Pick $N \in \mathbb{N}$ large enough such that $a, b \in \mathcal{O}_N$. For $M' \geq M \geq N$ we have $a, b \in \mathcal{O}_M \subseteq \mathcal{O}_{M'}$, $\mathcal{G}_{\mathcal{O}_M}(\cdot;a) - \mathcal{G}_{\mathcal{O}_{M'}}(\cdot;a) \leq 0$ and $\mathcal{G}_{\mathcal{O}_M}(\cdot;a) - \mathcal{G}_{\mathcal{O}}(\cdot;a) \leq 0$, as harmonic functions in $\mathcal{O}_N$. The inequalities follows from the Strong Maximum Principle (applied in $\mathcal{O}_M \supseteq \mathcal{O}_N$) and Proposition 16.14. Hence, $\{\mathcal{G}_{\mathcal{O}_M}(\cdot;a)\}_{M \geq N}$ is a monotone sequence of harmonic function in $\mathcal{O}_N \setminus \{a\}$. By Harnack's theorem (Theorem 12.13), the limit $\widetilde{\mathcal{G}}_{\mathcal{O}}(\cdot;a)$ is a (positive) harmonic function in $\mathcal{O}_N \setminus \{a\}$, dominated from above by $\mathcal{G}_{\mathcal{O}}(\cdot;a)$. Since we may take any larger N, the function $\widetilde{\mathcal{G}}_{\mathcal{O}}(\cdot;a)$ actually extends to a strictly positive harmonic function in $\mathcal{O} \setminus \{a\}$, dominated from above by $\mathcal{G}_{\mathcal{O}}(\cdot;a)$. Since $\mathcal{G}_{\mathcal{O}}(\cdot;a)$ is a Green's function, it vanishes at $\partial\mathcal{O}$, and – consequently – so does $\widetilde{\mathcal{G}}_{\mathcal{O}}(\cdot;a)$.

To conclude that $\widetilde{\mathcal{G}}_{\mathcal{O}}(\cdot;a)$ is a Green's function as well, it remains to check its behavior near $z = a$. Let $R > 0$ be such that $B[a,R] \subseteq \mathcal{O}$. For N large enough, we have $B[a,R] \subseteq \mathcal{O}_N$ and

$$\sup_{0 < |z-a| \leq R} |\mathcal{G}_{\mathcal{O}_N}(z;a) + \mathrm{Log}(|z-a|)|$$

$$= \sup_{|z-a|=R} |\mathcal{G}_{\mathcal{O}_N}(z;a) + \mathrm{Log}(|z-a|)|$$

$$\leq \sup_{|z-a|=R} |\mathcal{G}_{\mathcal{O}}(z;a)| + |\mathrm{Log}(R)|,$$

where we used the Strong Maximum Principle. Note that $\mathcal{G}_{\mathcal{O}_N}(z;a) +$ Log$(|z - a|)$ extends to a harmonic function in a neighborhood of $B[a, R]$. Hence, for any $w \in B[a, R] \setminus \{a\}$, we have

$$|\widetilde{\mathcal{G}}_{\mathcal{O}}(w;a) + \text{Log}(|w - a|)| = \lim_{N \to \infty} |\mathcal{G}_{\mathcal{O}_N}(w;a) + \text{Log}(|w - a|)|$$

$$\leq \sup_{|z-a|=R} |\mathcal{G}_{\mathcal{O}}(z;a)| + |\text{Log}(R)| < \infty.$$

This establishes that $\widetilde{\mathcal{G}}(\,\cdot\,;a)$ is a Greens function and by uniqueness (Proposition 16.8), we must have $\widetilde{\mathcal{G}}_{\mathcal{O}}(\,\cdot\,;a) = \mathcal{G}_{\mathcal{O}}(\,\cdot\,;a)$ for all $a \in \mathcal{O}$. Since $\widetilde{\mathcal{G}}_{\mathcal{O}}$ is a pointwise limit of symmetric functions, we may conclude that $\mathcal{G}_{\mathcal{O}}(a;b) = \mathcal{G}_{\mathcal{O}}(b;a)$ for all $a, b \in \mathcal{O}$ with $a \neq b$. $\qquad\square$

Exercises

16.1. Let $\mathcal{O} \subseteq \mathbb{C}$ be an open and bounded set. Let $\varphi\colon \partial\mathcal{O} \to \mathbb{R}$ be continuous. Suppose that the Dirichlet Problem with boundary data φ admits a solution u. Show that $u = u_\varphi$, where u_φ is the associated Perron function.

16.2. In this exercise the reader is asked to solve the Dirichlet Problem in unbounded subsets of the complex plane. Let $\mathcal{O} \subseteq \mathbb{C}$ be such that no connected components of $\mathbb{C}_\infty \setminus \mathcal{O}$ are singletons. Denote by $\partial_\infty\mathcal{O}$ the boundary of $\mathcal{O}$ as a subset of $\mathbb{C}_\infty$ and by $\overline{c\mathcal{O}}^\infty$ the closure of $\mathcal{O}$ in $\mathbb{C}_\infty$. Finally, suppose $\varphi\colon \partial_\infty\mathcal{O} \to \mathbb{R}$.

(α) Denote by C the connected component of $\mathbb{C}_\infty \setminus \mathcal{O}$ containing ∞. Pick $w \in C \setminus \{\infty\}$ and denote by R a continuous square root of $z \to z - w$. Put $\widetilde{\mathcal{O}} = R(\mathcal{O})$ and argue that the construction $\widetilde{\varphi}(\zeta) = \varphi(w + \zeta^2)$ gives a well defined continuous function on $\partial_\infty\widetilde{\mathcal{O}}$.

(β) Argue that $\mathbb{C} \setminus \overline{\widetilde{\mathcal{O}}}^\infty$ has an interior point and use this to find a continuous function $\widetilde{u}\colon \overline{\widetilde{\mathcal{O}}}^\infty \to \mathbb{R}$ with $\widetilde{u}(\zeta) = \widetilde{\varphi}(\zeta)$ for $\zeta \in \partial_\infty\widetilde{\mathcal{O}}$ and such that $\widetilde{u}$ is harmonic in $\widetilde{\mathcal{O}}$.

Hint: Consult the last step of the proof of Theorem 16.13.

(γ) Finally, show that there exists a continuous function $u\colon \overline{\mathcal{O}}^\infty \to \mathbb{R}$, such that $u(w) = \varphi(w)$ for $w \in \partial_\infty\mathcal{O}$ and such that u is harmonic in $\mathcal{O}$.

16.3. Let $S_r = \{z \in \mathbb{C} | |z| = r\}$ for $r \geq 0$. Put $\mathcal{O} = B(0,1) \backslash (S_0 \cup \bigcup_{n=2}^{\infty} S_{1/n})$. Let $\varphi \in C(\partial \mathcal{O})$ be real-valued.

(α) Argue that $\mathcal{O}$ is open and does not satisfy the assumptions in Theorem 16.3.

(β) Nevertheless, show that there exists a function $u \in C(B[0,1])$, harmonic in $\mathcal{O}$ and coinciding with φ on $\partial \mathcal{O}$.

16.4. Let $\mathcal{O} = \{x + iy \mid x \in \mathbb{R}, -\pi/2 < y < \pi/2\}$ and $a \in \mathcal{O}$. Compute the Green's function $\mathcal{G}_{\mathcal{O}}(\,\cdot\,; a)$.

16.5. Show that there are no Green's function in the set $\mathcal{O} = \mathbb{C} \setminus \{0\}$.

Hint: Argue first that it suffices to show that $\mathcal{G}_{\mathcal{O}}(\,\cdot\,; 1)$ does not exist.

16.6. In this exercise the reader is asked to give a simpler proof of existence and symmetry of Green's functions for open bounded simply connected regions $\mathcal{O}$.

(α) Let $a \in \mathcal{O}$ and $f_a \colon B(0,1) \to \mathcal{O}$ be the Riemann map with $f_a(0) = a$ and $f_a'(0) > 0$. Denote by $g_a =: \mathcal{O} \to B(0,1)$ the inverse map. Argue that $\mathcal{G}_{\mathcal{O}}(z; a) = -\operatorname{Log}(|g_a(z)|)$ is the Green's function.

(β) Let $a, b \in \mathcal{O}$. Show that there exists a complex number $C_{a,b} \in B(0,1)$, such that $f_a(z) = f_b(A(C_{a,b}, z))$ for all $z \in B(0,1)$. Here $A(c; \cdot)$ is the Möbius transformation from Example 8.14.

(γ) Argue that $C_{a,b} = -C_{b,a}$ for all $a, b \in \mathcal{O}$.

(δ) Conclude that for all $a, b \in \mathcal{O}$ with $a \neq b$, we have $\mathcal{G}_{\mathcal{O}}(b; a) = \mathcal{G}_{\mathcal{O}}(a; b)$.

16.7. In this exercise the reader is asked to provide an alternative proof of Riemann's mapping theorem, following the original idea of Riemann. Let $\mathcal{O} \subsetneq \mathbb{C}$ be connected and simply connected.

(α) Let $a \in \mathcal{O}$ and argue that the Green's function $\mathcal{G}_{\mathcal{O}}(\,\cdot\,; a)$ exists.

(β) Argue that there exists a holomorphic function $h \colon \mathcal{O} \to \mathbb{C}$ such that $\mathcal{G}_{\mathcal{O}}(z; a) = -\operatorname{Log}(|z - a|) + \operatorname{Re} h(z)$ for all $z \in \mathcal{O} \setminus \{a\}$.

(γ) Show that $g(z) := e^{-h(z)}(z - a)$ defines a holomorphic function in $\mathcal{O}$ with $g(\mathcal{O}) \subseteq B(0,1)$.

(δ) Let $z_0 \in \partial \mathcal{O}$ (possibly $z_0 = \infty$) and $\{z_n\}_{n \in \mathbb{N}} \subseteq \mathcal{O}$ with $z_n \to z_0$. Show that $|g(z_n)| \to 1$.

(ε) Prove that for any $0 < r < 1$, the set $K_r = g^{-1}(B[0,r])$ is compact and $\partial K_r \subseteq \{z \in \mathcal{O} \mid |g(z)| = r\}$.

(ζ) Let $w \in B(0,1)$ and $|w| < r < 1$. Show that g and $g - w$ has the same number of zeroes in K_r.

(η) Conclude that $g \colon \mathcal{O} \to B(0,1)$ is bijective.

Chapter 17

Extending Riemann Maps to the Boundary

Recall from Definition 8.19 that a region is an open, connected and non-empty subset of $\mathbb{C}$. The Riemann Mapping Theorem, Theorem 8.21, tells us, in particular, that any bounded, simply connected region Ω is conformally equivalent to the unit disc. It is natural to ask when this conformal equivalence and/or its inverse can be extended by continuity to the boundary. In this chapter we consider that question, with the main results being Theorem 17.17 and Theorem 17.19 for the conformal equivalence and its inverse, respectively. A particularly well-known positive answer, attributed to Carathéodory, deals with the case where the boundary of Ω is given by a closed Jordan curve. This result is given as Exercise 17.9. For a more complete treatment of the subject, we refer to the book by Conway (1995). For a very interesting account of the history of the Riemann Mapping Theorem and its generalizations, see Gray (1994).

17.1 Some analytic preliminaries

The following result will be used repeatedly. It is given as Exercise 8.18.

Lemma 17.1. *Let Ω be an open connected subset of $\mathbb{C}$. Let $\{f_n\}$ be a sequence of univalent functions on Ω, and assume that $\{f_n\}$ converges locally uniformly to a function f. Then either f is constant or it is a univalent function.*

Theorem 17.2. *If f is a biholomorphic map between the open sets G and Ω, then*

$$\text{Area}(\Omega) = \int_G |f'(z)|^2 \, dxdy. \qquad (17.1)$$

In particular, if $f \colon B(0,1) \to \Omega$ *is a biholomorphic map with series expansion* $f(z) = \sum_{n=0}^{\infty} a_n z^n$, *then*

$$\text{Area}(\Omega) = \int_{B(0,1)} |f'(z)|^2 \, dxdy = \pi \sum_{n=0}^{\infty} n|a_n|^2. \qquad (17.2)$$

Proof. The formula (17.1) is a consequence of change of variables in the integral. The second formula (17.2) follows from (17.1) in the following way. By monotone convergence and (17.1) we have

$$\text{Area}(\Omega) = \int_{B(0,1)} |f'(z)|^2 \, dxdy = \lim_{r \nearrow 1} \int_{B(0,r)} |f'(z)|^2 \, dxdy$$

$$= \lim_{r \nearrow 1} r^2 \int_{B(0,1)} |f'(rw)|^2 \, dxdy. \qquad (17.3)$$

Now the series for $f(rw)$ converges uniformly on $B(0,1)$, so one can exchange the double summation of $|f'(rw)|^2$ with the integral and thereby obtain (17.2). $\qquad \square$

Lemma 17.3. *Let G be a bounded, simply connected region and let g be a biholomorphic map from $B(0,1)$ to G. Then there exists a subset $E \subseteq \partial B(0,1)$ of Lebesgue measure (length) 0 such that if $w_0 \notin E$, then the curve γ in G defined by $\gamma(t) = g(tw_0)$, $t \in [1/2, 1[$, is a Jordan half-interval.*

Proof. We start by estimating a double integral, using the Cauchy-Schwarz inequality (technical note: We extend $g'(z)$ by 0 from $B(0,1)$ to $B[0,1]$ to have a well-defined integrand)

$$\int_0^{2\pi} \left(\int_{1/2}^1 |g'(te^{i\theta})| \, dt \right)^2 d\theta \leq \int_0^{2\pi} \frac{1}{2} \int_{1/2}^1 |g'(te^{i\theta})|^2 \, dtd\theta$$

$$\leq \int_0^{2\pi} \int_{1/2}^1 |g'(te^{i\theta})|^2 t \, dtd\theta. \qquad (17.4)$$

The last expression is just a planar integral written in polar coordinates, so we see that

$$\int_0^{2\pi} \left(\int_{1/2}^1 |g'(te^{i\theta})| \, dt \right)^2 d\theta \leq \int_{B(0,1)} |g'(z)|^2 \, dxdy = \text{Area}(G), \qquad (17.5)$$

where we used Theorem 17.2 to get the last identity.

Since the double integral is finite, there exists $E \subseteq \partial B(0,1)$ of zero measure such that if $z = e^{i\theta} \notin E$, then $\int_{1/2}^{1} |g'(te^{i\theta})| \, dt < +\infty$. But notice (by the Fundamental Theorem of Calculus and $|z| = 1$) that

$$|g(t_1 z) - g(t_2 z)| \leq \int_{t_1}^{t_2} |g'(tz)| \, dt, \qquad (17.6)$$

so finiteness of the integral implies existence of $\lim_{t \to 1_-} g(tz)$. $\qquad \square$

Remark 17.4. From (the argument leading to) (17.5) we see that for all $\alpha < \beta < \alpha + 2\pi$, and all $\rho > 1/2$, we have

$$\int_{\alpha}^{\beta} \left(\int_{\rho}^{1} |g'(te^{i\theta})| \, dt \right)^2 d\theta \leq 2(1 - \rho) \, \text{Area}(G). \qquad (17.7)$$

So (by the Mean Value Theorem for Integrals) there exists a $\theta \in [\alpha, \beta] \setminus E$ such that

$$\int_{\rho}^{1} |g'(te^{i\theta})| \, dt \leq \sqrt{\frac{2(1 - \rho) \, \text{Area}(\Omega)}{\beta - \alpha}}. \qquad (17.8)$$

Lemma 17.5. Let g be a bounded univalent function on $B(0,1)$. Then

$$\lim_{r \nearrow 1} (1 - r) \sup\{|g'(z)| \mid |z| = r\} = 0. \qquad (17.9)$$

Proof. Let $\varepsilon > 0$. Let $g(z)$ have series expansion $g(z) = \sum_{n=0}^{\infty} a_n z^n$. Using (17.2), we can choose $M > 0$ such that $\sum_{n=M+1}^{\infty} n|a_n|^2 < \varepsilon^2$. For the fixed M there exists a constant $C > 0$ such that the polynomial $P_M(z) = \sum_{n=0}^{M} na_n z^n$ satisfies $|P_M(z)| \leq C$ for all $z \in B(0,1)$.

We now estimate

$$|g'(z)|^2 \leq 2C^2 + 2 \left| \sum_{n=M+1}^{\infty} na_n z^n \right|^2$$

$$\leq 2C^2 + 2 \left(\sum_{n=M+1}^{\infty} n|a_n|^2 \right) \left(\sum_{n=1}^{\infty} n|z|^{2n} \right), \qquad (17.10)$$

where the last inequality follows from the Cauchy-Schwarz inequality. By differentiating the geometric series we get for all $|w| < 1$,

$$(1 - w)^{-2} = \sum_{n=1}^{\infty} nw^{n-1}, \qquad (17.11)$$

so we can recognize the last term above as

$$\sum_{n=1}^{\infty} n|z|^{2n} = \frac{|z|^2}{(1-|z|^2)^2}.$$

Inserting this and using the choice of M, we get the estimate

$$(1-|z|)^2|g'(z)|^2 \le (1-|z|)^2 2C^2 + \frac{2|z|^2}{(1+|z|)^2}\varepsilon^2$$

$$\le 2\delta^2 C^2 + \frac{2}{(2-\delta)^2}\varepsilon^2, \qquad (17.12)$$

if $1 - \delta \le |z| < 1$. $\qquad\qquad\square$

17.2   Curvilinear half-intervals

Definition 17.6. *A curvilinear half-interval is a continuous, injective mapping γ from a half-open interval $[a, b[\subseteq \mathbb{R}$. We define*

$$E_\gamma = \{z \in \mathbb{C} \mid \exists\{t_n\}_n \subseteq [a, b[\text{ with } t_n \to b, \text{ and } \gamma(t_n) \to z\}, \qquad (17.13)$$

as the limit set of γ.

Recall from Definition 15.1 that γ is a Jordan half-interval if and only if $E_\gamma = \{\xi\}$ consists of a unique point and $\xi \notin \gamma^*$.

Lemma 17.7. *Assume that G and D are open subsets of $\mathbb{C}$ and assume that f is a conformal equivalence between G and D. Assume that $\{z_n\}_n$ is a sequence in G all of whose limit points are on ∂G. Then all limit points for the sequence $\{f(z_n)\}_n$ will lie on ∂D.*

Proof. Write $w_n = f(z_n) \in D$. Clearly, all limit points are contained in $\overline{D}$, so suppose a subsequence $\{w_{n_j}\}_j$ converges to an interior point $w \in D$. Then f^{-1} is continuous at w, so we find that

$$z_{n_j} = f^{-1}(w_{n_j}) \to f^{-1}(w),$$

as $j \to \infty$, which is a contradiction. $\qquad\qquad\square$

Lemma 17.8. *Assume that G and D are open subsets of $\mathbb{C}$ and assume that f is a conformal equivalence between G and D. Assume that γ is a curvilinear half-interval with $\gamma^* \subseteq G$ and limit set $E_\gamma \subseteq \partial G$. Then $\sigma := f(\gamma)$ defines a curvilinear half-interval, with $\sigma^* \subseteq D$ and limit set E_σ contained in ∂D. If G is bounded then $E_\gamma \neq \emptyset$ and $E_\sigma \neq \emptyset$.*

Proof. The proof follows easily from the definitions and from Lemma 17.7.
□

Theorem 17.9. *Let $\gamma \colon [a, b[\, \to B(0,1)$ be continuous. Assume that the limit set E_γ, defined by (17.13), satisfies $E_\gamma \subseteq \{|z| = 1\}$ and consists of more than one point. Suppose that g is holomorphic and bounded on $B(0,1)$ and that the following limit exists*

$$\lim_{t \to b_-} g(\gamma(t)) =: c.$$

Then $g(z) = c$ for all $z \in B(0,1)$.

Remark 17.10. In particular, Theorem 17.9 shows that if g is a biholomorphic map between $B(0,1)$ and a bounded set G, and if γ is a Jordan half-interval in G with endpoint on ∂G, then $\sigma := g^{-1} \circ \gamma$ is a Jordan half-interval in $B(0,1)$ with endpoint on the boundary.

Remark 17.11. Theorem 17.9 will give the existence and continuity part of one of the extension theorems (Theorem 17.17), below.

Proof. Without loss, we can assume that $c = 0$. We suppose that g is not constant. We start by reducing the proof to the case where $g(0) = 1$.

There exists $k \in \mathbb{N} \cup \{0\}$ and $a_k \in \mathbb{C} \setminus \{0\}$ such that

$$\phi(z) := \frac{g(z)}{a_k z^k},$$

defines a holomorphic function on $B(0,1)$ with $\phi(0) = 1$. Since the set of limit points E_γ lies on the unit circle, we clearly have $\lim_{t \to b_-} \phi(\gamma(t)) = 0$. So we only have to verify that ϕ is bounded.

Let $M > 0$ be such that $|g(z)| \leq M$ for all $z \in B(0,1)$. Using the Weak Maximum Principle (Corollary 4.16), on ϕ gives for arbitrary $r < 1$ that $|\phi(z)| \leq M/(|a_k| r^k)$ on $B(0, r)$. By letting r tend to 1 we therefore get

$$|\phi(z)| \leq M/|a_k|,$$

for all $z \in B(0,1)$. So ϕ satisfies the assumptions of the theorem and $\phi(0) = 1$.

It remains to prove Theorem 17.9 in the case $c = 0$ and $g(0) = 1$. Let $\alpha \neq \beta$ be two points in E_γ. By possibly performing a rotation we can assume that $\alpha = e^{i\theta_0}$, $\beta = e^{-i\theta_0}$ for some $\theta_0 \in]0, \pi/2]$. Let $\alpha_n = \gamma(t_n)$, $\beta_n = \gamma(\tau_n)$ be sequences of points in γ with

$$\alpha_n \to \alpha = e^{i\theta_0}, \quad \beta_n \to \beta = e^{-i\theta_0},$$

as $n \to \infty$. We may assume that $t_n < \tau_n < t_{n+1}$ and $t_n \to b$ as $n \to \infty$. Also, by possibly deleting the first part of the sequence, we can assume that $|\gamma(t)| > 1/2$ for all $t \geq t_1$. Choose $m \in \mathbb{N}$ so large that $\pi/m < \theta_0$. Define

$$e_1 := \{re^{i\theta} \mid 1/2 \leq r \leq 1, 0 < \theta < \pi/m\} \subseteq \mathbb{C},$$
$$e_2 := \{re^{i\theta} \mid 1/2 \leq r \leq 1, \pi - \pi/m < \theta < \pi\} \subseteq \mathbb{C}.$$

We can then choose N so large that α_n and β_n are outside $e_1 \cup e_2$ for all $n \geq N$. Let γ_n be parametrized by $\gamma(t)$, $t_n \leq t \leq \tau_n$. By continuity, for all $n \geq N$ there exist $t'_n, \tau'_n \in]t_n, \tau_n[$ with $t'_n < \tau'_n$ and so that either $e^{-i\pi/m}\gamma(t'_n), \gamma(\tau'_n) \in [1/2, 1]$ and $\gamma(t) \in e_1$ for all $t'_n < t < \tau'_n$, or $e^{i\pi/m}\gamma(t'_n), \gamma(\tau'_n) \in [-1, -1/2]$ and $\gamma(t) \in e_2$ for all $t'_n < t < \tau'_n$. Furthermore, at least one of the two cases occurs infinitely often. We assume that it is the first – the latter case being similar – and denote the indices for which it occurs by n_j. Thus, defining γ'_{n_j} by restricting γ to the t-interval $[t'_n, \tau'_n]$ we get a sequence of curves with image $(\gamma'_{n_j})^*$ contained in $\bar{e}_1$.

Since $g(\gamma(t)) \to 0$ as $t \to b_-$, we have that

$$\mu_j := \max_{z \in (\gamma'_{n_j})^*} |g(z)| \to 0, \qquad \text{as } j \to \infty.$$

By (uniform) continuity there exists $\varepsilon_j > 0$ such that $|g(z)| < \mu_j + \frac{1}{j} =: \tilde{\mu}_j$ if $\text{dist}(z, (\gamma'_{n_j})^*) < \varepsilon_j$. We may assume that $\varepsilon_j < 1/j$. By Lemma 15.25 and Lemma 15.26 there exists a cross cut $\tilde{\gamma}_j$ in $e_1 \cap \{z \in \mathbb{C} \mid \text{dist}(z, (\gamma'_{n_j})^*) < \varepsilon_j\}$ from $\gamma(t'_{n_j})$ to $\gamma(\tau'_{n_j})$.

Define the Jordan arc l_k to have image $l_k^* = (\tilde{\gamma}'_k)^* \cup \overline{(\tilde{\gamma}'_k)^*}$. The curve l_k starts in a point of the form $r_k e^{i\pi/m}$, ends at $r_k e^{-i\pi/m}$, and has argument contained in $]-\pi/m, \pi/m[$ for all other points on the curve. Therefore, we can define a closed Jordan curve L_k with $L_k^* \subseteq B(0,1)$ by successively following the rotated curves $l_k, e^{-2\pi i/m}l_k, \ldots$ (m times). Let D_k be the Jordan region enclosed by L_k. Clearly, $B(0, 1/4) \subseteq D_k \subseteq B(0,1)$.

Let $\psi(z) = g(z)\overline{g\,(\overline{z})}$, $z \in B(0,1)$. By assumption $|\psi| \le M\tilde{\mu}_k$ on l_k^*. We can now define the holomorphic function χ on $B(0,1)$ by

$$\chi(z) = \prod_{j=0}^{m-1} \psi(e^{ij2\pi/m}z). \qquad (17.14)$$

By the Weak Maximum Principle (Corollary 4.16), applied on D_k,

$$|\chi(z)| \le M^{2m-1}\tilde{\mu}_k, \qquad \text{for all } z \in D_k. \qquad (17.15)$$

By taking the limit $k \to \infty$ we get that $\chi = 0$ on $B(0,1/4)$, which is a contradiction since $g(0) = 1$. $\qquad\square$

Theorem 17.12. *Let $\sigma\colon [0,1] \to B[0,1]$ be a closed Jordan curve and let $G \subseteq B(0,1)$ be the Jordan region enclosed by σ. Suppose $\zeta = \sigma(1/2) \in \partial B(0,1)$ and $\sigma^* \setminus \{\zeta\} \subseteq B(0,1)$. Let g be holomorphic and bounded on $B(0,1)$ and satisfy that*

$$\lim_{t \to 1/2_+} g(\sigma(t)) = c_1, \qquad \lim_{t \to 1/2_-} g(\sigma(t)) = c_2. \qquad (17.16)$$

Then $c_1 = c_2 = c$ and

$$\lim_{\overline{G} \ni z \to \zeta} g(z) = c. \qquad (17.17)$$

Proof. We start by considering the case where $c_1 = c_2 = c$. By possibly replacing g by $g - c$, we may assume that $c = 0$, so we need to prove

$$\lim_{G \ni z \to \zeta} g(z) = 0. \qquad (17.18)$$

Since g is bounded, we can fix $M > 0$ such that $|g(z)| \le M$ for all z in $B(0,1)$. We can assume without loss of generality that $\zeta = i$. Let $\varepsilon > 0$. We may choose $\eta > 0$ so small that $|g(z)| \le \varepsilon$ for all $z \in \sigma^* \setminus \{i\}$ with $\operatorname{Im} z \ge 1 - \eta$. Let $z_0 \in G$ with $\operatorname{Im} z_0 = d > 1 - \eta$. We will prove the inequality

$$|g(z_0)|^2 \le M\varepsilon. \qquad (17.19)$$

Define $G_1 = \{z \in G \mid \operatorname{Im} z \ge d\}$ and let G_2 be the reflected domain

$$G_2 := \{z \in \mathbb{C} \mid \overline{z} + 2di \in G_1\}.$$

Since $z_0 \in G$ we see that z_0 is an interior point of $G' := G_1 \cup G_2 \subseteq B(0,1)$. Define the reflected function g^* by $g^*(z) = \overline{g(\overline{z} + 2di)}$. This function is defined on $\{z \in \mathbb{C} \mid \overline{z} + 2di \in B(0,1)\}$ which contains G'. We can now define a holomorphic function χ on G' by

$$\chi(z) = g(z)g^*(z). \tag{17.20}$$

This function satisfies $\chi(z_0) = |g(z_0)|^2$. Also $|\chi(z)| \le M\varepsilon$ on $\partial G'$. However, since we do not know that χ extends continuously to the boundary points $\zeta = i$ and its reflection $\zeta' = (2d-1)i$, we cannot yet apply the Weak Maximum Modulus Principle.

We can choose and fix a holomorphic logarithm on G' (since for all $z \in G'$ we have $z - \zeta, z - \zeta' \notin\;]-\infty, 0]$. We can now, for any $\mu > 0$, define

$$\psi(z) = (z - \zeta)^\mu (z - \zeta')^\mu g(z)g^*(z), \tag{17.21}$$

with $\psi(\zeta) = \psi(\zeta') = 0$. With this definition, ψ is holomorphic on G' and continuous on $\overline{G'}$ and satisfies

$$|\psi(z)| \le 4^\mu M\varepsilon \qquad \text{for all } z \in \partial G'. \tag{17.22}$$

By the Weak Maximum Principle (Corollary 4.16), we have

$$|(z_0 - \zeta)^\mu (z_0 - \zeta')^\mu g(z_0)g^*(z_0)| = |\psi(z_0)| \le 4^\mu M\varepsilon. \tag{17.23}$$

The inequality (17.19) now follows by taking the limit $\mu \to 0$.

We finally have to argue that the case $c_1 \ne c_2$ cannot happen. In this case we define $F(z) = (g(z) - c_1)(g(z) - c_2)$.

Define Jordan half-intervals $\sigma_1, \sigma_2 \colon [0, 1/2[\to B(0,1)$ by

$$\sigma_1(t) = \sigma(t), \quad \sigma_2(t) = \sigma(1-t).$$

Clearly, we can apply the particular case just proved to the function F. Suppose for contradiction that $|c_1 - c_2| = \rho > 0$. Choose $\varepsilon > 0$ so small that $|F(z)| < \rho^2/4$ for all $z \in \overline{G} \cap B(\zeta, \varepsilon)$ and $|g(z) - c_j| < \rho/2$ for all $z \in \sigma_j^* \cap B(\zeta, \varepsilon)$. Let $m(r)$ be the function from Proposition 15.17. Choose $z_j \in \sigma_j^* \cap B(\zeta, m(\varepsilon))$, $j = 1, 2$. By continuity we can find $z_j' \in G$, $j = 1, 2$ such that $|z_j' - \zeta| < m(\varepsilon)$ and $|g(z_j') - c_j| < \rho/2$. Now, by Proposition 15.17 we can join z_1' and z_2' by a continuous curve δ in $G \cap B(\zeta, \varepsilon)$. At some point z_3 on this curve, we must have $|g(z_3) - c_1| = \rho/2$, but then

$$|F(z_3)| = |g(z_3) - c_2|\rho/2 \ge (|c_1 - c_2| - |g(z_3) - c_1|)\rho/2 = \rho^2/4, \tag{17.24}$$

which is a contradiction. $\qquad\qquad\qquad\qquad\qquad\qquad\qquad\qquad\qquad\square$

17.3 Accessible boundary points

Let $G \subseteq \mathbb{C}$ be open and connected and have boundary Γ. Recall (from Definition 15.20, Definition 15.21 and Proposition 15.27) that an accessible boundary point of G is a pair (z, γ), where $z \in \Gamma$ and γ is a Jordan half-interval in G with end-point z. Recall also the equivalence relation $\sim$ on the set of accessible boundary points of G and that $\partial_A G$ is the set of equivalence classes of accessible boundary points. Thus $\partial_A G$ consists of equivalence classes $[(z, \gamma)]$ of accessible boundary points.

It is a simple exercise (Exercise 17.1), to verify that if $G = B(0, 1)$, then $\partial_A G = \partial G$.

Example 17.13. Let $G_1 = B(0, 1) \setminus [0, 1]$ be the slit disc. Then it is easy to verify that all boundary points give rise to accessible boundary points. However, a point $z \in]0, 1[$ does not define a unique boundary point since the vertical lines with end point z from above and from below are non-equivalent.

Example 17.14. Let G_2 be defined by (see Fig. 17.1)

$$G_2 = \{z \in \mathbb{C} \mid \operatorname{Re} z, \operatorname{Im} z \in]0, 1[\}$$
$$\setminus \left(\bigcup_{n=1}^{\infty} \left[\tfrac{1}{2n}, \tfrac{1}{2n} + \tfrac{3}{4}i \right] \cup \left[\tfrac{1}{2n+1} + \tfrac{1}{4}i, \tfrac{1}{2n+1} + i \right] \right).$$

Then $[0, i] \subseteq \partial G_2$, but there is no continuous curve in G_2 with end-point in $[0, i]$.

Theorem 17.15. *Let G be a bounded, simply-connected region and let f be a biholomorphic map between G and $B(0, 1)$. Let $(z_0, \gamma) \in \partial_A G$ where γ is a Jordan half-interval, then $f \circ \gamma$ defines a Jordan half-interval in $B(0, 1)$. Furthermore, if $(z_0, \gamma') \in [(z_0, \gamma)]$, then $f \circ \gamma'$ and $f \circ \gamma$ are Jordan half-intervals with the same end point.*

So f induces a function – also denoted by f – from $\partial_A G$ to $\partial B(0, 1)$. This induced function is injective and has dense image in $\partial B(0, 1)$.

Remark 17.16. Theorem 17.15 combined with Exercise 17.1 implies that the biholomorphic map f from G to $B(0, 1)$ induces a map from $\partial_A G$ to $\partial_A B(0, 1)$.

Proof. Let $\sigma = f \circ \gamma$. We may assume that $\sigma \colon [0, 1[\to \mathbb{C}$. By Lemma 17.8, $\emptyset \neq E_\sigma \subseteq \partial B(0, 1)$. Assume that σ is not a Jordan half-interval. Then E_σ contains at least two distinct points. But then we can apply Theorem 17.9

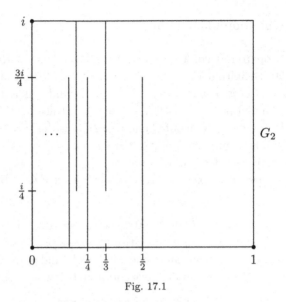

Fig. 17.1

to the function f^{-1}, because this is a bounded function by assumption and satisfies

$$\lim_{t \to 1_-} f^{-1}(\sigma(t)) = z_0.$$

By Theorem 17.9, f^{-1} is constant which is a contradiction. So σ must be a Jordan half-interval. Denote its end point by w_0.

Let now $(z_0, \gamma') \in [(z_0, \gamma)]$ and assume that $\sigma' := f \circ \gamma'$ has end point $w_1 \neq w_0$. Since $(z_0, \gamma') \sim (z_0, \gamma)$ there exists a continuous curve γ_3 with end point z_0 and which intersects both γ^* and $(\gamma')^*$ infinitely often in any neighborhood of z_0. Using Lemma 15.25 and Lemma 15.26 we may assume that γ_3 is a Jordan half-interval. But then

$$\{w_0, w_1\} \in E_{f \circ \gamma_3},$$

which is a contradiction to the first part of the proof.

Thus, we have established existence of the induced map from $\partial_A G$. That the map has dense image follows from Lemma 17.3 below. So we only have left to prove injectivity. Let γ_1, γ_2 be Jordan half-intervals in G with endpoints z_1, z_2 and such that $[(z_1, \gamma_1)]$ and $[(z_2, \gamma_2)]$ are inequivalent accessible boundary points. Assume that $\sigma_j = f \circ \gamma_j$, $j = 1, 2$, have the same end-point $w \in \partial B(0, 1)$.

Case 1: σ_1 and σ_2 intersect infinitely often in any neighborhood of w.
We can assume that both Jordan half-intervals are parametrized by $[0, 1[$. In this case we will have sequences $\{t_n\}, \{t'_n\} \subseteq [0, 1[$ such that $t_n, t'_n \to 1$ and $\gamma_1(t_n) = \gamma_2(t'_n)$. But then $z_1 = z_2 = z$ by taking the limit $n \to \infty$. Also, since their only limit point is z, the tails of γ_1, γ_2 will be contained in any given neighborhood of z and will intersect infinitely often, so the two curves define equivalent accessible boundary points.

Case 2: Suppose there exists $\varepsilon > 0$ such that $\sigma_1^* \cap \sigma_2^* \cap B(w, \varepsilon) = \emptyset$.
In this case we may – by possibly omitting the first part of the curves – assume that $\sigma_1^* \cap \sigma_2^* = \emptyset$, that $\sigma_1^*, \sigma_2^* \subseteq B(w, \varepsilon)$, and $\sigma_1(0), \sigma_2(0) \notin B(w, \varepsilon/2)$. Also, we assume that both σ_1 and σ_2 are parametrized by the interval $[0, 1[$.

Using Lemma 17.25 below we can construct a closed Jordan curve $\tilde{\sigma} : [0, 1] \to B[0, 1]$, with $\tilde{\sigma}(1/2) = w$, $\tilde{\sigma}^* \setminus \{w\} \subseteq B(0, 1)$ and such that $\sigma_1(t) = \tilde{\sigma}(2t), \sigma_2(t) = \tilde{\sigma}(1 - 2t)$, for $t \in [\frac{1}{2} - \delta, \frac{1}{2}[$ for some $\delta > 0$.

By Theorem 17.12 we have $z_1 = z_2 =: z$. However, we still need to establish that the two curves, γ_1 and γ_2 are equivalent as accessible boundary points.

Let G_1 be the Jordan region (in $B(0, 1)$) with boundary $\tilde{\sigma}^*$. We will construct a continuous curve $\sigma : [0, \infty[\to G_1$, with $\lim_{t\to\infty} \sigma(t) = w$ and $\sigma^* \subseteq \overline{G}_1$ and such that the tail of σ contains infinitely many points on both σ_1 and σ_2. By Theorem 17.12 we then have $\lim f^{-1}(\sigma(t)) = z$ and that $f^{-1} \circ \sigma$ provides an equivalence between γ_1 and γ_2. So it only remains to construct σ. The construction is very similar to the proof of Proposition 15.22.

Let the function m be the one from Proposition 15.17 applied to the domain G_1 with point w. Let $\{z_n\}_n \subseteq \overline{G}_1$ be a sequence with $|z_n - w| < \frac{m(1/n)}{2}$ and $z_n \in \sigma_1^*$ if n is even, $z_n \in \sigma_2^*$ if n is odd. Using that (by Theorem 15.19) z_n and z_{n+1} are accessible, there exists a continuous curve $\gamma_n : [0, 1] \to \overline{G}_1$ with $\gamma_n(0) = z_n, \gamma_n(1) = z_{n+1}$. By Proposition 15.17 we may assume that $\gamma_n^* \subseteq B(w, 1/n)$. We define σ by

$$\sigma(t) = \gamma_n(n + t), \qquad \text{for } t \in [n, n+1].$$

$\square$

17.4   The extension theorem

Recall the definition of simple boundary points from Definition 15.18 and that (by Proposition 15.27) any simple boundary point z_0 is also the end-point of a Jordan half-interval $\tilde{\gamma}$ in G, so $(z_0, \tilde{\gamma}) \in \partial_A G$.

Theorem 17.17. *Let $G \subsetneq \mathbb{C}$ be a simply connected region and let f be a biholomorphic map from G to $B(0,1)$. Let $S \subseteq \partial G$ be the collection of simple boundary points of G. Then f extends to a continuous, injective map defined on $G \cup S$.*

Proof. We first give the proof in the case where G is bounded. Let $z \in S$. We will prove that $f(z)$ is well-defined by

$$f(z) = \lim_{n \to \infty} f(z_n),$$

where $\{z_n\}_n \subseteq G$ is any sequence with $z_n \to z$.

So suppose $\{z_n\}_n$ is such a sequence and that the image sequence $w_n = f(z_n) \in B(0,1)$ has more than one accumulation point. Since z is simple there exists a continuous curve $\gamma \colon [0,1[\to G$ with endpoint z running through the points z_n. Then $\sigma = f \circ \gamma$ is a continuous curve in $B(0,1)$. The limit set $E_\sigma \subseteq \partial B(0,1)$ by Lemma 17.8 and contains at least 2 points by assumption. But this is a contradiction to Theorem 17.9.

Suppose now that $\{z_n\}_n \subseteq S \subseteq \partial G$ is a sequence with limit point $z \in S$. Using the above, we can for each n choose $z_n' \in G$ with $|z_n - z_n'| + |f(z_n) - f(z_n')| < 1/n$. It is clear that $z_n' \to z$ and by the previous argument $f(z_n') \to f(z)$. So it follows that $f(z_n) \to f(z)$.

It remains to prove that the extended function, still denoted by f, from $G \cup S$ to $B[0,1]$ is injective. By Theorem 17.15, f induces an injective function – which we for the purpose of this proof will denote by $\tilde{f}$ – from $\partial_A G$ to $\partial B(0,1)$, defined by

$$\tilde{f}((z_0, \gamma)) = \lim_{t \to 1_-} f(t),$$

if the curve γ is parametrized by $[0,1[$. So we see, by continuity of f, that if z_0 is a simple boundary point then $\tilde{f}((z_0, \gamma)) = f(z_0)$. Since $\tilde{f}$ is injective, this proves that $f \colon G \cup S \to B[0,1]$ is injective.

We now generalize to non-bounded sets G. This essentially follows from some steps of the proof of Theorem 8.21. Suppose $z_0 \in S$ and assume that $\{z_n\}_n \subseteq G$ satisfies $z_n \to z_0$ as $n \to \infty$. Define first ϕ to be a holomorphic square root of $z - z_0$, the existence is guaranteed on G, since G is simply connected. So $\phi(z)^2 = z - z_0$. Clearly, ϕ is a biholomorphic map to its open, simply connected image, $\phi(G)$. Furthermore, $\phi(G) \cap -\phi(G) = \emptyset$, so let $\zeta \in -\phi(G)$ and define $\Phi \colon G \to \mathbb{C}$ by $\Phi(z) = 1/(\phi(z) - \zeta)$. Then Φ is a biholomorphic map from G to $G' := \Phi(G)$. Since ζ is an interior point in $\mathbb{C} \setminus \phi(G)$, G' is bounded, as well as simply connected.

We will prove that $-1/\zeta$ is a simple boundary point of G'. It is clearly a boundary point, since $z_0 \in \partial G$. Suppose that $\{\eta_n\}_n$ is a sequence in G' with $\eta_n \to -1/\zeta$. Then $\Phi^{-1}(\eta_n) \in G$ and $\Phi^{-1}(\eta_n) \to z_0$, so there exists a continuous curve $\gamma \colon [0, 1[\to G$ with endpoint z_0 running through the points $\Phi^{-1}(\eta_n)$. Then $\sigma = \Phi \circ \gamma$ is a continuous curve in G'. Since $|\phi(z)| = \sqrt{|z - z_0|}$ we see that σ has endpoint $-1/\zeta$, proving that $-1/\zeta$ is a simple boundary point.

Since $-1/\zeta$ is a simple boundary point of G', the biholomorphic map $f \circ \Phi^{-1} \colon G' \to B(0, 1)$ extends by continuity to $-1/\zeta$. So we see that

$$\lim_{n \to \infty} f(z_n) = f \circ \Phi^{-1}(-1/\zeta).$$

Since the limit is independent of the sequence, this proves that $f(z_0)$ is well defined. The continuity of f extended to $G \cup S$ is proved as in the case of bounded G. $\qquad\square$

Corollary 17.18. *Let G be a bounded, simply connected region and let f be a biholomorphic map from G to $B(0, 1)$. Suppose all boundary points of G are simple. Then f extends to a homeomorphism between $\overline{G}$ and $B[0, 1]$.*

Proof. The existence and injectivity of the continuous extension follows from Theorem 17.17. So the extension of f will be a bijection between $\overline{G}$ and its image, but the image must be all of $B[0, 1]$ since it has to be compact. That the inverse function becomes continuous follows from a theorem in point set topology about continuous bijections between compact (Hausdorff) spaces (see Exercise 17.7). $\qquad\square$

In the other direction we have the following Theorem 17.19. The proof is more complicated and is postponed to the end of the chapter.

Theorem 17.19. *Let G be a bounded, simply connected region and let g be a biholomorphic map from $B(0, 1)$ to G. Then g extends continuously to $B[0, 1]$ if and only if the boundary of G is a continuous curve. Furthermore, the extended function is surjective from $B[0, 1]$ to $\overline{G}$.*

In the unbounded case, we only state the following simple result.

Corollary 17.20. *Suppose that $G, G' \subsetneq \mathbb{C}$ are simply connected regions and that $\mathbb{C} \setminus G'$ has non-empty interior. Suppose furthermore, either*

(a) *that G' is bounded and that the boundary of G' is a continuous curve,*

or

(b) *that G' is unbounded and that the boundary of G' in $\mathbb{C}_\infty$ is a continuous curve (with respect to the topology on $\mathbb{C}_\infty$).*

Let $\phi \colon G \to G'$ be a biholomorphic map. Then ϕ extends by continuity to the set of simple boundary points of G with values in $\mathbb{C}_\infty$.

Proof. We only consider the unbounded case. Let z_0 be an interior point of $\mathbb{C} \setminus G'$. Then $I(z) = (z - z_0)^{-1}$ is a homeomorphism of $\mathbb{C}_\infty$, so it is equivalent to prove the result with G' replaced by $G'' := I(G')$. But G'' is bounded and has boundary given by a continuous curve, so the result follows by combining Theorem 17.19 with Theorem 17.17. $\qquad\qquad\square$

We end this section with a result justifying the name *simple* boundary point.

Proposition 17.21. *Let G be a bounded, simply connected region and assume that the boundary of G is a continuous curve. Then the set S of simple boundary points coincides with*

$$S' := \{z_0 \in \partial G \mid \text{there exists a unique equivalence class} \\ [(z, \gamma)] \in \partial_A G \text{ with } z = z_0\}.$$

Proof. By Proposition 15.22, $S \subseteq S'$, so it suffices to prove the other inclusion. Let $f \colon B(0,1) \to G$ be a biholomorphic map, which exists by the Riemann Mapping Theorem (Theorem 8.21), with inverse g. Using Theorem 17.19, f extends to a continous surjective map, still denoted by f, from $B[0,1]$ to $\overline{G}$. Let $z_0 \in S'$ and let $\{z_n\}_n \subseteq G$ be a sequence, with $z_n \to z_0$. Denote $w_n = g(z_n) \in B(0,1)$. We will prove that $\{w_n\}$ has a limit $w \in \partial B[0,1]$.

Suppose not, then there exist subsequences, $\{w_{n_j}\}$, $\{w_{n'_j}\}$, with $w_{n_j} \to w_1$, $w_{n'_j} \to w_2$, with $w_1 \neq w_2$ and $w_1, w_2 \in \partial B(0,1)$ (by Lemma 17.7). Since w_1, w_2 are simple boundary points of $B(0,1)$, there exist continuous curves $\sigma_j \colon [0,1[\to B(0,1)$, with $\lim_{t \to 1_-} \sigma_j(t) = w_j$ and passing through the sequence $\{w_{n_j}\}$, respectively $\{w_{n'_j}\}$. By continuity of f up to the boundary, the curves $\gamma_1 = f \circ \sigma_1$ and $\gamma_2 = f \circ \sigma_2$ in G both have end point z_0. Therefore, they are equivalent by definition of S'. But then $w_1 = w_2$ by Theorem 17.15.

Let now $w_{n_j} = f(z_j)$ in the previous argument, i.e., the original sequence. The construction of γ_1 now proves that z_0 is simple. $\qquad\square$

We can now give the promised results concerning endpoints of Jordan arcs.

Proposition 17.22. *Let $\sigma\colon [0,1] \to \mathbb{C}$ be a Jordan arc. Suppose that $|\sigma(1)| = 1$ and $|\sigma(t)| < 1$ for all $t < 1$. Define $G = B(0,1) \setminus \sigma^*$. Then the boundary point $\sigma(0) \in \partial G$ is simple.*

Proof. We want to use Theorem 17.19, so we need to establish that G is a simply connected region. The simple connectedness follows from Corollary 9.12. To show that G is connected, let $a, b \in G$. By Corollary 15.5, there exists a continuous curve $\gamma\colon [0,1] \to \mathbb{C} \setminus \sigma^*$ with $\gamma(0) = a$, $\gamma(1) = b$. If γ stays inside $B(0,1)$ there is nothing to prove. If not, define

$$s_0 := \inf\{s \in [0,1] \mid |\gamma(s)| = 1\}, \quad t_0 := \sup\{t \in [0,1] \mid |\gamma(t)| = 1\}.$$

Clearly, $0 < s_0 \leq t_0 < 1$. Let $0 \leq \alpha_0 \leq \beta_0 \leq 2\pi$ be such that – as sets – $\{e^{i\alpha_0}, e^{i\beta_0}\} = \{\gamma(s_0), \gamma(t_0)\}$ and $\sigma(0) \notin e^{i[\alpha_0, \beta_0]}$. Then there exists $\varepsilon > 0$ such that σ^* is disjoint from

$$A := \{z = re^{i\theta} \mid 1 - \varepsilon < r < 1 \text{ and } \theta \in [\alpha_0 - \varepsilon, \beta_0 + \varepsilon]\,\}. \qquad (17.25)$$

Choose a point $a' = \gamma(s') \in A$, with $s' < s_0$ and a point $b' = \gamma(t')$, with $t' > t_0$. By connecting a' and b' by a continuous curve in A, we see that G is connected.

We first prove that $\sigma(0)$ is an accessible boundary point. By Theorem 17.19 any biholomorphic map $g\colon B(0,1) \to G$ extends by continuity to $B[0,1]$. We will also denote this extended map by g. Observe that $g(B[0,1]) = B[0,1]$; indeed, $g(B[0,1])$ is compact and therefore $\overline{G} \subseteq g(B[0,1]) \subseteq B[0,1]$ and the claim now follows from (the first part of) Theorem 15.24. We may now select $w_0 \in \partial B(0,1)$ such that $g(w_0) = \sigma(0)$. Define $\gamma(t) = g(tw_0)$ then γ defines a Jordan half-interval in G with endpoint $\sigma(0)$.

Suppose now that $\tilde{\gamma}$ is a continuous curve in G with endpoint $\sigma(0)$. Then if $[(\sigma(0), \tilde{\gamma})] \neq [(\sigma(0), \gamma)]$ as accessible boundary points, there exists $\varepsilon > 0$ such that γ and $\tilde{\gamma}$ do not intersect in $B(\sigma(0), \varepsilon)$. The tails of σ and γ define a cross-cut in $B(\sigma(0), \varepsilon)$, splitting it into two Jordan regions

(Proposition 15.13), which we call G_1 and G_2. Then $\tilde{\gamma}$ runs entirely in one of these, say G_1. Let g_1 be a homeomorphism between $B[0,1]$ and $\overline{G_1}$ (which exists by Corollary 17.18). Then it is easy to prove that $\tilde{\gamma}$ and γ are equivalent curves by constructing a curve establishing the equivalence in $B[0,1]$ and transporting it back using g_1. See Definition 15.21. By Proposition 17.21 this finishes the proof. $\square$

Corollary 17.23. *Let $G \subseteq \mathbb{C}$ be open and $a \in G$. Suppose $\gamma \colon [0,1] \to \mathbb{C}$ is a Jordan arc with $\gamma(0) = a$. Then a is a simple boundary point of the set $G \setminus \gamma^*$.*

Proof. Note first that by Theorem 15.24, a is in the boundary of $\mathbb{C} \setminus \gamma^*$. Hence, since a is an interior point of G, a is also in $\partial(G \setminus \gamma^*)$. It remains to argue that a is simple.

Let $\{z_n\} \subseteq G \setminus \gamma^*$ such that $z_n \to a$. Pick $r > 0$ such that $B[a,r] \subseteq G$ and $|a - \gamma(1)| > r$. Then $s = \inf\{t \in [0,1] \mid |\gamma(t) - a| = r\} \in \,]0,1[$. We may now apply Proposition 17.22 to $\gamma \colon [0,s] \to B[a,r]$ (suitably rescaled) to conclude that a is a simple boundary point of $B(a,r) \setminus \gamma([0,s])$. Hence, there exists an N and a curve $\sigma \colon [0,1[\, \to B(a,r) \setminus \gamma([0,s])$, such that $\lim_{t \to 1_-} \sigma(t) = a$ and σ passes through the points z_n, $n \geq N$ (in the right order). See Definition 15.18.

Let $r' = \inf_{t \in [s,1]} |\gamma(t) - a| > 0$ and pick $0 \leq t_0 < 1$ and $M > N$ such that $\sigma(t_0) = z_M$ and $|\sigma(t) - a| < r'$ for all $t \in [t_0, 1[$. Then the curve $\sigma \colon [t_0, 1[\, \to G \setminus \gamma^*$ demonstrates that s is simple boundary point for $G \setminus \gamma^*$. $\square$

17.5   Cross cuts and prime ends

The notion of prime ends is a useful tool to discuss continuity at the boundary. We start by some simple lemmas about curves.

Lemma 17.24. *Let G be open. Let $(z_1, \gamma_1), (z_2, \gamma_2)$ be inequivalent accessible boundary points in G. Then $\gamma_1^* \cap \gamma_2^*$ has no limit points in ∂G.*

Proof. Clearly, since both curves are Jordan half-intervals, if $\gamma_1^* \cap \gamma_2^*$ has a limit point in ∂G, the two half-intervals have the same endpoint $\{z_0\}$. Let $\{z_n\} \subseteq \gamma_1^* \cap \gamma_2^*$ be a sequence in G converging to z_0. By possibly extracting a subsequence, we may assume that

$$z_n = \gamma_1(t_n) = \gamma_2(\tau_n),$$

where both $\{t_n\}$ and $\{\tau_n\}$ are increasing. We then produce a curve γ_3 by connecting the points $\{z_n\}$ alternatively along γ_1 and γ_2. This curve γ_3 establishes the equivalence between γ_1 and γ_2. $\qquad\square$

Lemma 17.25. *Let G be open and connected. Let $\gamma_1, \gamma_2 \colon [0,1[\to G$ be Jordan half-intervals with end points $z_1, z_2 \in \partial G$ and satisfying $[(z_1, \gamma_1)] \neq [(z_2, \gamma_2)]$ as accessible boundary points. Then there exist $\varepsilon > 0$ and Jordan half-intervals $\widetilde{\gamma}_1, \widetilde{\gamma}_2 \colon [0,1[\to G$ with $\widetilde{\gamma}_j(t) = \gamma_j(t)$ for all $t \in [1-\varepsilon, 1[$, $j = 1, 2$ and such that $\widetilde{\gamma}_1(0) = \widetilde{\gamma}_2(0)$ and*

$$\widetilde{\gamma}_1^* \cap \widetilde{\gamma}_2^* = \{\widetilde{\gamma}_1(0)\}.$$

Furthermore, if λ is a continuous, piecewise linear curve in G connecting $\gamma_1(0)$ with $\gamma_2(0)$, then we can choose $\widetilde{\gamma}_1, \widetilde{\gamma}_2$ such that

$$\widetilde{\gamma}_1^* \cup \widetilde{\gamma}_1^* \subseteq \gamma_1^* \cup \lambda^* \cup \gamma_2^*.$$

Notice that since the tails of the curves $\widetilde{\gamma}_j^*$ agree with the original curves, they define the same accessible boundary points as them.

Proof. Using Lemma 17.24, we may assume – after possibly omitting the start of each curve – that $\gamma_1^* \cap \gamma_2^* = \emptyset$. Since G is open and connected we can choose $\lambda \colon [0,1] \to G$ to be a continuous, piecewise linear curve from $\gamma_1(0)$ to $\gamma_2(0)$. By Lemma 15.26, we may assume that λ is injective. Using Lemma 17.24 again, we can choose $a < 1$ so that $\gamma_1(a) \in \lambda^*$ and for all $t \in]a, 1[$,

$$\gamma_1(t) \notin \lambda^* \cup \gamma_2^*.$$

After reparametrizing γ_1 and λ and possibly omitting the first part of the curves, we can assume that $a = 0$ and $\gamma_1(a) = \lambda(0)$, $\lambda(1) = \gamma_2(0)$. We choose $b \in [0,1]$ as $b = \min\{t \in [0,1] \mid \lambda(t) \in \gamma_2^*\}$. We now make a Jordan half-interval $\widetilde{\gamma}_2$ by following λ from $\lambda(0)$ to $\lambda(b)$ and then continuing along γ_2. Up to reparametrization γ_1 and $\widetilde{\gamma}_2$ then satisfy the conclusion of the lemma. $\qquad\square$

Definition 17.26 (Cross Cut). *Let G be a bounded, simply connected region and $\sigma : [a, b] \to \mathbb{C}$ a curve. We call σ a cross cut in G if*

(a) $\sigma|_{]a,b[}$ *is injective with image in G.*

(b) *If $\gamma_1(t) := \sigma(\frac{a+b}{2} + t\frac{a-b}{2})$ and $\gamma_2(t) := \sigma(\frac{a+b}{2} + t\frac{b-a}{2})$, for $t \in [0, 1]$, then γ_1, γ_2 are Jordan half-intervals in G with end-points on ∂G and γ_1, γ_2 are non-equivalent in $\partial_A G$.*

It is sometimes convenient to write $\sigma = (\gamma_1, \gamma_2)$, where the γ_j's are given as above.

Remark 17.27. If G is a Jordan region, then any boundary point defines a unique accessible boundary point by Theorem 15.23, so the definition agrees with Definition 15.12.

Remark 17.28. Let γ_1 and γ_2 be Jordan half-intervals in G that determine different accessible boundary points and with end points $z_1, z_2 \in \partial G$. Using Lemma 17.25, we may assume (after possibly restricting to the tails of the curves and connecting the starting points) both curves to start in the same point $z_0 \in G$ and that $\gamma_1^* \cap \gamma_2^* = \{z_0\}$. That is, we can produce a cross cut having the two given accessible boundary points as tails.

Lemma 17.29. *Let G be a bounded, simply connected region and let $f : G \to B(0, 1)$ be a biholomorphic map. Let γ defined on $[a, b]$ be a cross cut in G. Then γ splits G in exactly two connected components G_1, G_2.*

Proof. By Theorem 17.15, $\sigma = f \circ \gamma$ defines a cross cut in $B(0, 1)$. The lemma now follows from Proposition 15.13. $\square$

Definition 17.30. *Let G be a bounded, simply connected region and let $f : G \to B(0, 1)$ be a biholomorphic map. Let $\{C_n\}_n = \{(\gamma_1^n, \gamma_2^n)\}_n$ be a sequence of cross cuts in G with end points z_1^n, z_2^n and let G_1^n, G_2^n be the corresponding connected components of $G \setminus ((\gamma_1^n)^* \cup (\gamma_2^n)^*)$.*

*If $\{C_n\}_n$ satisfies (a)–(d) below, we call the sequence a **prime end**. The point w_0 is called the **image point** (under f) and E the **base set** of the prime end. We define an equivalence relation on prime ends by saying that $\{C_n\}_n \sim \{C_n'\}_n$ if they have the same image point.*

The conditions are

(a) *Monotonicity 1: $G_1^{n+1} \subseteq G_1^n$ for all $n \geq 1$.*

(b) *Monotonicity 2:*

$$\{[(z_1^n, \gamma_1^n)], [(z_2^n, \gamma_2^n)]\} \cap \{[(z_1^{n+1}, \gamma_1^{n+1})], [(z_2^{n+1}, \gamma_2^{n+1})]\} = \emptyset.$$

(c) $E := \bigcap_{n=1}^{\infty} \overline{G_1^n} \subseteq \partial G.$

(d) *Let* $\Delta_n = f(G_1^n)$, *then* $\bigcap_{n=1}^{\infty} \overline{\Delta}_n$ *consists of exactly one point* $w_0 \in \partial B(0,1)$.

Remarks 17.31.

(A) Since the $\{\overline{G_1^n}\}_n$ are a decreasing sequence of compact sets it is immediate that $E \neq \emptyset$. The same applies to $\{\overline{\Delta}_n\}_n$. It is an exercise to prove that monotonicity and (c), implies that $\bigcap_{n=1}^{\infty} \overline{\Delta}_n \subseteq \partial B(0,1)$, so the only real assumption in (d) is that the set reduces to a single point.

(B) Theorem 17.36 below gives the existence of (an abundance of) prime ends.

(C) The notion of prime end is independent of the choice of biholomorphic function f (see Exercise 17.2).

Example 17.32. Let $G = B(0,1) \setminus [0,1]$ be the slit disc. Then points on the slit can be the base set of two inequivalents prime ends (see Fig. 17.2).

We will now prove that the base set $E \subseteq \partial G$ of a prime end is uniquely determined by the point $w_0 \in \partial B(0,1)$.

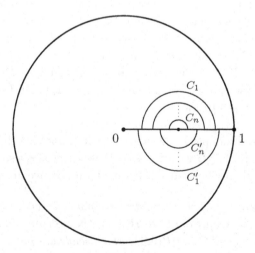

Fig. 17.2

Lemma 17.33. *Let $\{C_n\}_n$ and $\{C'_n\}_n$ be prime ends with the same image point $w_0 \in \partial B(0,1)$. Then*

$$E = E'.$$

Proof. By symmetry, we only need to prove that $E' \subseteq E$. Let $\{G_1^n\}_n$ be the monotone sequences of connected components corresponding to $\{C_n = (\gamma_1^n, \gamma_2^n)\}$ and $\{\widetilde{G}_1^n\}_n$ corresponding to $\{C'_n\}_n$. Let $n \in \mathbb{N}$. Using (b) of Definition 17.30, we have

$$\varepsilon := \operatorname{dist}(w_0, f(C_n^*)) > 0. \qquad (17.26)$$

Since $\bigcap_m \overline{f(\widetilde{G}_1^m)} = \{w_0\}$, we must have $f(\widetilde{G}_1^m) \subseteq B(w_0, \varepsilon) \cap B(0,1) \subseteq f(G_1^n)$ for all m sufficiently large. So $\widetilde{G}_1^m \subseteq G_1^n$ for all m sufficiently large, which finishes the proof. $\qquad \square$

Lemma 17.34. *Let $\{C_n\}_n$ be a prime end with image point w_0 and base set E. Then*

(a) *if $\{w_n\}_n \subseteq B(0,1)$ satisfies $w_n \to w_0$ as $n \to \infty$, then all limit points of $\{f^{-1}(w_n)\}_n$ will be contained in E.*

(b) *if $z_0 \in E$, then there exists a sequence of points $\{z_n\}_n \subseteq G$ such that $z_n \to z_0$ and $f(z_n) \to w_0$ as $n \to \infty$.*

Proof. Let G_1^n and $\Delta_n = f(G_1^n)$ be as in Definition 17.30. If $w_n \to w_0$, then by Lemma 17.7 we get that all limit points of $\{f^{-1}(w_n)\}_n$ are contained in ∂G. Furthermore, the same argument as in the proof of the previous lemma gives that $w_m \in \Delta_n$ for all m sufficiently large. So $f^{-1}(w_m) \in G_1^n$ for all m sufficiently large. This proves (a). To prove (b), let $z_0 \in E$. By definition of E we can choose $z_n \in G_1^n$ such that $|z - z_n| \leq 1/n$. By Definition 17.30, $w_n = f(z_n)$ converges to w_0. $\qquad \square$

Remark 17.35. Example 17.32 shows that it is not true that $f(z_n) \to w_0$ for *all* sequences $\{z_n\} \subseteq G$ with $z_n \to z_0$. However, if z_0 is a simple accessible boundary point then Theorem 17.17 provides a positive answer.

Theorem 17.36. *Let G be a bounded, simply connected region and let f be a biholomorphic map from G to $B(0,1)$. The mapping from equivalence classes of prime ends of G to $\partial B(0,1)$, which sends a prime end to its image point under f, is a bijection.*

Proof. By definition of the equivalence relation on prime ends, the mapping is injective, so we only need to establish surjectivity. So let $w_0 \in \partial B(0,1)$. For notational convenience only, we will assume that $w_0 = 1$. We will actually construct a sequence of cross cuts $\{C_n\}_n$ defining a prime end in G and such that the length of the curve C_n tends to 0 as $n \to \infty$.

Let g denote the inverse of the biholomorphic map f. We will construct the cross cuts in the unit disc $B(0,1)$ and then transport them back using the bounded holomorphic map g. Let $\{r_n\}_n$ be an increasing sequence with $r_n \nearrow 1$ as $n \to \infty$. Each cross cut $\gamma^{(n)}$ will be given as a union of 3 curves

$$\gamma_1^{(n)}(t) = r_n e^{it}, \qquad t \in [\alpha_n, \beta_n],$$
$$\gamma_2^{(n)}(t) = t e^{i\alpha_n}, \qquad t \in [r_n, 1[,$$
$$\gamma_3^{(n)}(t) = t e^{i\beta_n}, \qquad t \in [r_n, 1[.$$

Here we will choose $\alpha_n \nearrow 0$, $\beta_n \searrow 0$ as $n \to \infty$, so that the sets Δ_n defined by the cross cuts satisfy strict inclusions.

Given the r_n's, we have by Lemma 17.5 a sequence ε_n with $\varepsilon_n \to 0$ such that

$$\sup\{|g'(z)| \mid |z| = r_n\} \le \frac{\varepsilon_n}{1 - r_n}. \tag{17.27}$$

Choose now $\theta_n := (1 - r_n)/\sqrt{\varepsilon_n}$. By possibly making ε_n larger, we may assume that $\theta_n \to 0$. Using now Remark 17.4, we can choose $\alpha_n \in [-\theta_n, -\theta_n/2]$ such that

$$\int_{r_n}^1 |g'(te^{i\alpha_n})| \, dt \le \sqrt{\frac{2(1 - r_n) \operatorname{Area}(G)}{\theta_n/2}} = 2\sqrt{\sqrt{\varepsilon_n} \operatorname{Area}(G)} \to 0. \tag{17.28}$$

The choice of β_n in $[\theta_n/2, \theta_n]$ is similar. The finiteness of these integrals imply that $g \circ \gamma_2^{(n)}$ and $g \circ \gamma_3^{(n)}$ are Jordan half-intervals, so $f \circ \gamma^{(n)}$ will be a cross cut.

The length of the curve $g \circ \gamma_1^{(n)}$ is controlled by

$$\int_{\alpha_n}^{\beta_n} |g'(r_n e^{it})| \, dt \le 2\theta_n \frac{\varepsilon_n}{1 - r_n} \le 2\sqrt{\varepsilon_n} \to 0. \tag{17.29}$$

By possibly passing to a subsequence, we can assume that $\overline{\Delta_n} \cap B(0,1) \subseteq \Delta_{n-1}$. The corresponding C_n's define a prime end and furthermore, the length of C_n tends to zero. $\qquad \square$

We end this section by giving the proof of Theorem 17.19.

Proof of Theorem 17.19. It is clear that if the extention exists, it will be surjective.

Suppose first that g extends to a continuous map on $B[0,1]$. By Lemma 17.7 we then have $g(\partial B(0,1)) \subseteq \partial G$. But since $g(B[0,1])$ is closed we must have $g(B[0,1]) = \overline{G}$, so therefore $g(\partial B(0,1)) = \partial G$. So the restriction of g to the unit circle parametrizes ∂G as a continuous curve.

The other direction is much more complicated, so we divide the proof in several steps. Notice that by Lemma 17.34 it suffices to prove that the base set E of any prime end consists of exactly one point, i.e., that it satisfies

$$\text{diam}(E) = 0. \tag{17.30}$$

Assume that there is a continuous curve $h \colon [0,1] \to \mathbb{C}$ such that $h(0) = h(1)$ and $h([0,1]) = \partial G$.

Step 1: ∂G is locally connected.
We will prove that ∂G has the following property: For all $\varepsilon > 0$ exists $\delta > 0$ such that if $x, y \in \partial G$ with $|x - y| < \delta$, then there exists a connected set $A \subseteq \partial G$, with $x, y \in A$ and $|x - a| < \varepsilon$ for all $a \in A$.

Step 2: $\mathbb{C}_\infty \setminus G$ is locally connected.

Step 3: Proof of (17.30).

We prove each step individually.

Proof of Step 1. Suppose for contradiction that ∂G is not locally connected. Then there exists $\varepsilon > 0$ such that for all $n \in \mathbb{N}$ there exist $x_n, y_n \in \partial G$ with $|x_n - y_n| < 1/n$ but such that if $A \subseteq \partial G$ is connected and has $x_n, y_n \in A$, then there exists $a_n \in A$ with $|a_n - x_n| \geq \varepsilon$.

By possibly passing to a subsequence we may assume that $x_n \to \overline{x} \in \partial G$. It clearly follows that $y_n \to \overline{x}$. For convenience we will consider h as a periodic function defined on all of $\mathbb{R}$. Let

$$J = \{t_0 \in [0,1] \mid h(t_0) = \overline{x}\}.$$

Since h is not necessarily injective, J might be a large set. By the uniform continuity of h we can choose $\eta > 0$ such that $|h(s) - h(t)| < \varepsilon/2$ for all $s, t \in \mathbb{R}$ such that there exists $t_0 \in J$ with $|s - t_0|, |t - t_0| \leq \eta$. Define

$$A = \bigcup_{t_0 \in J} h([t_0 - \eta, t_0 + \eta]).$$

Clearly, $A \subseteq \partial G$ is connected. We will now prove that $x_n \in A$ for all n sufficiently large.

There exists $t_n \in [0,1]$ such that $x_n = h(t_n)$. Up to extracting a subsequence, we may assume that $t_n \to \bar{t} \in [0,1]$. By continuity $h(\bar{t}) = \bar{x}$, so $\bar{t} \in J$. But then $t_n \in [\bar{t} - \eta, \bar{t} + \eta]$ for all sufficiently large n and therefore $x_n \in A$.

The same applies to y_n so we have $x_n, y_n \in A$ and it is clear that $|x_n - a| < \varepsilon$ for all $a \in A$. This is a contradiction, so ∂G is locally connected.

Proof of Step 2. Let $\varepsilon > 0$. Since ∂G is locally connected, there exists $\delta_1 > 0$ such that if $x, y \in \partial G$ with $|x - y| < \delta_1$, then there exists a connected $A \subseteq \partial G$ with $x, y \in A$ and $|x - a| < \varepsilon/3$ for all $a \in A$. We may assume without loss that $\delta_1 < \varepsilon$.

Suppose now that $x \in \partial G$ and $y \in \mathbb{C} \setminus \overline{G}$ with $|x - y| < \delta_1/2$. Let $z \in \partial G$ be the point with $|z - y| = \min_{\zeta \in \partial G} |\zeta - y| < \delta_1/2$. Then $[z, y] \subseteq \mathbb{C} \setminus G$ and $\operatorname{dist}(x, [z, y]) < \delta_1$. By the definition of δ_1 there exists a connected $A_1 \subseteq \partial G$ such that $x, z \in A_1$ and $|x - a| < \varepsilon$ for all $a \in A_1$. We now take $A = A_1 \cup [z, y]$.

The same type of construction (with two line segments) works if $x, y \in \mathbb{C} \setminus \overline{G}$ with $\operatorname{dist}(x, \partial G), \operatorname{dist}(y, \partial G) < \delta_1/3$.

Finally, if $x \in \mathbb{C}_\infty \setminus \overline{G}$ with $\operatorname{dist}(x, \partial G) \geq \delta_1/3$, then clearly we can take $\delta = \delta_1/3$ and $A = B(x, \delta_1/3)$. This finishes the proof.

Proof of Step 3. Let $\{C_n\}_n$ be a prime end with base set $E = \bigcap_{n=1}^\infty \overline{G_1^n} \subseteq \partial G$ and image point w. By the proof of Theorem 17.36 we may assume that the length – and therefore the diameter – of C_n tends to zero with n. Let $\varepsilon > 0$. We will prove that

$$\limsup_n \operatorname{diam}(\overline{G_1^n}) < \varepsilon, \tag{17.31}$$

from which (17.30) follows immediately. Choose $\delta < \varepsilon$ corresponding to the given ε in the definition of local connectedness of $\mathbb{C}_\infty \setminus G$. Let $z_0 \in G$ be arbitrary. Choose n so large that $z_0 \notin \overline{G_1^n}$, $\operatorname{dist}(z_0, C_n) > \varepsilon$ and $\operatorname{diam}(C_n) < \delta$. Let $a_n, b_n \in \partial G$ be the endpoints of the cross cut C_n. Since $|a_n - b_n| < \delta$ there exists a connected $B_n \subseteq \mathbb{C}_\infty \setminus G$ with diameter less that ε and such that $a_n, b_n \in B_n$. By possibly taking the closure, we may assume that B_n is compact. Then $A_n := B_n \cup C_n \subseteq B(a_n, \varepsilon)$ is connected and compact. Let $z' \in G$ with $|a_n - z'| > \varepsilon$. We will prove that z_0 and z' can be connected by a continuous curve in G_2^n. This implies that $G_1^n \subseteq B[a_n, \varepsilon]$ and therefore (17.31).

Let Φ be a Möbius transform that sends z_0 to 0 and z' to ∞. Define $K_1 = \Phi(\mathbb{C}_\infty \setminus G)$, $K_2 = \Phi(A_n)$. Then K_1, K_2 are compact and connected and $0, \infty$ can be connected by continuous curves γ_1 in $\mathbb{C}_\infty \setminus K_1$ and γ_2 in $\mathbb{C}_\infty \setminus K_2$.

Since $K_1 \cap K_2$ is connected there is a connected component Ω of $\mathbb{C}_\infty \setminus (\gamma_1^* \cap \gamma_2^*)$ with $K_1 \cap K_2 \subseteq \Omega$. Let $K_j' := K_j \cap \Omega^c$. Then the K_j' are compact and disjoint, so there exist open and disjoint V_j's with $K_j' \subseteq V_j \subseteq \mathbb{C}_\infty \setminus \gamma_j^*$. Define $U = V_1 \cup V_2 \cup \Omega$ and $K = K_1 \cup K_2$. We apply Lemma 17.37 with the given U, K and with $z_0 = \infty$ to get a piecewise smooth Jordan curve σ in U with K contained in the bounded component of $\mathbb{C} \setminus \sigma^*$. Notice that $\mathbb{C} \setminus \gamma_j^*$ is simply connected by Corollary 9.12, so there exist continuous logarithms f_j on $\mathbb{C} \setminus \gamma_j^*$. Since Ω is connected, we may assume that $f_1 = f_2$ on Ω, so they combine to define a continuous logarithm f on U. Clearly, $f'(z) = 1/z$, so

$$\mathrm{Ind}_\sigma(0) = \frac{1}{2\pi i} \int_\sigma \frac{1}{z}\, dz = 0. \tag{17.32}$$

This proves that 0 lies in the unbounded component of $\mathbb{C} \setminus \sigma^*$, in particular, 0 and ∞ can be connected by a continuous curve not touching K. $\square$

Lemma 17.37. *Let U be an open subset of $\mathbb{C}$ and let K be a compact, connected subset of U and $z_0 \in \mathbb{C} \setminus U$. Then there exists a piecewise smooth, closed Jordan curve γ in U such that K and z lie in different connected components of $\mathbb{C} \setminus \gamma^*$.*

Proof. We first prove that there exists a piecewise smooth, closed Jordan curve γ in U such that K is in the bounded component of $\mathbb{C} \setminus \gamma^*$. Let $\gamma_1, \ldots, \gamma_n$ be the piecewise linear, closed paths constructed in Lemma 9.4. By Remark 9.5 these paths can be chosen as Jordan curves. We will prove that one of these curves can be used. Let $z_0 \in K$. By Theorem 15.16 and the connectedness of K it suffices to prove that there exists a j such that $\mathrm{Ind}_{\gamma_j}(z_0) \neq 0$. But Lemma 9.4 applied with $f(z) = 1$ gives

$$1 = \sum_{j=1}^n \mathrm{Ind}_{\gamma_j}(z_0).$$

So $\mathrm{Ind}_{\gamma_j}(z_0) \neq 0$ for at least one j.

To prove the lemma, we can now use the Möbius transformation $g(z) = (z - z_0)^{-1}$ which maps z_0 to ∞. Then $g(K)$ will be a compact and $g(U)$ an open subset of $\mathbb{C}$. At this point we can apply the first half of the proof, and go back again with the Möbius transformation to conclude the lemma. $\square$

Exercises

17.1. Let $G = B(0,1)$. Prove that if (z_0, γ) and (z_0, γ') are two accessible boundary points with the same end point, then $(z_0, \gamma) \sim (z_0, \gamma')$. Conclude that the accessible boundary points $\partial_A B(0,1)$ can be identified with $\partial B(0,1)$.

17.2. Consider a prime end defined via a biholomorphic map f between G and $B(0,1)$. Let f' be another biholomorphic map between G and $B(0,1)$. Use Example 8.14 to conclude that $f' \circ f^{-1}$ is a Möbius transformation which leaves $B[0,1]$ invariant. Use this fact to conclude that the notion of prime end is independent of the choice of biholomorphic function f involved in the definition. What can one say about the corresponding image points of the prime end?

17.3. Assume (a) and (c) of Definition 17.30. Prove that then $\bigcap_{n=1}^{\infty} \overline{\Delta}_n \subseteq \partial B(0,1)$.

Hint: If $w \in (\bigcap_{n=1}^{\infty} \overline{\Delta}_n) \cap B(0,1)$, then $z = f^{-1}(w) \in G$. Consider the decreasing sequence of compact sets $\overline{g_1^n} \cap \partial B(z, \varepsilon)$.

17.4. Prove (17.26).

17.5. Let $G = B(0,1) \setminus [0,1]$ be the slit disc.

(α) Prove that $z = \frac{1}{2}$ is not a simple boundary point.

Hint: Let $z_n = \frac{1}{2} + (-1)^n \frac{1}{n}$. Let γ be a continuous curve in G with $\gamma(t_n) = z_n$, $t_n < t_{n+1}$. Consider the intersections of γ with the real axis.

(β) Prove that 0 is a simple boundary point.

Hint: Construct γ by connecting consecutive points by say circular arcs followed by radial lines.

(γ) Conclude that the set S of simple boundary points of G satisfies

$$S = \{0\} \cup \{e^{i\theta} \mid \theta \in \,]0, 2\pi[\}.$$

17.6. Let G be defined by (see Fig. 17.3)

$$G = \{z \in \mathbb{C} \mid \operatorname{Re} z, \operatorname{Im} z \in \,]-1, 1[\}$$

$$\setminus \left([0, \tfrac{1}{2}] \cup [0, i] \cup \bigcup_{n=1}^{\infty} [\tfrac{1}{2n}, \tfrac{1}{2n} + \tfrac{3}{4}i] \cup [\tfrac{1}{2n+1} + \tfrac{1}{4}i, \tfrac{1}{2n+1} + i] \right).$$

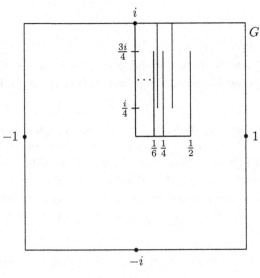

Fig. 17.3

Then any boundary point $z_0 \in [0, i]$ of G is "simple" in the sense that there exists a unique equivalence class $[z_0, \gamma] \in \partial_A G$ with z_0 as end-point. However, prove that if $z_0 \in [0, i]$, then z_0 is not a simple boundary point in the sense of Definition 15.18. Compare this with Proposition 17.21.

17.7. Let f be a continuous bijection between the compact metric spaces X and Y. Prove that $f^{-1} \colon Y \to X$ is continuous.

17.8. Suppose that $\Omega \subseteq \mathbb{C}$ is open and has the property that any closed Jordan curve γ with $\gamma^* \subseteq \Omega$ is null-homotopic in Ω.

(α) Let γ be a closed Jordan curve with $\gamma^* \subseteq \Omega$ and let $w \in \mathbb{C} \setminus \Omega$. Prove that $\mathrm{Ind}_\gamma(w) = 0$.

(β) Let σ be a closed curve with $\sigma^* \subseteq \Omega$. Show that there exists a closed Jordan curve γ with $\gamma^* \subseteq \Omega$ and such that σ^* is contained in the bounded component of $\mathbb{C} \setminus \gamma^*$.

    Hint: Use Lemma 17.37.

(γ) Show that Ω is simply connected.

    Hint: Use that a Jordan region is simply connected.

17.9. Let γ and σ be closed Jordan curves in $\mathbb{C}$ and let G and G' be the bounded connected components of $\mathbb{C} \setminus \gamma^*$ and $\mathbb{C} \setminus \sigma^*$. Prove that there exists a biholomorphic map $f: G \to G'$ and that f extends to a homeomorphism from $\overline{G}$ to $\overline{G'}$.

17.10. Suppose that $\gamma: [0, 1[\to \mathbb{C}$ is a simple continuous curve with $\gamma(0) = 0$, $\lim_{t \to \infty} |\gamma(t)| = \infty$. Define $D := \mathbb{C} \setminus \gamma^*$ and let $g: B(0, 1) \to D$ be a biholomorphic map. The aim of this exercise is to prove that g extends to a continuous map from $B[0, 1]$ to $\mathbb{C}_\infty$.

(α) Write $r(t) = |\gamma(t)|$. Show that there exists a continuous map $\theta:]0, 1[\to \mathbb{R}$ such that $\gamma(t) = r(t) e^{i\theta(t)}$.

(β) Define the following curves in $\mathbb{C}_\infty$:

$$\sigma_1(t) = \begin{cases} 0, & t = 0, \\ \sqrt{r(t)} \exp(i\theta(t)/2), & 0 < t < 1, \\ \infty, & t = 1, \end{cases}$$

$$\sigma_2(t) = \begin{cases} 0, & t = 0, \\ \sqrt{r(t)} \exp(i[\theta(t) + 2\pi]/2), & 0 < t < 1, \\ \infty, & t = 1, \end{cases}$$

and

$$\sigma(t) = \begin{cases} \sigma_1(t), & 0 \le t \le 1, \\ \sigma_2(2 - t), & 1 < t \le 2. \end{cases}$$

Prove that σ is a closed Jordan curve in $\mathbb{C}_\infty$.

(γ) Prove that $\mathbb{C}_\infty \setminus \sigma^* \ne \emptyset$.

Hint: The first lines of the proof of Theorem 15.24 are useful.

Prove furthermore that $\mathbb{C} \setminus \emptyset = U_1 \cup U_2$, where U_1, U_2 are disjoint, simply connected regions.

(δ) Let ψ be a holomorphic square root on D and let $w_0 \in D$. Assume that $\psi(w_0) \in U_1$. Prove that ψ maps D biholomorphically on U_1 (and $-\psi$ maps D biholomorphically on U_2).

(ε) Let $\tilde{g}$ be a biholomorphic map from $B(0, 1)$ to U_1. Prove that $\tilde{g}$ extends to a continuous map from $B[0, 1]$ with values in $\mathbb{C}_\infty$. Conclude.

Chapter 18

Carathéodory Convergence

Suppose that $F_n \colon B(0,1) \to \mathbb{C}$ are univalent with $\Omega_n = F_n(B(0,1))$. It is then natural to ask what geometric consequences will follow for the sets Ω_n is one supposes local uniform convergence of the sequence of (Riemann) maps $\{F_n\}$. These considerations result in the concept of Carathéodory Convergence.

18.1 The area theorem and consequences

The subject of univalent functions is very rich. We will only give a few results here. We refer to Conway (1995) for more details.

We begin with the Area Theorem, originally due to Gronwall (1914).

Theorem 18.1 (Area Theorem). *Suppose F is a univalent function on $\mathbb{C} \setminus B[0,1]$ with Laurent series*

$$F(z) = z + b_0 + \sum_{n=1}^{\infty} b_n z^{-n}. \tag{18.1}$$

Let K be the complement of the image of F. Then

$$\mathrm{Area}(K) = \pi(1 - \sum_{n=1}^{\infty} n|b_n|^2).$$

In particular,

$$\sum_{n=1}^{\infty} n|b_n|^2 \leq 1.$$

Proof. For $1 < r$, let $C_r = \partial B[0,r]$ be the circle of radius r with usual parametrization. Then $\gamma_r := F \circ C_r$ is a closed Jordan curve in $\mathbb{C}$. Let K_r

be the bounded component of $\mathbb{C} \setminus \gamma_r$. Using Green's theorem, we compute

$$\text{Area}(K_r) = \int_{K_r} 1 \, dx dy = \frac{1}{2i} \int_{\gamma_r} \bar{z} \, dz$$

$$= \frac{1}{2} \int_0^{2\pi} \overline{F(re^{it})} re^{it} F'(re^{it}) \, dt. \tag{18.2}$$

But

$$\overline{F(z)} z F'(z) = \left(\bar{z} + \bar{b}_0 + \sum_{k=1}^{\infty} \frac{\bar{b}_k}{\bar{z}^k} \right) \left(z - \sum_{k=1}^{\infty} \frac{k b_k}{z^k} \right)$$

$$= |z|^2 - \sum_{k=1}^{\infty} \frac{k |b_k|^2}{|z|^{2k}} + \sum_{\substack{k-\ell \neq 0 \\ k,\ell \geq 0}} \frac{c_{k,\ell}}{z^k \bar{z}^\ell},$$

where $c_{k,\ell}$ are suitable complex coefficients, and both sums are absolutely and uniformly convergent, for z with $|z| \geq r$. Inserting into (18.2), we find that

$$\text{Area}(K_r) = \pi \left(r^2 - \sum_{k=1}^{\infty} k |b_k|^2 r^{-2k} \right).$$

Taking the limit $r \to 1$ on both sides, using monotone convergence, yields Theorem 18.1. □

Lemma 18.2. *Let f be a univalent function on $B(0,1)$ with $f(0) = 0$ and $f'(0) = 1$. Then there exists an odd univalent function h with $h(0) = 0$, $h'(0) = 1$ and $h(z)^2 = f(z^2)$ for all $z \in B(0,1)$.*

Proof. Consider the function $\phi(z) = f(z)/z$. Clearly ϕ extends to a holomorphic function on $B(0,1)$ with $\phi(0) = 1$. Since f was injective, ϕ has no zeroes, so we can find a holomorphic square root $g(z)$ of $\phi(z)$ for $z \in B(0,1)$ with $g(0) = 1$. Define $h(z) = zg(z^2)$. Then it is immediate to check that h is holomorphic, odd and satisfies the required identities, in particular

$$h(z)^2 = f(z^2). \tag{18.3}$$

We only verify that h is injective. Suppose $h(z_1) = h(z_2)$. By (18.3) it is clear that $z_1 = \pm z_2$. But if $z_1 = -z_2$, then $h(z_1) = z_1 g(z_1^2) = -z_2 g(z_2^2) = -h(z_2)$. □

Theorem 18.3. *Let f be a univalent function on $B(0,1)$ with $f(0) = 0$ and $f'(0) = 1$. Write the power series of f as $f(z) = z + \sum_{n=2}^{\infty} a_n z^n$. Then $|a_2| \leq 2$.*

This result is due to Bieberbach (1916) and formed the starting point for the celebrated Bieberbach conjecture – now de Brange's theorem – that $|a_n| \leq n$. See de Branges (1985).

Proof. Let h be the odd function constructed in Lemma 18.2 with power series

$$h(z) = z + \sum_{\substack{n=3 \\ n \text{ odd}}} a'_n z^n \tag{18.4}$$

Since $h(z)^2 = f(z^2)$, we get that $a'_3 = a_2/2$. Let now $g(z) = \frac{1}{h(1/z)}$ on $\mathbb{C} \setminus B[0,1]$. Then g is univalent and odd with

$$\lim_{z \to \infty} \frac{g(z)}{z} = \lim_{w \to 0} \frac{w}{h(w)} = 1.$$

So we can apply the Area Theorem (Theorem 18.1), to g. Notice that g will have the Laurent series

$$g(z) = z + \sum_{\substack{n=1 \\ n \text{ odd}}} b_n z^{-n}, \tag{18.5}$$

with $b_1 = -a'_3 = -a_2/2$. So the Area Theorem gives

$$|a_2|^2/4 = |b_1|^2 \leq \sum_{n=1}^{\infty} n|b_n|^2 \leq 1.$$

$\square$

Theorem 18.4 (Koebe's 1/4-Theorem). *Let f be a univalent function on $B(0,1)$ with $f(0) = 0$ and $f'(0) = 1$. Then $B(0,1/4) \subseteq f(B(0,1))$.*

Proof. Let $f(z) = z + \sum_{n=2}^{\infty} a_n z^n$ be the power series of f. By Theorem 18.3 $|a_2| \leq 2$. Let $z_0 \in \mathbb{C} \setminus f(B(0,1))$. Define

$$g(z) = \frac{z_0 f(z)}{z_0 - f(z)}.$$

Then g is clearly a univalent map (being the composition of f by a Möbius transform) by the choice of z_0. A direct calculation gives that the power series of g starts like

$$g(z) = z + (a_2 + z_0^{-1})z^2 + \cdots$$

Theorem 18.3 applied to g gives

$$|a_2 + z_0^{-1}| \le 2,$$

so

$$|z_0^{-1}| \le |a_2| + |a_2 + z_0^{-1}| \le 4.$$

Since $z_0 \in \mathbb{C} \setminus f(B(0,1))$ was arbitrary, this finishes the proof. $\qquad\square$

Theorem 18.5. *Let f be a univalent function on $B(0,1)$ with $f(0) = 0$ and $f'(0) = 1$. Then*

$$\left| \frac{z f''(z)}{f'(z)} - \frac{2|z|^2}{1 - |z|^2} \right| \le \frac{4|z|}{1 - |z|^2}, \tag{18.6}$$

$$\frac{1 - |z|}{(1 + |z|)^3} \le |f'(z)| \le \frac{1 + |z|}{(1 - |z|)^3}, \tag{18.7}$$

and

$$\frac{|z|}{(1 + |z|)^2} \le |f(z)| \le \frac{|z|}{(1 - |z|)^2}, \tag{18.8}$$

for all $z \in B(0,1)$.

The result (18.7) goes under the name the *Distortion Theorem* and (18.8) is known as the *Growth Theorem*.

Proof. We prove the three estimates by calculations.

Proof of (18.6). Fix $z_0 \in B(0,1)$. Let $A(-z_0, \cdot)$ be the Möbius transformation from Example 8.14. Define g by

$$g(w) = \frac{f(A(-z_0, w)) - f(z_0)}{f'(z_0)(1 - |z_0|^2)}. \tag{18.9}$$

Then $g(0) = 0$, $g'(0) = 1$ and g is univalent on $B(0,1)$. The second term a_2 in the power series for g becomes

$$\frac{f''(z_0)(1 - |z_0|^2)}{2f'(z_0)} - \overline{z_0}$$

by direct calculation, so Theorem 18.3 yields

$$\left| \frac{f''(z_0)(1 - |z_0|^2)}{2f'(z_0)} - \overline{z_0} \right| \le 2.$$

This finishes the proof of inequality (18.6).

Proof of (18.7). Since f is univalent, f' never vanishes, so there exists a holomorphic function h on $B(0,1)$ with $h(0) = 0$ and $\exp(h(z)) = f'(z)$. In particular,

$$\operatorname{Re} h(z) = \log|f'(z)|. \tag{18.10}$$

By differentiation $h'(z) = \frac{f''(z)}{f'(z)}$, in particular, writing $z = re^{-i\theta}$,

$$r\frac{\partial}{\partial r} \operatorname{Re} h(z) = \operatorname{Re} \frac{z f''(z)}{f'(z)}. \tag{18.11}$$

So by (18.6),

$$\frac{2r^2 - 4r}{1 - r^2} \le r\frac{\partial}{\partial r} \log|f'(re^{i\theta})| \le \frac{4r + 2r^2}{1 - r^2}. \tag{18.12}$$

The estimate (18.7) follows (for $z = Re^{i\theta}$) by dividing by r and integrating from 0 to R.

Proof of (18.8). The upper bound follows directly from (18.7) by (radial) integration, so we only consider the lower bound. By the trivial inequality,

$$(1 + |z|)^2 = 1 + 2|z| + |z|^2 \ge 2|z| + 2|z|,$$

we can restrict to the case $|f(z)| < 1/4$. By replacing f by $e^{-i\theta} f(e^{i\theta} z)$, we may assume that $x := f(z) \in [0, 1/4[$. By Koebe's 1/4-theorem, Theorem 18.4, $[0, x] \subseteq f(B(0,1))$, so we can define $\gamma(t) = f^{-1}(t)$, for $t \in [0, x]$. Then

$$f(z) = x = \int_0^x 1\, dt = \int_0^x |f'(\gamma(t))||\gamma'(t)|\, dt \tag{18.13}$$

Notice that $|\gamma(t)|$ is differentiable for $t > 0$ and

$$\frac{d}{dt}|\gamma(t)| = |\gamma(t)|^{-1} \operatorname{Re}(\overline{\gamma(t)}\gamma'(t)) \le |\gamma'(t)| \tag{18.14}$$

Inserting this, and using (18.7), we get

$$\begin{aligned}
f(z) &\ge \int_0^x \frac{1 - |\gamma(t)|}{(1 + |\gamma(t)|)^3} \frac{d}{dt}|\gamma(t)|\, dt \\
&= \int_0^x \frac{d}{dt} \frac{|\gamma(t)|}{(1 + |\gamma(t)|)^2}\, dt,
\end{aligned} \tag{18.15}$$

which yields (18.8). $\square$

18.2   Carathéodory convergence

We start by defining the notion of Carathéodory convergence of sets. We consider a sequence $\{D_n\}_n$ where each $D_n \subseteq \mathbb{C}_\infty$ is open and $P \in D_n$ for all n.

Given the sequence $\{D_n\}$ and the base point P we can define a class of sets

$$\mathcal{K} = \big\{U \subseteq \mathbb{C}_\infty \mid U \text{ is open and connected with } P \in U \text{ and any}$$
$$\text{compact subset } C \subseteq U \text{ satisfies that there} \qquad (18.16)$$
$$\text{exists } N \geq 0 \colon C \subseteq D_n, \text{ for all } n \geq N\big\}.$$

Definition 18.6. *Let $\{D_n\}_n$ be a sequence of sets where each $D_n \subseteq \mathbb{C}_\infty$ is open and $P \in D_n$ for all n. The kernel $D = \ker(\{D_n\}_n, P)$ of the sequence with respect to the base point P is defined as follows.*

(a) *If P is not an interior point of $\bigcap_n D_n$ then $D = \{P\}$.*

(b) *If P is an interior point of $\bigcap_n D_n$ then*

$$D = \bigcup_{U \in \mathcal{K}} U. \qquad (18.17)$$

Remark 18.7. One can also characterize $\ker(\{D_n\}_n, P)$ as being the largest open, connected set O with $P \in O$ and such that if $C \subseteq O$ is compact, then $C \subseteq D_n$ for all but finitely many n. If no such open connected set exists, then $\ker(\{D_n\}_n, P) = \{P\}$. See also Exercise 18.3.

Definition 18.8 (Carathéodory Convergence). *The sequence of open sets $\{D_n\}_n$ converges to D in the sense of Carathéodory, with base point P if any subsequence of $\{D_n\}_n$ has D as its kernel with base point P. We write this as $D_n \to D$ with base point P.*

Example 18.9. Suppose that $\{D_n\}_n$ are open and connected with $P \in D_n$ for all n and $D_1 \subseteq D_2 \subseteq \cdots$. Then $D_n \to D$ in the sense of Carathéodory with base point P, where $D = \cup_{n=1}^{\infty} D_n$.

This can be seen as follows: Clearly D is open and connected with $P \in D$. Let $C \subseteq D$ be compact. Then $\{D_n\}_n$ is an open covering of C, so by compactness and monotonicity of the D_n's there exists $N \in \mathbb{N}$ such that $C \subseteq D_n$ for all $n \geq N$. This proves that $D \subseteq \ker(\{D_n\}_n, P)$ and the opposite inclusion is trivial. Furthermore, the same argument applies to any subsequence, which provides the convergence in the sense of Carathéodory.

Proposition 18.10. *Let* $f\colon \mathbb{C}_\infty \to \mathbb{C}_\infty$ *be a homeomorphism. Let* $\{\Omega_n\}_{n\geq 1}$ *be open sets containing the point* P *and let* $P \in \Omega \subseteq \mathbb{C}_\infty$. *Then*

$$\Omega_n \to \Omega, \qquad \text{with base point } P,$$

if and only if

$$f(\Omega_n) \to f(\Omega), \qquad \text{with base point } f(P).$$

The proof is left to the reader (Exercise 18.4).

Lemma 18.11. *Let* $\Omega \subseteq \mathbb{C}$ *be open and* $F, F_n\colon \Omega \to \mathbb{C}$ *univalent functions with* $F_n \to F$ *locally uniformly in* Ω. *Let* $a \in \Omega$. *There exists an open set* $\mathcal{O} \subseteq \mathbb{C}$ *and* $N \in \mathbb{N}$, *such that* $F(a) \in \mathcal{O} \subseteq F(\Omega)$, *and* $\mathcal{O} \subseteq F_n(\Omega)$, *for all* $n \geq N$.

Proof. Fix $a \in \Omega$ and and define $\zeta = F'(a)$ and $\zeta_n = F_n'(a)$, for all $n \geq 1$. Note that ζ and the ζ_n's are nonzero, since F and the F_n's are injective. The locally uniform convergence ensures, by Weierstrass (Theorem 4.10) that $\zeta_n \to \zeta$.

Let $R > 0$ be such that $B(a, R) \subseteq \Omega$. Define

$$H(z) = (R\zeta)^{-1}\bigl(F(Rz + a) - F(a)\bigr)$$
$$H_n(z) = (R\zeta_n)^{-1}\bigl(F_n(Rz + a) - F_n(a)\bigr).$$

Then H and the H_n's are holomorphic in $B(0, 1)$, maps zero to zero, and has derivative equal to 1 at zero. By Koebe's 1/4-theorem (Theorem 18.4), $B(0, 1/4) \subseteq H(\Omega)$ and $B(0, 1/4) \subseteq H_n(\Omega)$, for all $n \in \mathbb{N}$. Consequently,

$$B\bigl(F(a), R|\zeta|/4\bigr) \subseteq F(\Omega) \quad \text{and} \quad B\bigl(F_n(a), R|\zeta_n|/4\bigr) \subseteq F_n(\Omega).$$

The claim now follows, since $F_n(a) \to F(a)$ and $\zeta_n \to \zeta \neq 0$. $\square$

Theorem 18.12 (Carathéodory Convergence Theorem). *Let* $\{D_n\}_n$ *be a sequence of open, connected and simply connected sets with* $0 \in D_n$ *and* $D_n \subsetneq \mathbb{C}$ *for all* n. *Let* $f_n\colon B(0, 1) \to D_n$ *be the unique biholomorphic map with*

$$f_n(0) = 0, \quad f_n'(0) > 0, \qquad \text{for all } n.$$

Then the following two conditions are equivalent

(a) *The functions $\{f_n\}_n$ converge locally uniformly on $B(0,1)$ to a holomorphic function f.*

(b) *The sets $\{D_n\}_n$ converge in the sense of Carathéodory with base point 0 to a set D with $D \neq \mathbb{C}$.*

In either case we also have the following two conclusions.

(c) *$f \equiv 0$ if and only if $D = \{0\}$.*

(d) *If f is not constant, then f is biholomorphic between $B(0,1)$ and D and $f_n^{-1} \to f^{-1}$ locally uniformly on D.*

Furthermore, we also have

(e) *$f_n'(0) \to \infty$ if and only if $D_n \to \mathbb{C}$ in the sense of Carathéodory.*

Remark 18.13.

(A) Theorem 18.12 is stated in the case of base point $P = 0$. Of course, one can obtain a similar result with general base point by composing with a fixed Möbius transform mapping the base point P to 0.

(B) The local uniform convergence of the inverses f_n^{-1} in Theorem 18.12 deserves some clarification since they are not defined on the same set. However, by definition of the kernel D, any compact $C \subseteq D$ will be contained in D_n for all but finitely many n. Therefore, the tail of the sequence f_n^{-1} is defined on C and, possibly discarding finitely many n, one can ask whether $\sup_{w \in C} |f_n^{-1}(w) - f^{-1}(w)| \to 0$.

Proof. We first assume that (a) holds. By Lemma 17.1 the limiting function f is either a constant – in which case it is necessarily 0 – or f is univalent.

Case 1, $f \equiv 0$.
We will prove that in this case $D = \{0\}$. Suppose for contradiction that 0 is an interior point of $\bigcap_n D_n = \bigcap_n f_n(B(0,1))$. Then there exists a $\rho > 0$ such that $B(0,\rho) \subseteq D_n$ for all n. Let $\phi_n = f_n^{-1}$ defined on D_n. In particular, for all n, ϕ_n is defined on $B(0,\rho)$. By definition, $\phi_n(0) = 0, |\phi_n(z)| \leq 1$ for all $z \in B(0,\rho)$. Therefore, by Schwarz' lemma (Theorem 4.17),

$$|\phi_n'(0)| \leq 1/\rho.$$

But since for all n,

$$1 = |f_n'(0)||\phi_n'(0)|,$$

this is a contradiction to $|f'_n(0)| \to 0$ (by local uniform convergence). So the kernel $D = \{0\}$. The same argument applies to any subsequence, so we conclude that

$$D_n \to \{0\},$$

in the sense of Carathéodory.

Case 2, f not constant.
In this case f maps $B(0,1)$ biholomorphically to $G = f(B(0,1))$. We start by observing that $G \subseteq D = \ker(\{D_n\}_n, 0)$: Clearly, $0 \in G$ and G is open and connected. Let $C \subseteq G$ be compact and $c \in C$. By Lemma 18.11 there exists an open set $\mathcal{O}_c$ with $c \in \mathcal{O}_c \subseteq G$ such that $\mathcal{O}_c \subseteq D_n$ for all but finitely many n. Using the compactness of C on the open covering $\{\mathcal{O}_c\}_{c \in C}$ we find that $C \subseteq D_n$ for all but finitely many n.

We will now prove that $G = D$. Let $\{K_j\}_j$ be compact subsets of D as given by Proposition 9.10. Fix $j \geq 1$. Clearly, $\phi_n = f_n^{-1}$ is defined on $\operatorname{int} K_j$ for all n sufficiently large. By Montel's theorem (Theorem 8.15), there is a subsequence $\{\phi_{n_k}\}_k$ that converges locally uniformly to a holomorphic function defined on $\operatorname{int} K_j$. Since this can be done for any j, we get by a diagonal sequence argument the existence of ϕ, holomorphic on D and a subsequence, still denoted by $\{\phi_{n_k}\}_k$, that converges locally uniformly to ϕ. Since

$$0 < \frac{1}{f'(0)} = \lim_{k \to \infty} \frac{f'_{n_k}(0)\phi'_{n_k}(0)}{f'(0)} = \phi'(0), \tag{18.18}$$

we get from Lemma 17.1 that ϕ is univalent on D.

Let $z \in B(0,1)$ and define $w = f(z)$, $w_k = f_{n_k}(z)$. Choose a small disc $B[w, \varepsilon] \subseteq D$. Then $w_k \to w$ as $k \to \infty$, so the tail of the sequence is contained in $B[w, \varepsilon]$. Then

$$|\phi(w) - z| \leq |\phi(w) - \phi(w_k)| + |\phi(w_k) - \phi_{n_k}(w_k)| \to 0,$$

as $k \to \infty$ by the uniform convergence on $B[w, \varepsilon]$. This proves that $\phi = f^{-1}$ on $G = f(B(0,1)) \subseteq D$. Combining the convergence and the Open Mapping Theorem (Theorem 4.14), we get that $\phi(D) \subseteq B(0,1)$. Since ϕ is univalent we conclude that $G = D$.

The above argument applies not only to the entire sequence $\{\phi_n\}_n$ but to any subsequence. Since the kernel is always the same, this implies that $f_n(B(0,1)) \to f(B(0,1))$ in the sense of Carathéodory. Since the limiting function $\phi = f^{-1}$ is always the same, this proves that $\{\phi_n\}_n$ converges locally uniformly to f^{-1}.

We now prove the other direction, so we assume that $D_n \to D$ in the sense of Carathéodory.

Case 1, $D = \{0\}$.
We start by proving that in this case $f_n'(0) \to 0$. Suppose that is not true. Then there exists $\varepsilon > 0$ such that $f_{n_k}'(0) \geq \varepsilon$ along a subsequence. By Koebe's 1/4-theorem (Theorem 18.4), we have

$$B\big(0, \varepsilon/4\big) \subseteq B\big(0, f_{n_k}'(0)/4\big) \subseteq f_{n_k}\big(B(0,1)\big) = D_{n_k}.$$

So we conclude that $\ker(\{D_{n_k}\}_k) \neq \{0\}$, which is a contradiction. So $f_n'(0) \to 0$. By Theorem 18.5, this implies that $f_n \to 0$ uniformly on compact subsets of $B(0,1)$.

Case 2, $D \neq \{0\}$, $D \neq \mathbb{C}$.
In this case we must have that $\{f_n'(0)\}_n$ is a bounded sequence, since if not Koebe's 1/4-theorem would give that $D = \mathbb{C}$. The boundedness of $\{f_n'(0)\}$ gives by Theorem 18.5 that $\{f_n\}_n$ is uniformly bounded on compact subsets of $B(0,1)$. By Montel's theorem (Theorem 8.15), there exists a locally uniformly convergent subsequence $\{f_{n_k}\}_k$. By the first half of the proof (applied to the subsequence) the subsequence converges to f, the biholomorphic map between $B(0,1)$ and D with $f(0) = 0$, $f'(0) > 0$. Since this result can be applied to any subsequence of the original sequence we must have the local uniform convergence of the entire sequence $\{f_n\}_n$ to f.

The argument in the beginning of this paragraph also shows that $f_n'(0) \to \infty$ implies that $D_n \to \mathbb{C}$ in the sense of Carathéodory, i.e., one half of (e). On the other hand, if $D_n \to \mathbb{C}$ in the sense of Carathéodory. Then, for any $R > 0$, $B[0, R] \subseteq f_n(B(0,1))$, for all n sufficiently large. So the Schwarz lemma (Theorem 4.17), shows that $(f_n^{-1})'(0) \to 0$. This finishes the proof. $\square$

Exercises

18.1. Let $C \in \mathbb{R}$ and $F \colon \mathbb{C} \setminus B[0,1] \to \mathbb{C}$ be a univalent function with Laurent expansion

$$F(z) = e^C z + b_0 + \sum_{n=1}^{\infty} b_n z^{-n}.$$

Show that

$$\forall z \in \mathbb{C} \setminus B[0,1]: \quad |F(z) - e^C z - b_0| \leq \frac{e^C}{\sqrt{|z|^2 - 1}}.$$

Hint: The Area Theorem.

18.2. Let $\gamma_n(t) = e^{it}$ defined on $t \in [0, 2\pi - 1/n]$. Let $D_n = \mathbb{C} \setminus \gamma_n^*$. Prove that $D_n \to B(0,1)$ in the sense of Carathéodory with base point 0. Prove that $D_n \to \mathbb{C} \setminus B[0,1]$ in the sense of Carathéodory with base point 2.

18.3. Suppose 0 is an interior point of $\bigcap_n D_n$. Prove that then $D = \ker(\{D_n\}_n, 0) \in \mathcal{K}$, with $\mathcal{K}$ from (18.16).
Hint: Let $C \subseteq D$ be compact and let $c \in C$. Then there exists $\rho_c > 0$ and $U_c \in \mathcal{K}$, such that $B[c, \rho_c] \subseteq U_c$. From this construction one gets an open covering of the compact set C, from which it is easy to proceed.

18.4. Prove Proposition 18.10.

18.5. Suppose that $\{D_n\}_{n=1}^\infty$ is a sequence of open sets, $P \in D_n$ for all n.

(α) Notice that $\ker(\{D_n\}_n, P) \subseteq \ker(\{D_{n_k}\}_k, P)$, for any subsequence $\{D_{n_k}\}_k$.

(β) Prove that $D_n \to D$ in the sense of Carathéodory with base point P if and only if $D_{n_k} \to D$ for any subsequence $\{D_{n_k}\}_k$.

(γ) Prove that $D_n \to D$ in the sense of Carathéodory with base point P if and only if the following is true: For any subsequence $\{D_{n_k}\}_k$ one can extract a subsequence $\{D_{n_{k_l}}\}_l$ with $D_{n_{k_l}} \to D$.
Hint: The case $D \subseteq \ker(\{D_n\}_n, P)$ is easy. In the case $\ker(\{D_n\}_n, P) \subsetneq D$, choose $z \in D \cap \partial \ker(\{D_n\}_n, P)$ (why does such a z exist?) and $r > 0$ such that $B[z, r] \subseteq D$. Choose $\{n_j\}_j$ such that $B[z, r] \not\subseteq D_{n_j}$ and use the assumption to reach a contradiction.

18.6. Let $f: B(0,1) \to \mathbb{C}$ be a univalent map with $f(0) = 0$ and $f'(0) = 1$. Let a_2 denote the coefficient of z^2 in the power series expansion $f(z) = z + \sum_{n=2}^\infty a_n z^n$. If $|a_2| = 2$, argue that f must be the Koebe function $z \mapsto \frac{z}{(1-z)^2}$.

18.7. Let $\gamma_1^*, \ldots, \gamma_n^* \subseteq \mathbb{C}$ be a collection of Jordan arcs. Let f be univalent on $B(0,1)$ with $f(0) = 0$, $f'(0) = 1$. If $f(B(0,1)) = \mathbb{C} \setminus (\gamma_1^* \cup \cdots \cup \gamma_n^*)$, then f is called a *slit-map*. In the case $n = 1$, f is called a *single-slit map*.

(α) Prove that γ_j^* is unbounded for at least one $j \in \{1, \ldots, n\}$.

The aim of this exercise is to prove the following

> **Theorem.** *Let f be univalent on $B(0,1)$ with $f(0) = 0$, $f'(0) = 1$. Then there exists a sequence of single-slit mappings $\{f_n\}_n$ such that $f_n \to f$ locally uniformly on $B(0,1)$.*

(β) Argue that it suffices to prove that for all $\varepsilon > 0$ and $\rho \in \ ]0,1[$ exists a single slit map g such that

$$|f(z) - g(z)| \leq \varepsilon, \qquad \text{for all } |z| < \rho.$$

(γ) Define, for $0 < r < 1$, $f_r(z) = r^{-1}f(rz)$. Prove that $f_r \to f$ uniformly on $B[0,\rho]$ as $r \nearrow 1$ for any $\rho < 1$.

    Hint: Write $f(z) = z + \sum_{n=2}^{\infty} a_n z^n$. Prove the estimate $|f(z) - f_r(z)| \leq (1-r)\sum_{n=2}^{\infty}|a_n|(n-1)|z|^n$. What is the radius of convergence of this new series?

(δ) For fixed $r < 1$, let $w_0 = f_r(1) = r^{-1}f(r)$. Construct a Jordan arc Γ_n^* which consists of

    (a) a Jordan arc $\Gamma^{(\infty)}$ from ∞ to w_0,

    (b) the Jordan arc $\Gamma_n^{(1)}$ parametrized as $f_r(e^{i\theta})$ on $\theta \in [0, 2\pi - 1/n]$.

Let $D_n = \mathbb{C} \setminus \Gamma_n^*$ and prove that

$$D_n \to D := f_r(B(0,1)),$$

    in the sense of Carathéodory.

(ε) Conclude.

One important application of this result is in the proof of the Bieberbach conjecture. Notice that if it holds that $|f^{(n)}(0)| \leq C$ for all single-slit mappings, then the same bound is true for all univalent f on $B(0,1)$ with $f(0) = 0$, $f'(0) = 1$.

Chapter 19

Capacities

In this chapter we introduce two types of hulls, compact hulls and half-plane hulls. We furthermore define capacities associated with such hulls and study their properties.

Throughout this chapter, we will make frequent use of the following observation. If $U, V \subseteq \mathbb{C}_\infty$ are open sets and $f \colon U \to V$ a homeomorphism. Then U is (simply) connected if and only if V is (simply) connected. We leave the proof in the capable hands of the reader.

19.1 Logarithmic capacity

In this section we introduce the notions of compact hulls and logarithmic capacity, and we study their properties.

Definition 19.1. *A subset $K \subseteq \mathbb{C}$ is called a **compact hull**, if*

(a) *K contains at least two points.*

(b) *K is compact.*

(c) *$\mathbb{C}_\infty \setminus K$ is connected and simply connected.*

Easy examples of compact hulls are compact line segments, closed discs, ellipses and regular polygons. The following two examples provide us with two important classes of compact hulls.

Example 19.2. As a candidate for a compact hull, we consider the closure $K = \overline{G}$ of a Jordan region G. Let γ be a closed Jordan curve with $\gamma^* = \partial G$. That G contains at least two points is obvious.

Let $G'_\infty = \mathbb{C}_\infty \setminus K$. Jordan's curve theorem, Theorem 15.8, implies that $G'_\infty = G_\infty \cup \{\infty\}$, where G_∞ is the unbounded connected component of $\mathbb{C} \setminus \gamma^*$. This shows that G'_∞ is a connected subset of $\mathbb{C}_\infty$.

To see that G'_∞ is also simply connected as a subset of $\mathbb{C}_\infty$, we argue as follows. Fix a $w \in G$ and define a Möbius transformation $m \colon \mathbb{C}_\infty \to \mathbb{C}_\infty$ by setting $m(z) = (z - w)^{-1}$. Then $\sigma = m \circ \gamma \colon [0, 1] \to \mathbb{C}$ is also a closed Jordan curve and the bounded connected component of $\mathbb{C} \setminus \sigma^*$ is exactly $m(G'_\infty)$. Hence, by Remark 15.11, the set $m(G'_\infty)$ is simply connected in $\mathbb{C}$ and consequently, G'_∞ is simply connected in $\mathbb{C}_\infty$. This shows that K is a compact hull.

Example 19.3. Let $a < b$ be two real numbers and $\gamma \colon [a, b] \to \mathbb{C}$ a simple curve, i.e., a Jordan arc. We claim that $K = \gamma^*$ is a compact hull. Clearly, K is compact and contains at least two points, and it follows from Corollary 15.5 that $\mathbb{C}_\infty \setminus \gamma^*$ is connected.

To see that $\mathbb{C}_\infty \setminus K$ is simply connected, we argue as follows. Define $\sigma \colon [a, b] \to \mathbb{C}_\infty$ by setting $\sigma(t) = (\gamma(t) - \gamma(b))^{-1}$. Let $\Omega = \mathbb{C} \setminus \sigma^*$, which is an open subset of $\mathbb{C}$. Since $\mathbb{C}_\infty \setminus \Omega = \sigma^*$ is (path) connected, we conclude from Corollary 9.12 that Ω is simply connected. Hence, since $m(z) = (z - \gamma(b))^{-1}$ is a Möbius transformation, we also find that $\mathbb{C}_\infty \setminus \gamma^* = m^{-1}(\Omega)$ is simply connected, as a subset of the Riemann sphere. Consequently, $K = \gamma^*$ is a compact hull.

Before continuing, we pause to introduce some notations and constructions connected with compact hulls. We write $\Omega = \mathbb{C} \setminus K$ and $\Omega_\infty = \Omega \cup \{\infty\}$. Note that both Ω and Ω_∞ are open and connected, but only Ω_∞ is simply connected (as a subset of the Riemann sphere). For a $z_0 \in K$, we write

$$\mathcal{O} = \mathcal{O}(z_0) = \{w \in \mathbb{C} \mid w^{-1} + z_0 \in \Omega_\infty\},$$

which is a simply connected region containing $w = 0$. Recall from Definition 8.19 that a region is an open, nonempty and connected subset of $\mathbb{C}$. Since K contains at least two points, we must have $\mathcal{O} \neq \mathbb{C}$. The Riemann map $f_\mathcal{O} \colon B(0, 1) \to \mathcal{O}$, normalized by $f_\mathcal{O}(0) = 0$ and $f'_\mathcal{O}(0) > 0$, therefore has its existence (and unicity) ensured by the Riemann Mapping Theorem, Theorem 8.21. Going back to Ω, we get a univalent bijection $F_K \colon \mathbb{C} \setminus B[0, 1] \to \Omega$, by the construction

$$F_K(z) = \frac{1}{f_\mathcal{O}(1/z)} + z_0. \tag{19.1}$$

This function satisfies that

$$\lim_{z \to \infty} \frac{F_K(z)}{z} = \lim_{w \to 0} \frac{w}{f_\mathcal{O}(w)} = \frac{1}{f_\mathcal{O}'(0)}. \qquad (19.2)$$

We introduce the notation

$$G_K = F_K^{-1} \colon \Omega \to \mathbb{C} \setminus B[0,1]$$

for the biholomorphic function inverse to F_K. Figure 19.1 illustrates the setup. The notation established here will be used throughout the section without reference.

The above discussion yields the existence part of the following proposition.

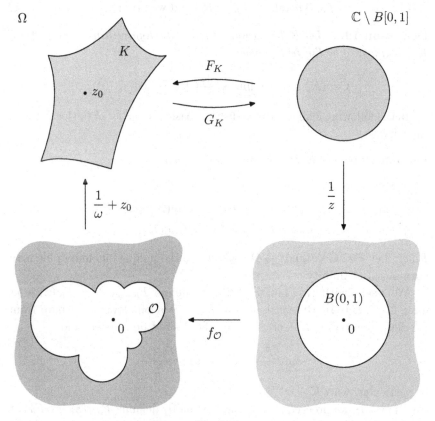

Fig. 19.1

Proposition 19.4. *Let K be a compact hull. There exists one and only one biholomorphic map $F_K \colon \mathbb{C} \setminus B[0,1] \to \Omega$, characterized by the property that the limit $\lim_{z \to \infty} F_K(z)/z$ exists and is (strictly) positive.*

Proof. We only have to establish uniqueness. Suppose $\widetilde{F}_K$ is another function, with the same properties as F_K. Then we can define $\tilde{f}_{\mathcal{O}} \colon B(0,1) \setminus \{0\} \to \mathcal{O} \setminus \{0\}$ by setting $\tilde{f}_{\mathcal{O}}(w) = 1/(\widetilde{F}_K(1/w) - z_0)$. By assumption, $\tilde{f}_{\mathcal{O}}$ is biholomorphic, with $\lim_{w \to 0} \tilde{f}_{\mathcal{O}}(w) = 0$. Consequently, 0 is a removable singularity. Putting $\tilde{f}_{\mathcal{O}}(0) = 0$ extends $\tilde{f}_{\mathcal{O}}$ to a biholomorphic map from $B(0,1)$ onto $\mathcal{O}$. Since $\tilde{f}_{\mathcal{O}}(w)/w = w^{-1}/(\widetilde{F}_K(w^{-1}) - z_0)$, for $w \in B(0,1) \setminus \{0\}$, we conclude using (19.2) that

$$\tilde{f}'_{\mathcal{O}}(0) = \left(\lim_{z \to \infty} \frac{\widetilde{F}_K(z)}{z} \right)^{-1} > 0,$$

and hence $\tilde{f}_{\mathcal{O}} = f_{\mathcal{O}}$. Therefore, $\widetilde{F}_K = F_K$ and we are done. $\qquad\square$

Definition 19.5. *Let K be a compact hull. The **logarithmic capacity** of K is defined to be the real number*

$$\mathrm{Cap}(K) = \mathrm{Log}\left(\lim_{z \to \infty} \frac{F_K(z)}{z} \right) = -\mathrm{Log}(f'_{\mathcal{O}}(0)).$$

In the following lemma we establish the basic properties of the logarithmic capacity.

Lemma 19.6. *Let K be a compact hull. Then*

(a) *For all $w \in \mathbb{C}$, we have $\mathrm{Cap}(K + w) = \mathrm{Cap}(K)$.*

(b) *For all $\alpha > 0$, we have $\mathrm{Cap}(\alpha K) = \mathrm{Cap}(K) + \mathrm{Log}(\alpha)$.*

(c) *For all $\theta \in \mathbb{R}$, we have $\mathrm{Cap}(e^{i\theta} K) = \mathrm{Cap}(K)$.*

Proof. Let $F_K \colon \mathbb{C} \setminus B[0,1] \to \Omega = \mathbb{C} \setminus K$ be the unique biholomorphic map normalized at infinity.

We begin with (a). Let $w \in \mathbb{C}$ and define $F_{K+w}(z) = F_K(z) + w$, for $z \in \mathbb{C} \setminus B[0,1]$. By construction, F_{K+w} is a biholomorphic map from $\mathbb{C} \setminus B[0,1]$ onto $\mathbb{C} \setminus (K + w)$. Clearly, the desired limit exists and

$$\lim_{z \to \infty} \frac{F_{K+w}(z)}{z} = \lim_{z \to \infty} \frac{F_K(z)}{z}.$$

In particular, $\mathrm{Cap}(K + w) = \mathrm{Cap}(K)$.

As for (b), we proceed in a similar fashion by defining $F_{\alpha K}(z) = \alpha F_K(z)$. Finally, to observe (c), we define $F_{e^{i\theta} K}(z) = e^{i\theta} F_K(e^{-i\theta} z)$. $\qquad\square$

Example 19.7. It is obvious that $K = B[0,1]$ is a compact hull with $F_K(z) = z$. Hence, $\mathrm{Cap}(B[0,1]) = 0$. Using the lemma above, we see that for any $w \in \mathbb{C}$ and $R > 0$, we have $\mathrm{Cap}(B[w,R]) = \mathrm{Log}(R)$.

Example 19.8. Let $K = [-4,0]$, which is a compact hull. The associated open simply connected set is $\mathcal{O} = \mathbb{C} \setminus]-\infty, -1/4]$. Here we chose $z_0 = 0 \in K$. The Riemann map $f_\mathcal{O} \colon B(0,1) \to \Omega$ is the Koebe function $f_\mathcal{O}(w) = f_{\mathrm{Koebe}}(w) = w/(1-w)^2$, which has $f'_{\mathrm{Koebe}}(0) = 1$. Consequently, $\mathrm{Cap}([-4,0]) = 0$ in this example as well. Note that we can compute

$$F_{[-4,0]}(z) = \frac{1}{f_{\mathrm{Koebe}}(z^{-1})} = \frac{(1-z^{-1})^2}{z^{-1}} = z - 2 + \frac{1}{z}.$$

If $K \subseteq \mathbb{C}$ is a compact line segment of length L, we conclude from Lemma 19.6 that $\mathrm{Cap}(K) = \mathrm{Log}(L/4)$.

Remark 19.9. Suppose K is a compact hull and $K = K^*$, that is; K is symmetric under reflection in the real axis. Then, since K^* is also a compact hull and $F_{K^*}(z) = \overline{F_K(\bar{z})}$, we must have $F_K(\bar{z}) = \overline{F_K(z)}$ for all $z \in \mathbb{C} \setminus K$. Consequently, $F_K(\mathbb{R} \setminus [-1,1]) = \mathbb{R} \setminus K$. We claim that this observation implies

$$F_K(\mathbb{H} \setminus B[0,1]) = \mathbb{H} \setminus K,$$

where $\mathbb{H} = \{z \in \mathbb{C} \mid \mathrm{Im}\, z > 0\}$ is the open upper half-plane. Indeed, let $z_1, z_2 \in F_K(\mathbb{H} \setminus B[0,1])$ and write $w_j = G_K(z_j) \in \mathbb{H} \setminus B[0,1]$. Denote by $\gamma \colon [0,1] \to \mathbb{C}$ a curve running from w_1 to w_2 with $\gamma^* \subseteq \mathbb{H} \setminus B[0,1]$. Since γ never meets the real line, neither does $\sigma = F_K \circ \gamma$, which implies that $\mathrm{Im}\, z_1$ and $\mathrm{Im}\, z_2$ has the same sign. To see that $\mathrm{Im}\, z_j > 0$, we just have to remark that the chosen normalization at infinity ensures that $\mathrm{Im}\, F_K(iy) > 0$, for $y > 0$ sufficiently large. We have now established that $F_K(\mathbb{H} \setminus B[0,1]) \subseteq \mathbb{H} \setminus K$. Equality now follows, since complex conjugation of this inclusion implies that $F_K(\mathbb{H}^* \setminus B[0,1]) \subseteq \mathbb{H}^* \setminus K$. Here we also used that $F_K(\mathbb{C} \setminus B[0,1]) = \mathbb{C} \setminus K$ and the notation $\mathbb{H}^* = \{z \in \mathbb{C} \mid \bar{z} \in \mathbb{H}\}$.

Theorem 19.10. *Let K_1 and K_2 be compact hulls with $K_1 \subseteq K_2$. Then $\mathrm{Cap}(K_1) \leq \mathrm{Cap}(K_2)$ with equality if and only if $K_1 = K_2$.*

Proof. By Lemma 19.6(a), we can without loss of generality assume that $0 \in K_1 \subseteq K_2$. The associated simply connected regions $\mathcal{O}_1$ and $\mathcal{O}_2$ are both defined with $z_0 = 0$. Write $f_j \colon B(0,1) \to \mathcal{O}_j$ for the associated normalized Riemann maps $f_{\mathcal{O}_j}$.

By assumption, the map $h = f_1^{-1} \circ f_2 \colon B(0,1) \to B(0,1)$ is a univalent function, with $0 \in \mathcal{U} := h(B(0,1))$. By Schwarz' lemma (Theorem 4.17), we must have $|h'(0)| \leq 1$. But

$$h'(0) = \frac{f_2'(0)}{f_1'(0)} = e^{\mathrm{Cap}(K_1) - \mathrm{Cap}(K_2)}.$$

This shows that $\mathrm{Cap}(K_1) \leq \mathrm{Cap}(K_2)$. If equality holds, we must have $h'(0) = 1$ and hence, again by Schwarz' lemma, $h(z) = z$. This implies that $f_1 = f_2$, and consequently; that $K_1 = K_2$. $\square$

It is in general a difficult task, if possible at all, to compute the logarithmic capacity of a compact hull. The following lemma will be of use in the exercises to this chapter and in the material to follow.

Lemma 19.11. *Let K_1 and K_2 be compact hulls. Suppose $f \colon \mathbb{C} \setminus K_1 \to \mathbb{C} \setminus K_2$ is biholomorphic and satisfies that $\limsup_{|z| \to \infty} |f(z) - z| < \infty$. Then $F_{K_2} = f \circ F_{K_1}$ and $\mathrm{Cap}(K_1) = \mathrm{Cap}(K_2)$.*

Proof. Let $F_{K_j} \colon \mathbb{C} \setminus B[0,1] \to \mathbb{C} \setminus K_j$, $j = 1, 2$, be the Riemann maps associated with K_1 and K_2. To see that $F_{K_2} = f \circ F_{K_1}$, we only have to verify that $f \circ F_{K_1}/z$ has a positive limit at infinity. See Proposition 19.4. We compute

$$\frac{f(F_{K_1}(z))}{z} = \frac{f(F_{K_1}(z)) - F_{K_1}(z)}{z} + \frac{F_{K_1}(z)}{z} \to e^{\mathrm{Cap}(K_1)},$$

where we used that $|F_{K_1}(z)| \to \infty$, as $|z| \to \infty$. This shows that $F_{K_2} = f \circ F_{K_1}$ and that $\mathrm{Cap}(K_2) = \mathrm{Cap}(K_1)$. $\square$

The logarithmic capacity and the radius of a compact hull are two closely related quantities, as clarified by the following proposition.

Proposition 19.12. *Let K be a compact hull and define*

$$\mathrm{rad}(K) = \tfrac{1}{2} \sup\{|z - w| \mid z, w \in K\}.$$

(a) *If $\mathrm{Cap}(K) = 0$, then $1 \leq \mathrm{rad}(K) \leq 4$, and $\mathrm{rad}(K) = 1$ if and only if K is a closed disc of radius one.*

(b) *More generally, for any compact hull K:*

$$e^{\mathrm{Cap}(K)} \leq \mathrm{rad}(K) \leq 4e^{\mathrm{Cap}(K)}. \tag{19.3}$$

Proof. The inequalities in (b) follow by applying the inequalities in (a) to the scaled compact hull $e^{-\text{Cap}(K)}K$. Hence, it suffices to prove (a).

Let K be a compact hull with $\text{Cap}(K) = 0$. Suppose towards a contradiction that $\text{rad}(K) < 1$. Hence, $K \subseteq B(a, 1)$ for some $a \in \mathbb{C}$. Recall that $\text{Cap}(B[a, 1]) = 0$. Using Theorem 19.10, we find the inequality $\text{Cap}(K) < \text{Cap}(B[a, 1]) = 0$. This contradicts the choice of K and we may therefore conclude that $\text{rad}(K) \geq 1$.

As for the other inequality, we may by Lemma 19.6(a) assume that $0 \in K$. Note that the associated Riemann map $f_{\mathcal{O}}$, where $\mathcal{O} = \{w \in \mathbb{C} \mid w^{-1} \in \mathbb{C}_\infty \setminus K\}$, satisfies that $f_{\mathcal{O}}(0) = 0$ and $f_{\mathcal{O}}'(0) = 1$. By Koebe's 1/4-theorem (Theorem 18.4), we find that $B(0, 1/4) \subseteq f_{\mathcal{O}}(B(0, 1))$. But this implies in turn that $\mathbb{C} \setminus B[0, 1/4] \subseteq F_K(\mathbb{C} \setminus B[0, 1])$. Therefore, we must have $K \subseteq B[0, 4]$, which concludes the proof of the inequality.

Assume now that $\text{Cap}(K) = 0$ and $\text{rad}(K) = 1$. Then $K \subseteq B[a, 1]$, for some a. Consequently, by Theorem 19.10, we must have $K = B[a, 1]$. This completes the proof of (a). $\square$

We end this section with bounds on G_K and F_K uniform in compact hulls contained in a disc of radius R.

Lemma 19.13. *Let $R > 0$. Then, for any compact hull $K \subseteq B[0, R]$, we have*

$$\forall z \in \mathbb{C} \setminus B(0, 1): \quad |F_K(z) - e^{\text{Cap}(K)}z| \leq 2R,$$

$$\forall z \in \mathbb{C} \setminus K: \quad |G_K(z) - e^{-\text{Cap}(K)}z| \leq \frac{4R}{\text{rad}(K)} + 4.$$

Proof. Let $K \subseteq B[0, R]$ be a compact hull.

Define for $w \in B(0, 1) \setminus \{0\}$:

$$f(w) = F_K(1/w) - e^{\text{Cap}(K)}/w.$$

Note that f is uniformly bounded and therefore 0 is a removable singularity (Theorem 4.3). The extension to $B(0, 1)$ is denoted by f as well. If $f = c$ is constant, we must have $F_K(z) = e^{\text{Cap}(K)}z + c$, which implies that $K = e^{\text{Cap}(K)}B[0, 1] + c$. Since $K \subseteq B[0, R]$, we must have $|c| \leq R$. This implies the claimed bound on F_K.

If f is not constant, then by the Maximum Modulus Principle, $|f|$ does not attain its supremum. Hence, $\sup_{w \in B(0,1)}|f(w)| = \limsup_{|w| \to 1_-}|f(w)|$. Since $\limsup_{|z| \to 1_+}|F_K(z)| = \sup_{z \in K}|z|$, we conclude that

$$\forall w \in B(0, 1): \quad |f(w)| \leq R + e^{\text{Cap}(K)} \leq R + \text{rad}(K),$$

where we used Proposition 19.12 to conclude the last inequality. This implies the claimed bound on F_K.

Let now $z = G_K(\zeta)$ for a $\zeta \in \mathbb{C} \setminus K$. Inserting z into the estimate just derived for F_K yields $|\zeta - e^{\mathrm{Cap}(K)} G_K(\zeta)| \leq R + \mathrm{rad}(K)$. Hence, by Proposition 19.12, we get the claimed bound for G_K. $\square$

19.2   Convergence of compact hulls

In the following we wish to allow for K's which are singletons, i.e., of the form $K = \{a\}$ where $a \in \mathbb{C}$. We assign to such a set the constant function $F_{\{a\}}(z) = a$ for $|z| > 1$, and the logarithmic capacity $\mathrm{Cap}(\{a\}) = -\infty$.

Recall from Definition 18.8 the notion of Carathéodory convergence of open sets.

Definition 19.14. *Let* $\{K_n\}_{n=1}^{\infty}$ *be a sequence of compact hulls or singletons. The sequence is said to converge in Carathéodory sense, if* $\{\mathbb{C}_{\infty} \setminus K_n\}_{n=1}^{\infty}$ *converges in Carathéodory sense, as subsets of* $\mathbb{C}_{\infty}$ *with base point* ∞. *If the sequence converges, we write* $K_n \to K$, *where* $K = \mathbb{C}_{\infty} \setminus \lim_{n \to \infty}(\mathbb{C}_{\infty} \setminus K_n)$.

Remark 19.15. Let us make three remarks:

(A) Suppose $\{K_n\}_{n=1}^{\infty}$ is a sequence of compact hulls or singletons converging in Carathéodory sense to a set K. It is an easy consequence of Theorem 18.12 that the limiting set $K \subseteq \mathbb{C}$ is either a singleton, a compact hull, or equal to $\mathbb{C}$.

(B) Suppose we have compact hulls or singletons K and K_n, with $K_n \to K$ in Carathéodory sense. Since ∞ is the base point for convergence, there must exist an $R > 0$ such that $\mathbb{C} \setminus B(0, R) \subseteq \Omega_n \cap \Omega$, for all $n \in \mathbb{N}$. This implies that $K_n, K \subseteq B(0, R)$ for all $n \in \mathbb{N}$, i.e., all the K_n's must be contained in a common ball.

(C) Suppose the limit set K is a compact hull. Then, eventually the K_n's are all compact hulls and it makes sense to ask if $G_{K_n} = F_{K_n}^{-1} \colon \mathbb{C} \setminus K_n \to \mathbb{C} \setminus B[0, 1]$ converges locally uniformly in the set $\mathbb{C} \setminus K$ to the function $G_K = F_K^{-1} \colon \mathbb{C} \setminus K \to \mathbb{C} \setminus B[0, 1]$. See also Remark 18.13(B).

Before we formulate our main result concerning Carathéodory convergence of compact hulls, we state and prove a useful topological lemma.

Lemma 19.16. *Let K_1 and K_2 be two connected compact subsets of $\mathbb{C}$, and $\Psi\colon \mathbb{C}_\infty \setminus K_1 \to \mathbb{C}_\infty \setminus K_2$ a homeomorphism with $\Psi(\infty) = \infty$. Let $\gamma_1\colon [0,1] \to \mathbb{C} \setminus K_1$ be a closed Jordan curve, and define another closed Jordan curve by setting $\gamma_2 = \Psi \circ \gamma_1\colon [0,1] \to \mathbb{C} \setminus K_2$. Denote by G_j the bounded connected component of $\mathbb{C} \setminus \gamma_j^*$ and by H_j the unbounded connected component, $j = 1, 2$. Then we must have: Either*

(a) $K_j \subseteq G_j$, *for* $j = 1, 2$, $\Psi(G_1 \setminus K_1) = G_2 \setminus K_2$ *and* $\Psi(H_1) = H_2$,

or

(b) $K_j \subseteq H_j$, *for* $j = 1, 2$, $\Psi(G_1) = G_2$ *and* $\Psi(H_1 \setminus K_1) = H_2 \setminus K_2$.

Proof. Since K_1 is connected, it is either contained in G_1 or in H_1. Suppose $K_1 \subseteq H_1$. Then $\Psi(G_1)$ is a Jordan region with boundary $\Psi(\gamma_1^*) = \gamma_2^*$. Hence $G_2 = \Psi(G_1)$. Consequently, we must also have $\Psi(H_1 \setminus K_1) = H_2 \setminus K_2$.

Suppose now that $K_1 \subseteq G_1$. Let $R_1 > 0$ be such that $\gamma_1^* \subseteq B[0, R_1]$. By the first step above applied with the roles of K_1 and K_2 reversed, and Ψ replaced by Ψ^{-1}, we must have $K_2 \subseteq G_2$.

Let $w \in H_1$, and denote by $\sigma_1\colon [0, \infty[\to H_1$ a curve with $\sigma(0) = w$ and $\lim_{t\to\infty} |\sigma(t)| = \infty$. Such a curve exists, since H_1 is open, connected and contains $\mathbb{C} \setminus B[0, R_1]$. Hence, $\sigma_2 = \Psi \circ \sigma_1\colon [0, \infty[\to \mathbb{C} \setminus \gamma_2^*$, and – by assumption – $\lim_{t\to\infty} |\sigma_2(t)| = \infty$. This shows that $\sigma_2(0) = \Psi(w) \in H_2$, the unbounded connected component of $\mathbb{C} \setminus \gamma_2^*$. Consequently, $\Psi(H_1) \subseteq H_2$. Since the same argument done in reverse shows that $\Psi^{-1}(H_2) \subseteq H_1$, we conclude that $\Psi(H_1) = H_2$. This implies in addition that $\Psi(G_1 \setminus K_1) = G_2 \setminus K_2$. $\square$

Theorem 19.17. *Let K and $\{K_n\}_{n\in\mathbb{N}}$ be compact hulls or singletons. The two following properties are equivalent.*

(a) $K_n \to K$ *in Carathéodory sense.*

(b) $\Gamma_{K_n} \to F_K$ *locally uniformly in* $\mathbb{C} \setminus B[0,1]$.

If one, and hence both, of the above properties holds true, then $\mathrm{Cap}(K_n) \to \mathrm{Cap}(K)$. Furthermore, if K is a compact hull, then $G_{K_n} \to G_K$ locally uniformly in $\mathbb{C} \setminus K$.

Remark 19.18. A glance at Theorem 18.12 (c) may tempt one to claim that $K_n \to \mathbb{C}$ in Carathéodory sense if and only if $|F_{K_n}(z)| \to \infty$ locally uniformly in $\mathbb{C} \setminus B[0,1]$. However, the sequence $K_n = B[2n, n] \to \mathbb{C}$ with the Riemann maps $F_{K_n}(z) = n(z + 2)$ provides a counterexample to one of the implications. The reader is in Exercise 19.10 asked to validate the other implication.

Proof. Here we consider only the case where K is a compact hull, in which case we may assume that all the K_n's are compact hulls as well. The case where K is a singleton, we address at the end of the section.

We begin with (a) $\Rightarrow$ (b), which we divide into three steps.

Step 1: Let $\zeta \in \partial K$. We claim that there exists a sequence of points $\zeta_n \in K_n$, such that $\zeta_n \to \zeta$, for $n \to \infty$. Indeed, let $\zeta_n \in K_n$ be such that the function $K_n \ni w \to |\zeta - w|$ is minimized. Such a ζ_n exists since K_n is compact. Assume – towards a contradiction – that $\zeta_n \not\to \zeta$. Then there exists $\varepsilon > 0$ and a subsequence $\{\zeta_{n_k}\}_{k=1}^{\infty}$, such that $|\zeta - \zeta_{n_k}| \geq \varepsilon$, for all k. Hence $B(\zeta, \varepsilon) \cap K_{n_k} = \emptyset$ for all k, where we used the construction of the ζ_n's as minimizing the distance to ζ.

Put $U = (\mathbb{C}_\infty \setminus K) \cup B(\zeta, \varepsilon)$. Then U is open and contains ∞. By the choice of z_0 in the boundary of K, hence of $\mathbb{C}_\infty \setminus K$, the set U is connected. The property that $B(\zeta, \varepsilon) \cap K_{n_k} = \emptyset$ for all k now ensures that $U \subseteq \ker(\{\mathbb{C}_\infty \setminus K_{n_k}\}_{k \in \mathbb{N}}, \infty)$. By the assumed property (a), we may conclude that $U \subseteq \mathbb{C}_\infty \setminus K$, which absurdly implies that ζ is an interior point of $\mathbb{C}_\infty \setminus K$.

The claim would in general be false for an interior point of K.

Step 2: We translate the original compact hulls and define $\widetilde{K}_n = K_n - \zeta_n$ and $\widetilde{K} = K - \zeta$. Then $\widetilde{K}_n \to \widetilde{K}$ in Carathéodory sense. Note that $F_{\widetilde{K}_n}(z) = F_{K_n}(z) - \zeta_n$ and $F_{\widetilde{K}}(z) = F_K(z) - \zeta$. In particular, it suffices to argue that $F_{\widetilde{K}_n} \to F_{\widetilde{K}}$ locally uniformly in $\mathbb{C} \setminus B[0,1]$. We now have $0 \in \widetilde{K}$ and $0 \in \widetilde{K}_n$ for all $n \in \mathbb{N}$.

Step 3: Let, as usual, $\mathcal{O} = \{z \in \mathbb{C} \mid z^{-1} \in \widetilde{\Omega}_\infty\}$ and $\mathcal{O}_n = \{z \in \mathbb{C} \mid z^{-1} \in \widetilde{\Omega}_{\infty,n}\}$ be the associated simply connected regions. Here $\widetilde{\Omega}_\infty = \mathbb{C}_\infty \setminus \widetilde{K}$ and $\widetilde{\Omega}_{\infty,n} = \mathbb{C}_\infty \setminus \widetilde{K}_n$. It follows from Proposition 18.10 that $\mathcal{O}_n \to \mathcal{O}$ in Carathéodory sense (with base point 0) for $n \to \infty$.

By Carathéodory's convergence theorem (Theorem 18.12), the associated Riemann maps $f_{\mathcal{O}_n} \colon B(0,1) \to \mathcal{O}_n$ converge locally uniformly to $f_{\mathcal{O}} \colon B(0,1) \to \mathcal{O}$. Hence, $F_{\widetilde{K}_n}(z) = f_{\mathcal{O}}(1/z)^{-1}$ converge locally uniformly in $\mathbb{C} \setminus B[0,1]$ to $F_{\widetilde{K}}(z) = f_{\mathcal{O}}(1/z)^{-1}$.

We now turn to the converse implication (b) $\Rightarrow$ (a), which we divide into five steps.

Step 1': We claim that the K_n's are uniformly bounded. Let N be such that $|F_K(w) - F_{K_n}(w)| \leq 1$ for all $w \in \partial B(0,2)$ and $n \geq N$, which is possible by (b). For each n, define a closed Jordan curve $\gamma_n \colon [0,1] \to \mathbb{C} \setminus K_n$ by $\gamma_n(t) = F_{K_n} \circ \sigma(t)$, where $\sigma(t) = 2e^{i2\pi t}$. Then, for all $n \geq N$ and $t \in [0,1]$:

$$|\gamma_n(t)| \leq |F_K(\sigma(t))| + |F_K(\sigma_t) - F_{K_n}(\sigma(t))|$$
$$\leq \sup\{|F_K(w)| \,|\, |w| = 2\} =: R < \infty.$$

We saw in Lemma 19.16 that the K_n's must be contained in the bounded connected components of $\mathbb{C} \setminus \gamma_n^*$. (Recall Jordan's curve theorem, Theorem 15.8.) Since $\gamma_n^* \subseteq B[0,R]$, we conclude that $K_n \subseteq B[0,R]$ for all $n \geq N$. This shows that the K_n's are uniformly bounded as claimed.

Step 2': For each n pick an $\eta_n \in K_n$. By Step I'. there exists a subsequence $\{\eta_{n_k}\}_{k=1}^\infty$ and a point η such that $\eta_{n_k} \to \eta$. We claim that $\eta \in K$. If not, let $r > 0$ be such that $B(\eta, r) \cap K = \emptyset$ and put $\Omega = G_K(B(\eta, r)) \subseteq \mathbb{C} \setminus B[0,1]$. Since $F_{K_{n_k}} \to F_K$ locally uniformly in Ω (by property (b)), we may invoke Lemma 18.11 with $a = G_K(\eta)$ to conclude that there exists $r' > 0$ with $B(\eta, r') \cap K_{n_k} = \emptyset$ for all k. This contradicts the choice of η as limit point for the η_{n_k}'s.

Step 3': Replace $F_{K_{n_k}}$ by $\widetilde{F}_{K_{n_k}} = F_{K_{n_k}} - \eta_{n_k}$ and F_K by $\widetilde{F}_K = F_K - \eta$, which correspond to compact hulls $\widetilde{K} = K - \eta$ and $\widetilde{K}_{n_k} = K_{n_k} - \eta_{n_k}$, respectively. Note that $0 \notin \widetilde{K}$ and $0 \notin \widetilde{K}_{n_k}$ for any k.

Step 4': Let $\mathcal{O}_{n_k}, \mathcal{O}, f_{\mathcal{O}_{n_k}} \colon B(0,1) \to \mathcal{O}_n$ and $f_{\mathcal{O}} \colon B(0,1) \to \mathcal{O}$ denote the open sets and Riemann maps from Step III above, associated with $\widetilde{F}_{K_{n_k}}$ and $\widetilde{F}_K$, respectively.

To see that $f_{\mathcal{O}_{n_k}} \to f_{\mathcal{O}}$ locally uniformly in $B(0,1)$, we argue as follows. Let $C \subseteq B(0,1) \setminus \{0\}$ be a compact set. Then $C^{-1} = \{z \in \mathbb{C}_\infty \,|\, z^{-1} \in C\}$ is a compact subset of $\mathbb{C} \setminus B[0,1]$. By assumption, $F_{\widetilde{K}_{n_k}} \to F_K$ uniformly on C^{-1}, and consequently, $f_{\mathcal{O}_{n_k}} \to f_{\mathcal{O}}$ uniformly on C. By the Weak Maximum Principle, or Cauchy's integral formula, this implies that $f_{\mathcal{O}_{n_k}} \to f_{\mathcal{O}}$ locally uniformly in $B(0,1)$.

By Theorem 18.12, we conclude that $\mathcal{O}_{n_k} \to \mathcal{O}$ in Carathéodory sense with base point 0. As in Step III above, this implies that $\mathbb{C}_\infty \setminus K_{n_k} \to \mathbb{C}_\infty \setminus K$ in Carathéodory sense with base point ∞.

Step 5': We have argued that $\{K_n\}_{n \in \mathbb{N}}$ has a subsequence converging to K in Carathéodory sense. We could equally well have started our chain of arguments with a subsequence $\{F_{K_{n_k}}\}$, in which case we would have extracted a further subsequence $\{K_{n_{k_\ell}}\}_{\ell \in \mathbb{N}}$ with $K_{n_{k_\ell}} \to K$ in Carathéodory sense. By Exercise 18.5 (applied to the complements in $\mathbb{C}_\infty$), this implies (a).

Having established the equivalence of (a) and (b), we derive the claimed consequences of having one (hence both) of the two properties satisfied.

As argued in Step 1 and 2 above, we may assume that $0 \in K$ and that $0 \in K_n$ for all n. (Recall also Lemma 19.6(a).) We recall from Step 3 that $F_{K_n} \to F_K$ locally uniformly in $\mathbb{C} \setminus K$, implies that $f_{\mathcal{O}_n} \to f_\mathcal{O}$ locally uniformly in $B(0,1)$. Hence, by Theorem 18.12, we have $g_{\mathcal{O}_n} \to g_\mathcal{O}$ locally uniformly in $\mathcal{O}$. Since $G_K(z) = g_\mathcal{O}(1/z)^{-1}$ and $G_{K_n}(z) = g_{\mathcal{O}_n}(1/z)^{-1}$, we conclude that $G_{K_n} \to G_K$ locally uniformly in $\mathbb{C} \setminus K$.

Since $\mathrm{Cap}(K) = -\mathrm{Log}(f'_\mathcal{O}(0))$ and $\mathrm{Cap}(K_n) = -\mathrm{Log}(f'_{\mathcal{O}_n}(0))$, the convergence $\mathrm{Cap}(K_n) \to \mathrm{Cap}(K)$ follows from Weierstrass (Theorem 4.10). $\square$

In the following lemma we approximate, in Carathéodory sense, an arbitrary compact hull from the outside by closures of Jordan regions. Recall from Example 19.2 that the closure of a Jordan region is a compact hull.

Lemma 19.19. *Let K be a compact hull with associated normalized Riemann map $F_K \colon \mathbb{C} \setminus B[0,1] \to \mathbb{C} \setminus K$.*

(a) *For $r > 1$, $\gamma_r \colon [0, 2\pi] \to \mathbb{C} \setminus K$, defined by $\gamma_r(t) = F_K(re^{it})$, defines a closed smooth Jordan curve. The associated Jordan regions G_r, are strictly decreasing ($\overline{G}_r \subseteq G_s$, for $1 < r < s$) and all contain K.*

(b) *Put $K_n = \overline{G}_{1+n^{-1}}$. Then $K_n \to K$ in Carathéodory sense.*

(c) $\mathrm{Cap}(K_n) \to \mathrm{Cap}(K)$.

Proof. We begin with (a). Let $r > 1$. Denote by $\sigma_r(t) = re^{it}$, for $t \in [0, 2\pi]$, the closed Jordan curve running counter clockwise around the origin once, on the circle of radius r. Then $\sigma_r^* \subseteq \mathbb{C} \setminus B[0,1]$. The image curve $\gamma_r = F_K \circ \sigma_r$ is a again a closed Jordan curve.

Let G_r denote the bounded connected component of $\mathbb{C} \setminus \gamma_r^*$. It is a direct consequence of Lemma 19.16 that if $1 < r < s$, we must have $K \subseteq G_r \subseteq \overline{G}_r \subseteq G_s$.

We now turn to (b). For each $n \in \mathbb{N}$, let K_n denote the compact hull arising from $\gamma_{1+n^{-1}}$ as the closure of the associated Jordan region $G_{1+n^{-1}}$. Then $K \subseteq K_{n+1} \subseteq K_n$. That $\mathbb{C}_\infty \setminus K_n \to \mathbb{C}_\infty \setminus K$ in Carathéodory sense, follows from the monotonicity of $\{K_n\}$, and the fact that $\bigcap_{n=1}^{\infty} K_n = K$. See Example 18.9. The claim (c) now follows from Theorem 19.17. $\square$

We begin preparations to establish Theorem 19.17 in the case where K is a singleton.

Lemma 19.20. *Let $a \in \mathbb{C}$ and $\{K_n\}_{n\in\mathbb{N}}$ be a sequence of compact hulls. The following are equivalent.*

(a) $K_n \to \{a\}$ *in Carathéodory sense.*

(b) *There exists a sequence of radii $r_n \to 0$, such that $K_n \subseteq B(a, r_n)$.*

(c) $\mathrm{Cap}(K_n) \to -\infty$, *for $n \to \infty$, and there exist $k_n \in K_n$, $n \in \mathbb{N}$, such that $k_n \to a$.*

Proof. To see that (a) $\Rightarrow$ (b), let $\varepsilon > 0$ and observe that $C = \mathbb{C}_\infty \setminus B(a, \varepsilon)$ is a compact subset of $\mathbb{C}_\infty \setminus \{a\}$. Hence, by assumption, there exists $N \in \mathbb{N}$ such that $C \subseteq \mathbb{C}_\infty \setminus K_n$, for $n \geq N$. Consequently, $K_n \subseteq B[a, \varepsilon]$, for $n \geq N$. This implies (b). Conversely, it is obvious that (b) $\Rightarrow$ (a).

It follows easily from (19.3) that (b) $\Leftrightarrow$ (c). $\square$

We note that for a compact hull K, the associated normalized map F_K has an expansion around infinity, a Laurent series, absolutely convergent throughout $\mathbb{C} \setminus B[0, 1]$, and of the form

$$F_K(z) = e^{\mathrm{Cap}(K)} z + b_0 + \sum_{k=1}^{\infty} \frac{b_k}{z^k}. \tag{19.4}$$

That the sum converges absolutely for all z, with $|z| > 1$, follows from Theorem 6.3.

Remark 19.21. Note that if K is compact hull with $\mathrm{Cap}(K) = 0$, we may feed the Area Theorem (Theorem 18.1) with the function F_K. Hence, we conclude the inequality $\sum_{k=1}^{\infty} k|b_k|^2 \leq 1$.

Proof of Theorem 19.17, singleton case. Here $K = \{a\}$. To prove (a) $\Rightarrow$ (b), we may assume that the K_n's are compact hulls, i.e., not singletons, with $K_n \to \{a\}$ in Carathéodory sense. (Along a subsequence $\{K_{n_k}\}_{k\in\mathbb{N}}$ of singletons $K_{n_k} = \{a_{n_k}\}$, which also converges in Carathéodory sense, we would trivially get convergence of the constant functions $F_{\{a_{n_k}\}} = a_{n_k}$ to $F_{\{a\}} = a$.) By Lemma 19.20(b), there exist radii $r_n > 0$ with $\lim_{n\to\infty} r_n = 0$

and $K_n \subseteq B(a, r_n)$. We can assign to the K_n's their logarithmic capacity $c_n := \mathrm{Cap}(K_n)$. From Lemma 19.20, we find that

$$\lim_{n \to \infty} c_n = -\infty, \tag{19.5}$$

which equals $\mathrm{Cap}(\{a\})$.

We rescale the K_n's to get compact hulls $\widetilde{K}_n = e^{-c_n} K_n$, with $\mathrm{Cap}(\widetilde{K}_n) = 0$. Denote by $b_k^{(n)}$ the expansion coefficients of F_{K_n}, such that $e^{-c_n} b_k^{(n)}$ are the expansion coefficient of $F_{\widetilde{K}_n}$. We can now apply the Area Theorem, cf. Remark 19.21, to the $\widetilde{K}_n$'s, and conclude that

$$\forall k \geq 1 \; \forall n \in \mathbb{N}: \quad |b_k^{(n)}| \leq \frac{e^{c_n}}{\sqrt{k}}.$$

Consequently, we can estimate for z, with $|z| > 1$,

$$|F_{K_n}(z) - b_0^{(n)}| \leq e^{c_n}|z| + \sum_{k=1}^{\infty} \frac{|b_k^{(n)}|}{|z|^k} \leq e^{c_n}\left(|z| + \sum_{k=1}^{\infty} \frac{1}{\sqrt{k}|z|^k}\right).$$

From this bound and (19.5), we conclude that $|F_{K_n}(z) - b_0^{(n)}| \to 0$ locally uniformly in $\mathbb{C} \setminus B[0,1]$.

It remains to argue that $b_0^{(n)} \to a$, in the limit $n \to \infty$. Let $\sigma : [0, 2\pi] \to \mathbb{C} \setminus B[0,1]$ be the closed Jordan curve $\sigma(t) = 2e^{it}$. Composing with F_{K_n}, we get a sequence of closed Jordan curves $\gamma_n = F_{K_n} \circ \sigma$ with K_n contained in the bounded component of $\mathbb{C} \setminus \gamma_n^*$. See Lemma 19.16. We have

$$\sup_{t \in [0, 2\pi]} |\gamma_n(t) - b_0^{(n)}| \to 0 \qquad \text{for } n \to \infty.$$

Hence

$$\sup_{s, t \in [0, 2\pi]} |\gamma_n(s) - \gamma_n(t)|$$

$$\leq \sup_{s \in [0, 2\pi]} |\gamma_n(s) - b_0^{(n)}| + \sup_{t \in [0, 2\pi]} |\gamma_n(t) - b_0^{(n)}| \to 0,$$

for $n \to \infty$. Let $k_n \in K_n$. We can now estimate

$$|a - b_0^{(n)}| \leq |a - k_n| + |k_n - \gamma_n(t)| + |\gamma_n(t) - b_0^{(n)}|$$

$$\leq r_n + \sup_{s \in [0, 2\pi]} |\gamma_n(s) - \gamma_n(t)| + |\gamma_n(t) - b_0^{(n)}| \to 0,$$

for $n \to \infty$. This completes the argument for (a) $\Rightarrow$ (b).

To establish that (b) $\Rightarrow$ (a), we employ the same curves σ and γ_n, for $n \in \mathbb{N}$, as above. Recall that K_n is inside the bounded component of $\mathbb{C} \setminus \gamma_n^*$. By assumption, $r_n := \sup_{t \in [0,2\pi]} |\gamma_n(t) - a| \to 0$, for $n \to \infty$. Hence, we must have $K_n \subseteq B(a, r_n)$, and consequently, that $K_n \to \{a\}$ in Carathéodory sense. See Lemma 19.20. $\qquad\square$

19.3 Half-plane capacity

We now turn to the study of half-plane hulls and their half-plane capacity. We introduce the notation $\mathbb{H} = \{z \in \mathbb{C} \mid \operatorname{Im} z > 0\}$ for the open upper half-plane.

Definition 19.22. *A subset $A \subseteq \mathbb{H}$ is called a half-plane hull, if*

(a) $\overline{A} \cap \mathbb{H} = A$.

(b) *A is a bounded set.*

(c) *$\mathbb{H} \setminus A$ is a simply connected region.*

For a set $U \subseteq \mathbb{C}$, we write $U^* = \{\bar{z} \mid z \in U\}$ for the reflection of U in the real axis. Note that $-U^*$ is the reflection of U in the imaginary axis. (We reserve the notation $\overline{U}$ for the closure of a set U.)

Remark 19.23. Apart from the obvious half-plane hulls, there are two classes of half-plane hulls that are central to our investigation.

One class is the range of simple curves growing from the real axis up into the upper half-plane. That is, sets of the form $\gamma(]0, R])$, where $\gamma \colon [0, R] \to \overline{\mathbb{H}}$ is a simple curve with $\gamma(0) \in \mathbb{R}$ and $\gamma(t) \in \mathbb{H}$ for $t \in]0, R]$.

The second class are Jordan regions with a segment of the real axis forming part of the boundary. If $\gamma \colon [0, 1] \to \overline{\mathbb{H}}$ is a simple curve with $\gamma(0) < \gamma(1)$ real numbers and $\gamma(t) \in \mathbb{H}$ for all $t \in]0, 1[$. Then we can form a closed Jordan curve by adding the segment $[\gamma(1), \gamma(0)]$ with the indicated orientation. Then $A = \overline{G} \cap \mathbb{H}$ will be a half-plane hull, where G is the bounded connected component of the complement of the closed Jordan curve.

We refer the reader to Example 19.34 below for an argument that the sets A just constructed are in fact half-plane hulls.

For a $z_0 = a + ib \in \mathbb{H}$, we define the Möbius transform

$$C_{z_0}(z) = \frac{z - z_0}{z - \bar{z}_0}. \tag{19.6}$$

Note that $\mathrm{Cay} = C_i$ is the Cayley transform, cf. Example 8.13, and that

$$C_{z_0}(z) = \mathrm{Cay}(z/b - a/b).$$

Since $z \to z/b - a/b$ is an automorphism of the upper half-plane $(b > 0)$, we conclude from Example 8.13 that

$$C_{z_0}(\mathbb{H}) = B(0,1) \quad \text{and} \quad C_{z_0}(z_0) = 0.$$

In the following, A denotes a half-plane hull and $\Omega = \mathbb{H} \setminus A$, its simply connected open complement inside the open upper half-plane. Let $z_0 \in \Omega$ and abbreviate $\mathcal{O} = C_{z_0}(\Omega) \subseteq B(0,1)$, which is a simply connected region.

Let $R > 0$ be such that $A \subseteq B[0, R]$. Denote by $J_R = S^1 \setminus C_{z_0}([-R, R])$, an open subarc of the unit circle $S^1 = \partial B(0,1)$. In fact, J_R is an open neighborhood in S^1 of $1 = C_{z_0}(\infty)$.

Denote by $\tilde{f}_{\mathcal{O}} \colon B(0,1) \to \mathcal{O}$ the associated Riemann map with $\tilde{f}_{\mathcal{O}}(0) = 0$ and $\tilde{f}'_{\mathcal{O}}(0) > 0$, and write $\tilde{g}_{\mathcal{O}} = \tilde{f}_{\mathcal{O}}^{-1} \colon \mathcal{O} \to B(0,1)$ for its inverse. Since J_R consists of simple accessible boundary points for $\mathcal{O}$, we can extend $\tilde{g}_{\mathcal{O}}$ to a continuous injective function $\tilde{g}_{\mathcal{O}} \colon \mathcal{O} \cup J_R \to B[0,1]$ (Theorem 17.17). The image $\tilde{g}_{\mathcal{O}}(J_R)$ is again an open connected subarc of S^1, containing $\tilde{g}_{\mathcal{O}}(1) =: e^{i\theta}$. Define $g_{\mathcal{O}} = e^{-i\theta} \tilde{g}_{\mathcal{O}}$, which is a Riemann map from $\mathcal{O}$ onto $B(0,1)$, extending continuously to $J_R \ni 1$, mapping J_R into the open subarc $B_R = g_{\mathcal{O}}(J_R) \ni 1$ of S^1.

Going back to Ω, we can define

$$\widehat{G}_A = \mathrm{Cay}^{-1} \circ g_{\mathcal{O}} \circ C_{z_0} \colon \Omega \to \mathbb{H}. \tag{19.7}$$

The choice of $g_{\mathcal{O}}$ ensures that $\lim_{|z| \to \infty, z \in \Omega} |\widehat{G}_A(z)| = \infty$, and that $\widehat{G}_A$ has an extension by continuity to $\mathbb{R} \setminus [-R, R]$ with $\widehat{G}_A \colon \mathbb{R} \setminus [-R, R] \to \mathrm{Cay}^{-1}(B_R) \setminus \{\infty\}$ an injective mapping. The construction is depicted in Fig. 19.2.

As a consequence of the above construction, $\widehat{G}_A$ extends by Schwarz reflection (see Exercise 4.27) to a univalent map $\widehat{G}_A \colon \mathbb{C} \setminus B[0, R] \to \mathbb{C}$ with $\lim_{|z| \to \infty} |\widehat{G}_A(z)| = \infty$. Hence, the function $h(w) = 1/\widehat{G}_A(w^{-1})$ is univalent in $B(0, 1/R) \setminus \{0\}$ with a removable singularity at $w = 0$. Setting $h(0) = 0$ yields a univalent extension to $B(0, 1/R)$. Being univalent, we must have $h'(0) \neq 0$. Consequently, the function $\tilde{h}(w) = h(w)/w$ also has a removable singularity at $w = 0$, and extends to a non-vanishing holomorphic function in $B(0, 1/R)$. Note that $\tilde{h}(0) = h'(0) \neq 0$. Hence, we have an expansion

$$\frac{1}{\tilde{h}(w)} = \sum_{n=0}^{\infty} a_n w^n$$

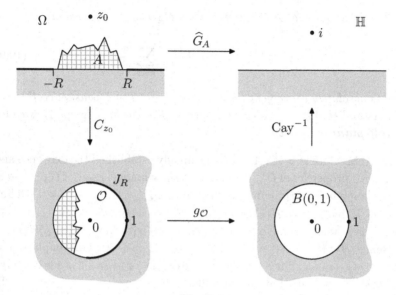

Fig. 19.2

with $a_0 = 1/h'(0) \neq 0$, and radius of convergence (at least) $1/R$. Therefore

$$\widehat{G}_A(z) = \frac{z}{\bar{h}(1/z)} = a_0 z + a_1 + \sum_{k=1}^{\infty} \frac{a_k}{z^k}, \tag{19.8}$$

as an expansion absolutely convergent for $z \in \mathbb{C}$ with $|z| > R$. Since $\widehat{G}_A(x)$ is real for x real (and sufficiently large), we must have $a_k \in \mathbb{R}$ for all $k \in \mathbb{N}_0$. Furthermore, setting $z = iy$, and taking y large, we observe that $a_0 > 0$. We can now define

$$G_A(z) = (\widehat{G}_A(z) - a_1)/a_0,$$

which is again a biholomorphic map from Ω onto $\mathbb{H}$. It has the expansion $G_A(z) = z + b_0/z + O(|z|^{-2})$ at infinity.

The above construction does not perhaps appear canonical in that it depends on the complex number z_0 and the radius $R > 0$. The following proposition establishes that it is in fact canonical.

Proposition 19.24. *Let A be a half-plane hull. There exists exactly one holomorphic bijection $G_A \colon \mathbb{H} \setminus A \to \mathbb{H}$, satisfying*

$$\lim_{\substack{|z| \to \infty \\ z \in \Omega}} G_A(z) - z = 0.$$

Let $R > 0$ be such that $A \subseteq B[0, R]$. Then G_A has the expansion

$$G_A(z) = z + \frac{\mathrm{hcap}(A)}{z} + \sum_{k=2}^{\infty} \frac{b_k}{z^k}, \qquad (19.9)$$

which is absolutely convergent for z with $|z| > R$. The expansion coefficients $\mathrm{hcap}(A)$ and $\{b_k\}_{k=2}^{\infty}$ are all real numbers. The constant $\mathrm{hcap}(A)$ is called the **half-plane capacity** of A.

Proof. Abbreviate $\Omega = \mathbb{H} \setminus A$. We have already established the existence and the claimed properties of G_A. It remains to show uniqueness. Let $\widetilde{G}_A : \Omega \to \mathbb{H}$ be another biholomorphic map, with $\lim_{|z| \to \infty, z \in \Omega} \widetilde{G}_A(z) - z = 0$. Define $H : \mathbb{H} \to \mathbb{H}$ by setting $H(z) = \widetilde{G}_A(z) \circ F_A(z)$, where $F_A = G_A^{-1} : \mathbb{H} \to \Omega$. Then H is a holomorphic automorphism of $\mathbb{H}$, with $\lim_{|z| \to \infty, z \in \Omega} |H(z)| = \infty$. Hence $H(z) = az + x$, where $a > 0$ and $x \in \mathbb{R}$. This implies that $\widetilde{G}_A(z) = aG_A(z) + x$. Since $\widetilde{G}_A(z) - z = a(G_A(z) - z) + (a - 1)z + x$ converge to zero at ∞, we must have $a = 1$ and $x = 0$. $\qquad \square$

In the following we write G_A for the associated biholomorphic map from $\Omega = \mathbb{H} \setminus A$ onto $\mathbb{H}$, normalized at infinity. We remark that this normalization is sometimes called hydrodynamic normalization. The notation $F_A : \mathbb{H} \to \Omega$ is reserved for the inverse map G_A^{-1}. We hope the reader can tell from the indexing subset, when a function pertains to a half-plane hull and when it pertains to a compact hull. Note that we must necessarily have

$$F_A(z) = z - \frac{\mathrm{hcap}(A)}{z} + O(z^{-2})$$

for large z.

Remark 19.25. Before continuing, we pause to discuss a connection with the normalized Riemann maps associated with compact hulls. Let A be a half-plane hull and $z_0 \in \Omega = \mathbb{H} \setminus A$. Put

$$K = B[0, 1] \cup C_{\bar{z}_0}(A), \qquad (19.10)$$

where the Möbius transform $C_{\bar{z}_0} = 1/C_{z_0}$ was defined in (19.6). The reader is asked in Exercise 19.11 to show that K is a compact hull. We write $G_K : \mathbb{C} \setminus K \to \mathbb{C} \setminus B[0, 1]$ for the associated normalized biholomorphic map, the existence of which is ensured by Proposition 19.4.

Since the Möbius transform $C_{\bar{z}_0}$ maps ∞ to 1, there is an $r > 0$ small enough such that $J = \{e^{i\theta} \mid |\theta| < r\} \subseteq \partial K$. By Corollary 17.20, applied with $\phi(z) = 1/G_K(1/z)$, G_K extends by continuity to $(\mathbb{C} \setminus K) \cup J$ and $G_K(z) \in \partial B(0,1)$ for $z \in J$.

We can use Schwarz reflection to extend G_K across the arc J to an analytic (in fact univalent) function on the set $(\mathbb{C} \setminus K) \cup J \cup \Theta(\mathbb{C} \setminus K)$. Here $\Theta(w) = \mathrm{Cay}(\mathrm{Cay}^{-1}(w)) = 1/\overline{w}$ is a reflection in the unit circle. Note that $\Theta \circ \Theta = \mathrm{id}$. The construction of the extension is $G_K(w) = \Theta(G_K(\Theta(w)))$, for $w \in \Theta(\mathbb{C} \setminus K)$. In particular, all derivatives $G_K^{(k)}(1)$ exists and $G_K'(1) \neq 0$. See also Exercise 4.27.

We may now define $\check{G}_A : \Omega \to \mathbb{H}$ by

$$\check{G}_A(z) = C_{-i}^{-1}(G_K(C_{\bar{z}_0}(z))/G_K(1)), \tag{19.11}$$

Then $\check{G}_A$ is a biholomorphic function with $\lim_{z \to \infty, z \in \Omega} |\check{G}_A(z)| = \infty$.

Expanding around $w = 1$, we find that

$$G_K(w) = G_K(1) + G_K'(1)(w-1) + \tfrac{1}{2}G_K''(1)(w-1)^2 + O((w-1)^3). \tag{19.12}$$

As for the Möbius transform $C_{\bar{z}_0}$, we find that

$$C_{\bar{z}_0}(z) = \frac{z - \bar{z}_0}{z - z_0} = 1 + \frac{2i \,\mathrm{Im}\, z_0}{z} + \frac{2iz_0 \,\mathrm{Im}\, z_0}{z^2} + O(z^{-3}). \tag{19.13}$$

Using (19.12) and the computation $C_{-i}^{-1}(w) = i + 2i/(w-1)$, we expand

$$\check{G}_A(z) = i + \frac{2i}{G_K(C_{\bar{z}_0}(z))/G_K(1) - 1}$$

$$= i + \frac{2i}{\frac{G_K'(1)}{G_K(1)}(C_{\bar{z}_0}(z) - 1) + \frac{1}{2}\frac{G_K''(1)}{G_K(1)}(C_{\bar{z}_0}(z) - 1)^2 + O(C_{\bar{z}_0}(z) - 1)^3)}.$$

Finally, we exploit (19.13) to arrive at the expansion

$$\check{G}_A(z)$$

$$= i + \frac{2i}{\frac{2i \,\mathrm{Im}\, z_0 G_K'(1)}{G_K(1)} z^{-1} + \left(\frac{2iz_0 \,\mathrm{Im}\, z_0 G_K'(1)}{G_K(1)} - \frac{2(\mathrm{Im}\, z_0)^2 G_K''(1)}{G_K(1)}\right) z^{-2} + O(z^{-3})}$$

$$= i + \frac{G_K(1)}{\mathrm{Im}\, z_0 G_K'(1)} \cdot \frac{z}{1 + \left(z_0 + \frac{i \,\mathrm{Im}\, z_0 G_K''(1)}{G_K'(1)}\right) z^{-1} + O(z^{-2})}$$

$$= \frac{G_K(1)}{\mathrm{Im}\, z_0 G_K'(1)} z + i - \frac{G_K(1)}{G_K'(1)} \left(\frac{z_0}{\mathrm{Im}\, z_0} + \frac{i G_K''(1)}{G_K'(1)}\right) + O(z^{-1}).$$

Since $\check{G}_A$ maps into the upper half-plane, we must have

$$a := \frac{G_K(1)}{G_K'(1)} > 0 \quad \text{and} \quad \mathrm{Re}\, \frac{G_K''(1)}{G_K'(1)} = -1 + \frac{1}{a}. \tag{19.14}$$

These somewhat strange looking facts may also (almost) be derived from the observation that $|G_K(e^{i\theta})|^2 = 1$ for small θ. See Exercise 19.12.

Putting

$$b := \operatorname{Re} z_0 - \operatorname{Im} \frac{G_K''(1)}{G_K'(1)} \operatorname{Im} z_0 \tag{19.15}$$

we find, using Proposition 19.24, that

$$G_A(z) = \frac{\operatorname{Im} z_0}{a} \check{G}_A(z) + b. \tag{19.16}$$

One may also view this remark as an alternative existence proof.

Lemma 19.26. *Let A be a half-plane hull. Then*

(a) *For all $x \in \mathbb{R}$, we have* $\operatorname{hcap}(A + x) = \operatorname{hcap}(A)$.

(b) *For all $\alpha > 0$, we have* $\operatorname{hcap}(\alpha A) = \alpha^2 \operatorname{hcap}(A)$.

(c) *We have* $\operatorname{hcap}(-A^*) = \operatorname{hcap}(A)$.

Proof. We remark that $A + x$, αA and $-A^*$ are again half-plane hulls, provided $x \in \mathbb{R}$ and $\alpha > 0$.

Let G_A denote the normalized Riemann map associated with the original half-plane hull A. Then we can construct Riemann maps $G_{A+x}(z) = G_A(z - x) + x$, $G_{\alpha A}(z) = \alpha G_A(z/\alpha)$ and $G_{-A^*}(z) = -\overline{G_A(-\bar{z})}$. We expand

$$G_{A+x}(z) = z + \frac{\operatorname{hcap}(A)}{z - x} + O(|z - x|^{-2}) = z + \frac{\operatorname{hcap}(A)}{z} + O(|z|^{-2}),$$

$$G_{\alpha A}(z) = z + \alpha \frac{\operatorname{hcap}(A)}{z/\alpha} + O(|z/a|^{-2}) = z + \frac{\alpha^2 \operatorname{hcap}(A)}{z} + O(|z|^{-2}),$$

$$G_{-A^*}(z) = z - \frac{\overline{\operatorname{hcap}(A)}}{-\bar{z}} + O(|\bar{z}|^{-2}) = z + \frac{\operatorname{hcap}(A)}{z} + +O(|z|^{-2}),$$

which implies the lemma. $\qquad\square$

Example 19.27. Let $A = \{iy \mid y \in \,]0, R]\}$. Clearly, A is a half-plane hull and we proceed to construct a normalized Riemann map onto the upper half-plane. The map $z \to z^2$ maps Ω bijectively onto $\mathbb{C} \setminus [-R^2, \infty[$. Composing with $z \to z + R^2$ maps bijectively onto $\mathbb{C} \setminus [0, \infty)$. Finally, we employ the square root $re^{i\theta} \to \sqrt{r}e^{i\theta/2}$, where $\theta \in \,]0, 2\pi[$, to get back up into the upper half-plane. This gives the holomorphic bijection

$$G_A(z) = \sqrt{z^2 + R^2}.$$

We expand $G_A(z) = z\sqrt{1 + R^2/z^2} = z + \frac{R^2}{2z} + O(|z|^{-2})$, and conclude that G_A is normalized correctly at infinity, and we read off the half-plane capacity $\text{hcap}(A) = R^2/2$.

Example 19.28. Let $A = \mathbb{H} \cap B[0,1]$. It is clear that A is a half-plane hull. To construct G_A we compose

$$M_1(z) = \frac{z-1}{z+1},$$

the function $M_2(z) = z^2$, and finally

$$M_3(z) = \frac{z+1}{1-z}.$$

The composition maps ∞ to ∞ and preserves the upper half plane. The first map takes the half-disc to the first quadrant, taking infinity to 1, the second maps onto the upper half-plane, preserving 1, and the third is an automorphism of the upper half-plane mapping 1 back out to infinity.

We compose and find

$$M_3(M_1(z)^2) = \frac{\left(\frac{z-1}{z+1}\right)^2 + 1}{1 - \left(\frac{z-1}{z+1}\right)^2} = \frac{(z-1)^2 + (z+1)^2}{(z+1)^2 - (z-1)^2}$$

$$= \frac{2z^2 + 2}{4z} = \tfrac{1}{2}\left(z + \tfrac{1}{z}\right).$$

Multiplying by 2 yields the normalized Riemann map $G_A(z) = z + 1/z$, and hence the half-plane capacity $\text{hcap}(A) = 1$.

Example 19.29. Let C denote the circle in the complex plane passing through the three points $\pm a$ and $i\varepsilon$ with $a > 0$ and $\varepsilon > 0$. Then $C = \partial B[c,r]$, where

$$r = \frac{a^2 + \varepsilon^2}{2\varepsilon} \quad \text{and} \quad c = -i\frac{a^2 - \varepsilon^2}{2\varepsilon}. \tag{19.17}$$

As a compact hull we take $A = B[c,r] \cap \mathbb{H}$. We construct G_A as a composition of three maps, M_1, M_2 and M_3, plus a final normalization. The role of the three maps is illustrated in Fig. 19.3. The reader may track the three distinguished points $-a$, a and ∞ clockwise through the figure.

We wish to construct a Möbius transformation of the upper half-plane taking $C \cap \mathbb{H}$ to a half line starting at 0. That is, the map should take $a \to 0$ and $-a \to \infty$. Up to a (positive) multiplicative factor such a map is

$$M_1(z) = \frac{z-a}{z+a}.$$

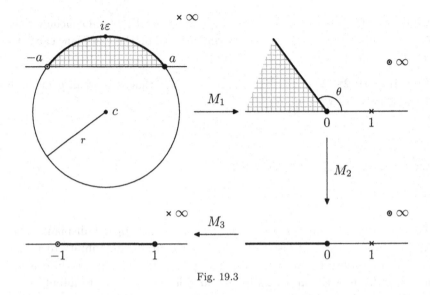

Fig. 19.3

To identify the line that $C \cap \mathbb{H}$ is mapped into, we compute

$$M_1(i\varepsilon) = \frac{i\varepsilon - a}{i\varepsilon + a} = -\frac{a^2 - \varepsilon^2}{a^2 + \varepsilon^2} + i\frac{2\varepsilon a}{a^2 + \varepsilon^2}.$$

When $\varepsilon = a$ this is the imaginary axis, as in the previous example.

In the following we assume that $\varepsilon < a$ such that the center c of the circle C sits in the lower half-plane. The principal argument of $M_1(i\varepsilon)$ is

$$\theta = \mathrm{Arg}(M_1(i\varepsilon)) = \pi - \arctan\left(\frac{2\varepsilon a}{a^2 - \varepsilon^2}\right). \qquad (19.18)$$

Then $M_1(C \cap \mathbb{H}) = \{Re^\theta \,|\, R > 0\}$ and $M_1(\mathbb{H} \setminus A) = \{z \in \mathbb{H} \,|\, 0 < \mathrm{Arg}(z) < \theta\}$. Put

$$p = \frac{\pi}{\theta} = \frac{\pi}{\pi - \arctan\left(\frac{2\varepsilon a}{a^2 - \varepsilon^2}\right)} \qquad (19.19)$$

and define the second map in the chain to be

$$M_2(z) = z^p = e^{p\,\mathrm{Log}(z)},$$

where $\mathrm{Log}(z)$ is the principal logarithm of $z \in \mathbb{C} \setminus \,]-\infty, 0]$. Note that $M_2 \circ M_1(\mathbb{H} \setminus A) = \mathbb{H}$.

We stress that we are not done yet, since $M_2 \circ M_1(\infty) = 1$. We need to map 1 back to infinity. This we do with

$$M_3(z) = \frac{z + 1}{1 - z},$$

which is an automorphism of the upper half-plane. We now have a biholomorphic map $M_3 \circ M_2 \circ M_1 \colon \mathbb{H} \setminus A \to \mathbb{H}$ taking ∞ to ∞.

It remains to identify the correct normalization. For this we expand the functions one at a time. Expand at infinity to second order

$$M_1(z) = \frac{1 - \frac{a}{z}}{1 + \frac{a}{z}} = \left(1 - \frac{a}{z}\right)\left(1 - \frac{a}{z} + \frac{a^2}{z^2} - \frac{a^3}{z^3}\right) + O(|z|^{-4})$$

$$= 1 - \frac{2a}{z} + \frac{2a^2}{z^2} - \frac{2a^3}{z^3} + O(|z|^{-4}).$$

Since $M_1(\infty) = 1$, we must expand M_2 around $z = 1$. Using the expansion $\mathrm{Log}(1 + w) = w - w^2/2 + w^3/3 + O(|w|^4)$, we get

$$\mathrm{Log}(M_1(z)) = -\frac{2a}{z} + \frac{2a^2}{z^2} - \frac{2a^3}{z^3} - \frac{1}{2}\left(-\frac{2a}{z} + \frac{2a^2}{z^2}\right)^2 - \frac{8a^3}{3z^3} + O(|z|^{-4})$$

$$= -\frac{2a}{z} - \frac{(6 - 12 + 8)a^3}{3z^3} + O(|z|^{-4}).$$

Consequently,

$$e^{p\,\mathrm{Log}(M_1(z))} = 1 - \frac{2pa}{z} + \frac{2p^2a^2}{z^2} - \frac{(4p^2 + 2)pa^3}{3z^3} + O(|z|^{-4}).$$

Finally, we expand

$M_3(M_1(z)^p)$

$$= \frac{2 - \frac{2pa}{z} + \frac{2p^2a^2}{z^2}}{\frac{2pa}{z} - \frac{2p^2a^2}{z^2} - \frac{(4p^2+2)pa^3}{3z^3}} + O(|z|^{-2})$$

$$= \frac{z}{pa}\frac{1 - \frac{pa}{z} + \frac{p^2a^2}{z^2}}{1 - \frac{pa}{z} + \frac{(2p^2+1)a^2}{3z^2}} + O(|z|^{-2})$$

$$= \frac{z}{pa}\left(1 - \frac{pa}{z} + \frac{p^2a^2}{z^2}\right)\left(1 + \frac{pa}{z} - \frac{(2p^2 + 1)a^2}{3z^2} + \frac{p^2a^2}{z^2}\right) + O(|z|^{-2})$$

$$= \frac{z}{pa}\left(1 - \frac{pa}{z} + \frac{p^2a^2}{z^2}\right)\left(1 + \frac{pa}{z} + \frac{(p^2 - 1)a^2}{3z^2}\right) + O(|z|^{-2})$$

$$= \frac{z}{pa}\left(1 + \frac{(p^2 - 1)a^2}{3z^2}\right) + O(|z|^{-2}).$$

We multiplying with pa to arrive at the normalized Riemann map

$$G_A(z) = pa M_3 \circ M_2 \circ M_1(z)$$

$$= z - \frac{(p^2 - 1)a^2}{3z} + O(|z|^{-2}). \tag{19.20}$$

We can thus read off the half-plane capacity to be

$$\operatorname{hcap}(A) = \tfrac{1}{3}(p^2 - 1)a^2. \tag{19.21}$$

If $\varepsilon = a$, we have $p = 2$ and we arrive at a^2, which fits with the previous example and the scaling law from Lemma 19.26(b).

Proposition 19.30. *Let A be a half-plane hull and define $C = \overline{A \cup A^*}$.*

(a) *The set C is compact with $\mathbb{C}_\infty \setminus C$ connected.*

(b) *The function G_A extends to a univalent map defined in $\mathbb{C} \setminus C$ with $G_A(\mathbb{R} \setminus C) \subseteq \mathbb{R}$ and $J := \mathbb{C} \setminus G_A(\mathbb{C} \setminus C)$ a compact subset of $\mathbb{R}$.*

(c) *C is a compact hull if and only if J is a compact interval.*[1]

Proof. That C is compact is obvious. Abbreviate $\Omega = \mathbb{H} \setminus A$.

To see that $\mathbb{C}_\infty \setminus C$ is connected it suffices to argue that it is path connected. In fact, it suffices to argue that any point $z \in \mathbb{C} \setminus C$ can be connected to ∞ by a curve running inside $\mathbb{C}_\infty \setminus C$. If $z \in \Omega$ or $z \in \Omega^*$, then this is possible due to boundedness of A and connectedness of Ω (and hence of Ω^*). If $z \in \mathbb{R}$, then we may use closedness of C to argue for the existence of a $c > 0$, such that the line segment $[z, z + ic]$ is contained inside $\mathbb{C} \setminus C$. Since $z + ic \in \Omega$, we may as before connect it to ∞ through Ω. This completes the proof of (a).

As for (b), we observe first that $\mathbb{R} \setminus C$ consists of simple boundary points for $\mathbb{H} \setminus A$. This follows easily, since for any $x \in \mathbb{R} \setminus C$ there exists $r > 0$ such that $B(x, r) \cap \mathbb{H} \subseteq \mathbb{H} \setminus A$. By Theorem 17.17 and an application of the Cayley transform (formula (19.6) with $z_0 = i$), the function G_A extends by continuity to $\mathbb{R} \setminus C$ and the extension is injective and takes real values on the real axis. The existence of the extension to $\mathbb{C} \setminus C$ now follows from Schwarz' reflection principle, cf. Exercise 4.27 and the extension is univalent.

That $\mathbb{C} \setminus G_A(\mathbb{C} \setminus C)$ is a compact subset of $\mathbb{R}$, follows from $G_A(\mathbb{H} \setminus A) = \mathbb{H}$ and $G_A(\mathbb{C}_\infty \setminus C)$ being an open neighborhood of ∞.

The remaining claim (c) follows from (a) and (b), since $\mathbb{C}_\infty \setminus J$ is simply connected if and only if J is an interval, a singleton or empty. Note that J is nonempty if and only if C is nonempty and C (hence also J) cannot be a singleton. $\square$

Let us illustrate the above proposition with two examples.

---

[1] By convention, an interval is neither empty nor a singleton.

Example 19.31. Let us return to the examples with the half-plane hulls $A_1 = {]0, Ri]}$ and $A_2 = \mathbb{H} \cap B[0,1]$. See Examples 19.27 and 19.28. Here $C_1 = [-Ri, Ri]$ and $C_2 = B[0,1]$ are the associated compact sets from Proposition 19.30, $G_{A_1}(z) = \sqrt{z^2 + R^2}$ and $G_{A_2}(z) = z + 1/z$.

Due to the choice of square root, G_{A_1} extends by continuity to $\sqrt{x^2 + R^2}$ for $x > 0$ and to $-\sqrt{x^2 + R^2}$ for $x < 0$. From this observation we conclude that for A_1 we have $J_1 = [-R, R]$. The Schwarz reflected map G_{A_1} maps $\mathbb{C} \setminus [-Ri, Ri]$ onto $\mathbb{C} \setminus [-R, R]$. This is consistent with Lemma 19.11, which implies that the compact hulls $C_1 = [-Ri, Ri]$ and $J_1 = [-R, R]$ have the same logarithmic capacities. See also Lemma 19.6(c).

As for G_{A_2}'s Schwarz extension, it is given by the same expression and is already defined for all nonzero complex numbers. We simply observe that $J_2 = [G_{A_2}(-1), G_{A_2}(1)] = [-2, 2]$, which, just like $C_2 = B[0,1]$, has logarithmic capacity 0.

Example 19.32. Let us revisit the more complicated example where $A = B[c, r] \cap \mathbb{H}$ is the half-plane hull sitting below the arc running from $-a$ to a via $i\varepsilon$, as discussed in Example 19.29. Here $C = B[c, r] \cap B[\bar{c}, r]$ is a compact hull. Recall from (19.17) how the center c and the radius r depend on $0 < \varepsilon < a$.

We leave it to the reader to observe that the extension of G_A, cf. (19.20), to the real axis (and the Schwarz reflection) is simply given by the same expression with the complex root $M_2(z) = z^p$ reducing to the standard real root x^p for $x > 0$.

We can now compute $J = [\alpha, \beta]$, where $\alpha = \lim_{x \to -a_-} G_A(x) = -pa$ and $\beta = \lim_{x \to a_+} G_A(x) = pa$. Since the logarithmic capacity of $[-pa, pa]$ is $\mathrm{Log}(pa/2)$, we conclude from Lemma 19.11 that the logarithmic capacity of C equals $\mathrm{Log}(pa/2)$ as well. The number $p = p(a, \varepsilon)$ is defined in (19.19). What we have described here is in fact a strategy to compute logarithmic capacities of compact hulls that are symmetric under reflection in the real axis.

We have the following partial converse to Proposition 19.30.

Lemma 19.33. *Let $A \subseteq \mathbb{H}$. If the set $\overline{A \cup A^*}$ is a compact hull, then A is half-plane hull.*

Proof. Let $K = \overline{A \cup A^*}$. Since K is bounded, so is A. Furthermore, using that K is closed, $A \subseteq \overline{A} \cap \mathbb{H} \subseteq K \cap \mathbb{H} = A$, which proves that $\overline{A} \cap \mathbb{H} = A$.

Recall from Remark 19.9 that F_K induces a homeomorphism from $\mathbb{H} \setminus B[0,1]$ to $\mathbb{H} \setminus K = \mathbb{H} \setminus A$. Here F_K is a the normalized Riemann map coming from Proposition 19.4, associated with the compact hull K. Since $\mathbb{H} \setminus B[0,1]$ is both connected and simply connected, we may conclude that $\Omega = \mathbb{H} \setminus K$ is both connected and simply connected as well. Hence, A is a half-plane hull. $\square$

Example 19.34. We can now give a short argument why the sets A from Remark 19.23 are half-plane hulls.

Let $A = \gamma(]0, R]) \subseteq \mathbb{H}$, where $\gamma \colon [0, R] \to \overline{\mathbb{H}}$ is a simple curve and $\gamma(0) \in \mathbb{R}$. Then $\overline{A \cup A^*} = \sigma^*$, where $\sigma \colon [-R, R] \to \mathbb{C}$ is given by $\sigma(t) = \gamma(t)$ and $\sigma(-t) = \overline{\gamma(t)}$ for $t \in [0, R]$. By Example 19.3, K is a compact hull and we may therefore conclude by Lemma 19.33 that A is a half-plane hull.

As for the other case, let $\gamma \colon [0, 1] \to \overline{\mathbb{H}}$ be a simple curve with $\gamma(0) < \gamma(1)$ real numbers and $\gamma(t) \in \mathbb{H}$ for $t \in \,]0, 1[$. Form a closed Jordan curve $\sigma \colon [-1, 1] \to \mathbb{C}$ by setting $\sigma(t) = \gamma(t)$ and $\sigma(-t) = \overline{\gamma(t)}$ for $t \in [0, 1]$. Denote by U the associated Jordan region. Then $A = \overline{U} \cap \mathbb{H}$ is the set from Remark 19.23 and $\overline{A \cup A^*} = \overline{U}$. By Example 19.2, $\overline{U}$ is a compact hull and consequently, by Lemma 19.33, the set A is a half-plane hull.

Lemma 19.35. *Let A be a half-plane hull. Then*

(a) $\mathrm{hcap}(A) \geq 0$.

(b) $\mathrm{hcap}(A) = 0$ *if and only if $A = \emptyset$.*

Proof. Suppose A is a half-plane hull. Let $I = [a, b]$ with $a < b$ be such that $K = I \cup A \cup A^*$ is a compact hull, cf. Exercise 19.13.

Write $J = \mathbb{C} \setminus G_A(\mathbb{C} \setminus K) \subseteq \mathbb{R}$. Since K is a compact hull, J is a compact interval or a singleton, where we exploited that G_A is a homeomorphism between $\mathbb{C} \setminus K$ and $\mathbb{C} \setminus J$. Note that J cannot be a singleton. Indeed, if $J = \{\alpha\}$ is a singleton, then $F_A \colon \mathbb{C} \setminus \{\alpha\} \to \mathbb{C} \setminus K$ has a removable singularity at α, which would imply that K itself is a singleton. This contradicts the choice of I with $a < b$.

Write $J = [d - c, d + c]$, where $d \in \mathbb{R}$ and $c > 0$.

Note that the function $F_{[-2,2]}(z) = z + z^{-1}$ maps $\mathbb{C} \setminus B[0,1] \to \mathbb{C} \setminus [-2, 2]$. By translating $A \to A - d$, and hence $G_A \to G_{A+d}(z) = G_A(z + d) - d$, we may assume that $J = [-c, c]$. Finally, scaling $A \to 2A/c$, and replacing G_A by $G_{2A/c}(z) = 2G_A(cz/2)/c$, we may furthermore assume that $J = [-2, 2]$. According to Lemma 19.26, these operations do not change the sign of $\mathrm{hcap}(A)$.

The inverse map $F_A \colon \mathbb{C} \setminus [-2, 2] \to \mathbb{C} \setminus K$ has the expansion $F_A(z) = z - \mathrm{hcap}(A)/z + O(|z|^{-2})$ at infinity. Composing with $F_{[-2,2]}$, we get a new biholomorphic map $F_K \colon \mathbb{C} \setminus B[0,1] \to \mathbb{C} \setminus K$:

$$F_K(z) = F_A \circ F_{[-2,2]}(z) = z + \frac{1 - \mathrm{hcap}(A)}{z} + O(|z|^{-2}).$$

By the Area Theorem (Theorem 18.1), we must have $1 - \mathrm{hcap}(A) \le 1$ and hence $\mathrm{hcap}(A) \ge 0$. If $\mathrm{hcap}(A) = 0$, the Area Theorem[2] furthermore implies that $F_K(z) = z + z^{-1} = F_{[-2,2]}(z)$. Consequently, $F_A(z) = z$ and hence, $K = [-2,2]$ and therefore, $A = \emptyset$. $\qquad\square$

Proposition 19.36. *The half-plane capacity is a strictly monotone function of half-plane hulls, i.e.; for two half-plane hulls $A_1 \subseteq A_2$ we have $\mathrm{hcap}(A_1) \le \mathrm{hcap}(A_2)$ with equality if and only if $A_1 = A_2$.*

Proof. Denote by G_{A_1} and G_{A_2} the two Riemann maps associated with the simply connected regions $\Omega_1 = \mathbb{C} \setminus A_1$ and $\Omega_2 = \mathbb{H} \setminus A_2$. Put $A = G_{A_1}(A_2 \setminus A_1) \subseteq \mathbb{H}$. The set A is itself a half-plane hull, which we now proceed to argue.

We begin by arguing that A is closed as a subset of the open upper half-plane, which is equivalent to showing that $\Omega = \mathbb{H} \setminus A$ is open. Since G_{A_1} is a bijection, we must have $\Omega = G_{A_1}(\Omega_2)$. But G_{A_1}, by the Open Mapping Theorem (Theorem 4.14), maps open sets to open sets, and hence we are done.

To see that A is bounded, we recall that G_{A_1} is analytic in a neighborhood of infinity (after Schwarz reflection) and hence it maps an open neighborhood of infinity to an open neighborhood of infinity. Since G_{A_1} is injective and A_2 is bounded, we conclude that $G_{A_1}(A_2 \setminus A_1)$ must be bounded as well.

That Ω is connected and simply connected follows from $G_{A_1} \colon \Omega_2 \to \Omega$ being a homeomorphism. Here we used that Ω_2 is connected and simply connected, since A_2 is a half-plane hull.

Having established that A is a half-plane hull, we get a normalized Riemann map $G_A \colon \Omega \to \mathbb{H}$. We claim that $G_{A_2} = G_A \circ G_{A_1}$. This follows from an expansion at infinity

$$
\begin{aligned}
G_A \circ G_{A_1}(z) &= G_{A_1}(z) + \frac{\mathrm{hcap}(A)}{G_{A_1}(z)} + O(|z|^{-2}) \\
&= z + \frac{\mathrm{hcap}(A_1)}{z} + \frac{\mathrm{hcap}(A)}{z + O(|z|^{-1})} \\
&= z + \frac{\mathrm{hcap}(A_1) + \mathrm{hcap}(A)}{z} + O(|z|^{-2}).
\end{aligned}
$$

---

[2] Using the Area Theorem to prove that only the empty half-plane hull has zero half-plane capacity was suggested to us by Subhojoy Gupta.

This expansion verifies the claim and establishes the identity

$$\text{hcap}(A_2) = \text{hcap}(A_1) + \text{hcap}(A). \tag{19.22}$$

The proposition now follows from Lemma 19.35. $\qquad\qquad\square$

19.4 Convergence of half-plane hulls

We now turn to the study of Carathéodory convergence of half-plane hulls.

Definition 19.37. *Let $\{A_n\}_{n\in\mathbb{N}}$ be a sequence of half-plane hulls. We say that $A_n \to A$, with A a half-plane hull, in Carathéodory sense if:*

(a) *There exists $R > 0$ such that $A, A_n \subseteq B(0, R)$ for all n.*

(b) *The simply connected sets $\Omega_n = \mathbb{H} \setminus A_n$ converge to $\Omega = \mathbb{H} \setminus A$ in Carathéodory sense with base point $2Ri$.*

Theorem 19.38. *Let A be a half-plane hull and $\{A_n\}_{n\in\mathbb{N}}$ a sequence of half-plane hulls. Suppose there exists $R > 0$ such that $A, A_n \subseteq B(0, R)$, for all $n \in \mathbb{N}$. The two following properties are equivalent*

(a) *$A_n \to A$ in Carathéodory sense.*

(b) *$F_{A_n} \to F_A$ locally uniformly in the upper half-plane.*

If one of the two properties above are satisfied, we furthermore have

(c) *$G_{A_n} \to G_A$ locally uniformly in $\mathbb{H} \setminus A$.*

(d) *$\text{hcap}(A_n) \to \text{hcap}(A)$.*

(e) *$b_k^{(n)} \to b_k$, for $k \geq 2$. Here the b_k's and the $b_k^{(n)}$'s are the expansion coefficients from (19.9), associated with A and A_n, respectively.*

Proof. We begin with (a) $\Rightarrow$ (b).

The key idea is to exploit the relation between the half-plane hulls A_n and A and the compact hulls $K = C_{\bar{z}_0}(A) \cup B[0,1]$ and $K_n = C_{\bar{z}_0}(A_n) \cup B[0,1]$ from Remark 19.25. Here we abbreviated $z_0 = 2Ri$, a point in $\Omega = \mathbb{H} \setminus A$ as well as in $\Omega_n = \mathbb{H} \setminus A_n$ for all n.

By the definition of Carathéodory convergence for compact hulls, we find that $K_n \to K$ in Carathéodory sense. See Proposition 18.10 and Definition 19.14.

Let G_K, F_K and G_{K_n}, F_{K_n} denote the normalized Riemann maps associated with the compact hulls K and K_n, respectively.

We remind the reader that G_{A_n} and G_A may be expressed in terms of G_{K_n}, G_K and the derivatives $G_{K_n}^{(k)}(1)$ and $G_K^{(k)}(1)$ for $k = 0, 1, 2$, where G_{K_n} and G_K has been extended by Schwarz reflection in an arc to a neighborhood of 1. See (19.11), (19.14), (19.15) and (19.16). Consequently, F_A and F_{A_n} may also be expressed in terms of F_{K_n} and F_K together with the same derivatives of the G's at 1.

By Theorem 19.17, we know that $F_{K_n} \to F_K$ locally uniformly in $\mathbb{C} \setminus B[0, 1]$. By Remark 19.25, it remains to argue that $G_{K_n}^{(k)}(1) \to G_K^{(k)}(1)$ for $n \to \infty$, $k = 0, 1, 2$. It also follows from Theorem 19.17 that

$$G_{K_n} \to G_K \tag{19.23}$$

locally uniformly in $\mathbb{C} \setminus K$. Hence, the Schwarz reflected G_{K_n}'s, cf. Remark 19.25, converge locally uniformly to the Schwarz reflected G_K in $(\mathbb{C} \setminus K) \cup \Theta(\mathbb{C} \setminus K)$. Note that since $A_n, A \subseteq B(0, R)$ for all R, the arc $J \subseteq \partial B(0, 1)$ across which we reflect, may be chosen independent of n.

It follows from Lemma 19.13 that G_{K_n} is uniformly bounded in a disc of the form $B[1, r] \subseteq (\mathbb{C} \setminus K) \cup J \cup \Theta(\mathbb{C} \setminus K)$ for some $r > 0$. By Vitali-Porter's theorem (Exercise 8.10), we conclude that $G_{K_n} \to G_K$ locally uniformly in $B(1, r)$. Consequently, by Weierstrass' theorem (Theorem 4.10),

$$\text{for } k = 0, 1, 2, \dots : \quad G_{K_n}^{(k)}(1) \to G_K^{(k)}(1). \tag{19.24}$$

This completes the proof of (a) $\Rightarrow$ (b).

We now proceed to argue that (b) $\Rightarrow$ (a). Define holomorphic functions $\tilde{h} = F_A \circ \text{Cay}^{-1} \colon B(0, 1) \to \Omega$, and $\tilde{h}_n = F_{A_n} \circ \text{Cay}^{-1} \colon B(0, 1) \to \Omega_n$. By assumption, $\tilde{h}_n \to \tilde{h}$ locally uniformly in $B(0, 1)$. Abbreviate $w = \tilde{h}^{-1}(2Ri) \in B(0, 1)$ and $w_n = \tilde{h}_n^{-1}(2Ri) \in B(0, 1)$. Since $B(2Ri, R/2) \subseteq F_A(\mathbb{H})$ and $B(2Ri, R/2) \subseteq F_{A_n}(\mathbb{H})$ for all n, we observe that $\tilde{h}^{-1}$ and $\tilde{h}_n^{-1}$ are all defined in the common disc $B(2Ri, R/2)$. Repeating the argument for Case 2 in Theorem 18.12 (which relied on Montel's theorem), one may argue that $\tilde{h}_n^{-1} \to \tilde{h}^{-1}$ locally uniformly in $B(2Ri, R/2)$. In particular, $w_n \to w$ which is what we shall need.

Define now $h(z) = \tilde{h}(A(-w; z))$ and $h_n(z) = \tilde{h}_n(A(-w_n; z))$. Here $A(\zeta; \cdot)$ are the automorphisms of the unit disc from Example 8.14. Recall that $A(-\zeta, 0) = \zeta$. Then $h(0) = h_n(0) = 2Ri$, $h_n \to h$ locally uniformly in $B(0, 1)$. By Weierstrass' convergence theorem, cf. Theorem 4.10, we conclude that $h_n'(0) \to h'(0)$. Since h and the h_n's are univalent, we must have $h'(0) \neq 0$ and $h_n'(0) \neq 0$.

Abbreviate $u = h'(0)/|h'(0)|$ and $h_n = h'_n(0)/|h'_n(0)|$ and define Riemann maps $f(z) = h(\bar{u}z)$ and $f_n(z) = h_n(\bar{u}_n z)$. These univalent functions satisfy that $f(0) = f_n(0) = 2Ri$, $f'(0) > 0$ and $f'_n(0) > 0$. We furthermore have $f_n \to f$ locally uniformly in $B(0,1)$, $f(B(0,1)) = \Omega$ and $f_n(B(0,1)) = \Omega_n$. Hence, invoking Theorem 18.12, we conclude that $\Omega_n \to \Omega$, in Carathéodory sense with base point $2Ri$.

It remains to establish the claimed consequences of having one, hence both, of the two equivalent properties satisfied for a bounded sequence of half-plane hulls.

We may express G_A and the G_{A_n}'s in terms of G_K and the G_{K_n}'s from above together with $G_K^{(k)}(1)$ and $G_{K_n}^{(k)}(1)$ for $k = 0,1,2$. It now follows from (19.23) and (19.24) together with Remark 19.25 that $G_{A_n} \to G_A$ locally uniformly in $\mathbb{H} \setminus A$.

Our next task is to establish convergence of expansion coefficients. As usual, we recycle the notation G_A and G_{A_n} for the Riemann maps Schwarz reflected across $\mathbb{R} \setminus [-R, R]$ to holomorphic functions in $\mathbb{C} \setminus B(0, R)$. Let $H(w) = G_A(w^{-1}) - w^{-1}$ and $H_n(w) = G_{A_n}(w^{-1}) - w^{-1}$, which are holomorphic functions in $B(0, 1/R)$, when extended to zero at zero. See Proposition 19.24. It follows that

$$H(w) = \mathrm{hcap}(A)w + \sum_{k=2}^{\infty} b_k w^k$$

and

$$H_n(w) = \mathrm{hcap}(A_n)w + \sum_{k=2}^{\infty} b_k^{(n)} w^k,$$

with radii of convergence $1/R$. Put $b_1 = \mathrm{hcap}(A)$ and $b_1^{(n)} = \mathrm{hcap}(A_n)$. On compact subsets $C \subseteq B(0, 1/R) \setminus \{0\}$, we have uniform convergence $H_n \to H$. By Cauchy's integral formula, we now conclude that for $k = 1, 2, 3 \ldots$ we have

$$b_k = \frac{1}{2\pi i} \int_{|w|=(2R)^{-1}} \frac{H(w)}{w^k} \, dw$$

$$= \lim_{n \to \infty} \frac{1}{2\pi i} \int_{|w|=(2R)^{-1}} \frac{H_n(w)}{w^k} \, dw = \lim_{n \to \infty} b_k^{(n)},$$

such that all the expansion coefficients in fact converge. In particular, $\mathrm{hcap}(A_n) \to \mathrm{hcap}(A)$. $\square$

The demand in Definition 19.37 that the sequence of half-plane hulls be bounded appears to be of a technical nature, and we suspect that it may be replaced by a weaker requirement. The following example explores consequences of weakening the boundedness assumption.

Example 19.39. Let $A_n^\kappa = \mathbb{H} \cap B[n, n^\kappa]$, where $n \in \mathbb{N}$ and $\kappa \in \mathbb{R}$. Observe that A_n does not converge in Carathéodory sense for any κ, since the sequence of half-plane hulls fails to be confined inside a common disc. In this example we examine convergence of $\Omega_n = \mathbb{H} \setminus A_n$, $G_n = G_{A_n}$, $F_n = F_{A_n}$ and hcap(A_n) for different values of κ.

Let us first compute the objects of interest. We have

$$\text{hcap}(A_n) = n^{2\kappa}$$

and

$$G_{A_n}(z) = z + \frac{n^{2\kappa}}{z - n} \quad \text{and} \quad F_{A_n}(z) = \tfrac{1}{2}(z + n) + \tfrac{1}{2}\sqrt{(z - n)^2 - 4n^{2\kappa}},$$

where $\sqrt{\cdot} : \mathbb{C} \setminus [0, +\infty[\to \mathbb{H}$ is the branch of the square root with positive imaginary part. Expanding around $z = \infty$ yields

$$G_{A_n}(z) = z + \frac{n^{2\kappa}}{z} + \sum_{k=2}^{\infty} \frac{b_k^{(n)}}{z^k}, \qquad \text{where } b_k^{(n)} = n^{2\kappa + k - 1}.$$

$\kappa > 1$: Here none of the objects, Ω_n, G_n, F_n or hcap(A_n) converge.

$\kappa = 1$: Here $\Omega_n \to \{z \in \mathbb{H} \mid \text{Re}\, z < 0\}$ (with base point $i - 1$), but G_n, F_n and hcap(A_n) do not converge.

$1/2 \le \kappa < 1$: Here $\Omega_n \to \mathbb{H}$ (with base point i), but G_n, F_n and hcap(A_n) do not converge.

$0 < \kappa < 1/2$: Here $\Omega_n \to \mathbb{H}$ (with base point i), $G_n \to G_\emptyset = z$, $F_n \to F_\emptyset = z$ (locally uniformly in $\mathbb{H}$) but hcap(A_n) diverges.

$\kappa = 0$: Here $\Omega_n \to \mathbb{H}$ (with base point i), $G_n \to G_\emptyset = z$, $F_n \to F_\emptyset = z$ (locally uniformly in $\mathbb{H}$) but hcap$(A_n) = 1 \neq \text{hcap}(\emptyset)$.

$\kappa < 0$: Here $\Omega_n \to \mathbb{H}$ (with base point i), $G_n \to G_\emptyset = z$, $F_n \to F_\emptyset = z$ (locally uniformly in $\mathbb{H}$) and hcap$(A_n) \to \text{hcap}(\emptyset) = 0$.

In particular we observe that if we do not impose an a priori assumption on boundedness of the sequence of half-plane hulls in Theorem 19.38, the equivalence of (a) and (b) in the theorem would fail to hold true.

The example however indicates that the demand in Definition 19.37 that the half-plane hulls be confined to a disc, should perhaps be replaced by a weaker demand that the A_n's be located below a curve of the form $\mathrm{Im}\, z = R|\mathrm{Re}\, z|^\kappa$ for some $\kappa < 0$ and $R > 0$.

Note that given $k_0 \in \mathbb{N}$ one may choose $\kappa < 0$ appropriately, such that we get convergence $b_k^{(n)} \to 0$ for $k \leq k_0$, but with $b_k^{(n)}$ divergent for $k > k_0$. Hence, if one were to modify Definition 19.37 as suggested in the preceding paragraph, then one would loose the convergence of all the expansion coefficients $b_k^{(n)}$, which was one of the consequences in Theorem 19.38.

The following lemma exemplifies the conclusions of Theorem 19.38 and is proved by an explicit construction.

Lemma 19.40. *Let A be a nonempty half-plane hull, $I = [\inf(\overline{A} \cap \mathbb{R}), \sup(\overline{A} \cap \mathbb{R})]$ and $K = A \cup A^* \cup I$. There exists a sequence of simple curves/Jordan arcs $\gamma_n \colon [0,1] \to \overline{\mathbb{H}}$ with*

(a) *$\gamma_n(0) < \gamma_{n+1}(0) < \gamma_{n+1}(1) < \gamma_n(1)$ real numbers, and $\gamma_n(]0,1[) \subseteq \mathbb{H}$.*

(b) *$A \subseteq \overline{U}_{n+1} \cap \mathbb{H} \subseteq U_n$ for all $n \in \mathbb{N}$, where U_n is the Jordan region with boundary $\gamma_n^* \cap [\gamma_n(0), \gamma_n(1)]$.*

(c) *Let $A_n = \overline{U}_n \cap \mathbb{H}$. Then $A_n \to A$ in Carathéodory sense as half-plane hulls[3] and $\mathrm{hcap}(A_n) \to \mathrm{hcap}(A)$.*

(d) *$K_n = \overline{A_n \cup A_n^*} \to K$ in Carathéodory sense as compact hulls.*

(e) *$G_{A_n} \to G_A$ locally uniformly in $\mathbb{C} \setminus K$.*

Proof. We shall make use of the constructions and results from Proposition 19.30, which we ask the reader to recall before continuing. Let $a = \inf(\overline{A} \cap \mathbb{R})$ and $b = \sup(\overline{A} \cap \mathbb{R})$.

Let $G_A \colon \mathbb{H} \setminus A \to \mathbb{H}$ be the normalized Riemann map associated to A. Recall that G_A can be extended by Schwarz reflection across the real axis to a univalent map, again denoted by G_A, defined in $\mathbb{C} \setminus C$. If we restrict G_A to $\mathbb{C} \setminus K$, we again get a univalent map and $G_A(\mathbb{C} \setminus K) = \mathbb{C} \setminus \tilde{J}$, where $\tilde{J} = [\alpha, \beta]$, $\alpha = \inf J$ and $\beta = \sup J$. Here we used that G_A is injective, in fact strictly increasing, on $]\infty, a[$ and $]b, \infty[$.

We may assume that $\tilde{J} = [-1, 1]$. Indeed, by translating A along the real axis, a distance $(\alpha + \beta)/2$, and rescaling $A \to cA$ with $c = 2/(\beta - \alpha)$, we can bring ourselves to this situation. See (the proof of) Lemma 19.26.

We now exploit the construction from Example 19.29. Let $a_n = 1 + 1/n$ and $\varepsilon_n = 1/n$. Denote by B_n the half-plane hull from Example 19.29,

---

[3] Recall from Remark 19.23 that the A_n's are half-plane hulls.

the region between the real line and the circular arc passing through $\pm a_n$ and $i\varepsilon_n$. Let r_n and c_n be the associated radius and center of the circle on which the n'th arc lies. See (19.17).

The angular extension of the arc from a_n to $-a_n$ is $[\pi/2 - \tau_n, \pi/2 + \tau_n]$, where

$$\tau_n = \arctan\left(\frac{2\varepsilon_n a_n}{a_n^2 - \varepsilon_n^2}\right).$$

See (19.18) and Fig. 19.4.

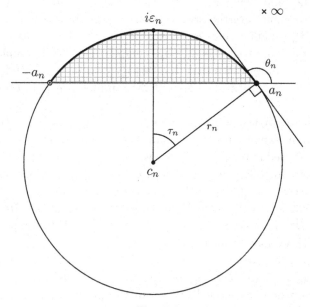

Fig. 19.4

Define closed Jordan curves $\sigma_n \colon [-1, 1] \to \mathbb{H}$ by setting, for $t \in [0, 1]$, $\sigma_n(t) = c_n + i r_n e^{i\tau_n(1-2t)}$ and subsequently $\sigma_n(-t) = \overline{\sigma_n(t)}$. The boundary of the associated Jordan region is $\sigma^* = \partial\overline{B}_n \cup B_n^*$.

We may now define our choice of Jordan arc to be

$$\gamma_n = F_A \circ \sigma_n.$$

Since $\sigma_n(0) = -a_n$, $\sigma_n(1) = a_n$ and F_A is strictly increasing on $]-\infty, -1[\cup]1, \infty[$, we have ensured the property (a). Note that the γ_n's in the lemma are simply the restriction of the γ_n's above to $[0, 1]$.

Let V_n denote the bounded connected component of $\mathbb{C} \setminus \gamma_n^*$. By Lemma 19.16, we have $K \subseteq V_n$ for all n. Since $U_n = V_n \cap \mathbb{H}$, we conclude that $A \subseteq U_n$ for all n. That $\overline{V}_{n+1} \subseteq V_n$ also follows from Lemma 19.16 and therefore we conclude that $\overline{U}_{n+1} \cap \mathbb{H} \subseteq U_n$ for all n as well. Note that K and the $\overline{V}_n$'s are connected sets. We have now established the property (b).

In order to verify (c), we observe first the easy inclusions $A \subseteq \bigcap_{n=1}^{\infty} U_n \subseteq \bigcap_{n=1}^{\infty} A_n$. Let $z \in \mathbb{H} \setminus A$. Then $G_A(z) \in \mathbb{H}$. Pick n such that $\varepsilon_n < \operatorname{Im} G_A(z)$. With this choice of n we have $G_A(z) \notin B_n$ and therefore $z \notin A_n$. This show that $\mathbb{H} \setminus A \subseteq \bigcup_{n=1}^{\infty} \mathbb{H} \setminus A_n$, implying the converse inclusion $\bigcup_{n=1}^{\infty} A_n \subseteq A$. That $A_n \to A$ in Carathéodory sense now follows from Example 18.9. Although the limit $\operatorname{hcap}(A_n) \to \operatorname{hcap}(A)$ is now a consequence of Theorem 19.38, we shall nevertheless give a direct proof of this fact below.

As for (d), we abbreviate $K_n = \overline{A_n \cup A_n^*} = \overline{V}_n$, the closure of the n'th Jordan region. Note that the K_n's are compact hulls by Example 19.2. It follows from Example 18.9 that $K_n \to \bigcap_{n=1}^{\infty} K_n$ in Carathéodory sense. That $\bigcap_{n=1}^{\infty} K_n = K$ is a repetition of the argument used to establish (c) above.

We turn to our last item (e). Note first that $G_{A_n} = G_{B_n} \circ G_A$, which follows from an expansion around infinity and Proposition 19.24. A similar argument was used in the proof of Proposition 19.36. Here $G_{B_n} : \mathbb{H} \setminus B_n \to \mathbb{H}$ is the normalized Riemann map associated with the half-plane hull B_n.

Observe that $G_{B_n}(z) \to z$ locally uniformly in $\mathbb{C} \setminus [-1, 1]$. See also Example 19.32 for a discussion of the explicit extension of G_{B_n} from $\mathbb{H} \setminus B_n$ to $\mathbb{C} \setminus \overline{B_n \cup B_n^*}$. Since G_A maps compact subsets of $\mathbb{C} \setminus K$ to compact subsets of $\mathbb{C} \setminus [-1, 1]$, we may conclude (e).

It remains to give a direct argument for $\operatorname{hcap}(A_n) \to \operatorname{hcap}(A)$. The identity $G_{A_n} = G_A \circ G_{B_n}$ implies the formula $\operatorname{hcap}(A_n) = \operatorname{hcap}(A) + \operatorname{hcap}(B_n)$ as when we derived (19.22) in the proof of Proposition 19.36. From (19.19) and (19.21), we observe that $\operatorname{hcap}(B_n) \to 0$, which implies the result.   □

As a consequence, we have the following proposition, which – apart from being of independent interest – plays a subtle role in our final theorem of the chapter.

Proposition 19.41. *Let A be a half-plane hull and $x > \max(\overline{A} \cap \mathbb{R})$ ($x < \min(\overline{A} \cap \mathbb{R})$), a real number. Then $G_A(x) \geq x$ ($G_A(x) \leq x$), where G_A has been extended by continuity to $\mathbb{R} \setminus \overline{A}$.*

Proof. Assume first that there exists a Jordan arc $\gamma \colon [0, 1] \to \overline{\mathbb{H}}$, such that $\gamma(0) < \gamma(1)$ are real numbers, $\gamma(]0, 1[) \subseteq \mathbb{H}$ and $\overline{A}$ is the closure of the Jordan region enclosed by $\gamma^* \cup [\gamma(0), \gamma(1)]$. Let $I = [\gamma(0), \gamma(1)]$ and put $J = \mathbb{R} \setminus G_A(\mathbb{R} \setminus I)$, which is a compact interval.

By Corollary 17.20, the map $F_A \colon \mathbb{H} \to \mathbb{H} \setminus A$ extends by continuity to $\overline{\mathbb{H}} \to \overline{\mathbb{H} \setminus A}$, which we also denote by F_A. note that $F_A(x) \in \mathbb{R}$ for $x \in \mathbb{R} \setminus J$. We can now use the half-plane version of Schwarz' lemma (Theorem 14.7) (with $f(z) = -i(F_A(z) - z)$) to compute for $w \in \mathbb{H}$

$$F_A(w) - w = \frac{1}{\pi} \int_{\mathbb{R}} \frac{\operatorname{Im}(F_A(s) - s)}{s - w} \, ds = \frac{1}{\pi} \int_J \frac{\operatorname{Im} F_A(s)}{s - w} \, ds.$$

Let $x \in \mathbb{R} \setminus I$. Both sides of the equation above extends by continuity to $w = G_A(x)$, which yields the identity

$$\forall x \in \mathbb{R} \setminus I : \quad x - G_A(x) = \frac{1}{\pi} \int_J \frac{\operatorname{Im} F_A(s)}{s - G_A(x)} \, ds.$$

If $x > \max I$, then $G_A(x) > \max J$ and the right-hand side is negative, since $\operatorname{Im} F_A(s) \geq 0$ for all $s \in \mathbb{R}$. Hence, $G_A(x) \geq x$. Similarly, if $x < \min I$, we must have $G_A(x) < \min J$ and consequently, $G_A(x) < x$.

The proposition now follows from Lemma 19.40. $\qquad\square$

For a stronger result along the same lines as the above proposition, we refer the reader to Exercise 20.5.

Our next task is to study convergence of G_{A_n} and F_{A_n} Schwarz reflected across the real line to $\mathbb{C} \setminus C_n$ and $\mathbb{C} \setminus J_n$, respectively. The sets C_n and J_n were introduced in Proposition 19.30. Suppose we have a sequence of half-plane hulls A_n, $n \in \mathbb{N}$, and we form the associated compact sets $C_n = \overline{A_n \cup A_n^*}$. One may naturally ask what implications Carathéodory convergence of the A_n's have on the C_n's. Lemma 19.40 provides a positive class of examples, whereas the following example is negative:

Example 19.42. Let $A_n = \,]0, i/n] \cup [-1 + i/n, 1 + i/n]$. We leave it to the reader to check that this is indeed a sequence of half-plane hulls and that $A_n \to A = \emptyset$ in Carathéodory sense. However, the associated compact sets, which are in fact compact hulls, are $C_n = [-i/n, i/n] \cup [-1 + i/n, 1 + i/n] \cup [-1 - i/n, 1 - i/n]$. The C_n's do in fact converge in Carathéodory sense but $C_n \to [-1, 1]$, which is not naturally related to $A = \emptyset$.

Matters may be even worse. Let A_{2n+1} be as above, but for even indices we replace with $A_{2n} = \,]0, i/n] \cup [-2 + i/n, 2 + i/n]$. We still have $A_n \to \emptyset$, but now the associated compact hulls do not converge at all in Carathéodory sense.

Before we continue, we establish the following half-plane analogue of Lemma 19.13, which is where we need Proposition 19.41.

Lemma 19.43. *Let $R > 0$. For any half-plane hull $A \subseteq B(0, R)$, the following holds:*

(a) *Writing $C = \overline{A \cup A^*}$ and $J = \mathbb{C} \setminus G_A(\mathbb{C} \setminus C)$, we have $J \subseteq [-4R, 4R]$,*

$$\sup_{z \in B[0,R] \setminus C} |G_A(z)| \leq 4R$$

and

$$\forall z \in \mathbb{C} \setminus B(0, R): \quad |G_A(z) - z| \leq \frac{5R^2}{|z|}.$$

(b) *For any nonempty half-plane hull A, with $A \subseteq B(0, R)$, we have*

$$\sup_{w \in B[0,4R] \setminus J} |F_A(w)| \leq 5R$$

and

$$\forall w \in \mathbb{C} \setminus B(0, 4R): \quad |F_A(w) - w| \leq \frac{36R^2}{|w|}.$$

Proof. Recall from Proposition 19.30 that G_A extends (by Schwarz reflection) to a univalent map from $\mathbb{C} \setminus C$ into $\mathbb{C}$ and the complement of its range, $J = \mathbb{C} \setminus G_A(\mathbb{C} \setminus C)$, is a compact subset of the real line.

We begin with (a). The extended G_A is defined outside of $B(0, R)$, for any half-plane hull $A \subseteq B(0, R)$. Furthermore, by Proposition 19.41, G_A does not vanish outside $B(0, R)$ and hence, the inverted map

$$g_A(w) = R G_A(R w^{-1})^{-1}$$

is holomorphic in $B(0, 1)$. A priori away from $w = 0$, but since $|G_A(z)| \to \infty$, as $|z| \to \infty$, the point 0 is a removable singularity. The extension is univalent and satisfies $g_A(0) = 0$. The scaling is furthermore chosen such that $g'_A(0) = 1$. By Koebe's 1/4-theorem (Theorem 18.4), we must have $B(0, 1/4) \subseteq \mathrm{Range}(g_A)$. Consequently, $\mathbb{C} \setminus B[0, 4R] \subseteq G_A(\mathbb{C} \setminus B[0, R])) \subseteq G_A(\mathbb{C} \setminus C)$. Since G_A is univalent, we conclude $J \subseteq [-4R, 4R]$ and

$$G_A(B[0, R] \setminus C) \subseteq B[0, 4R]. \tag{19.25}$$

To estimate $G_A(z) - z$ outside of $B(0, R)$, we define

$$h_A(w) = G_A(w^{-1}) - w^{-1},$$

which is holomorphic in $B(0, 1/R)$ and continuous on $B[0, 1/R]$. Here we again extend by 0 across the removable singularity at $w = 0$. By the Weak

Maximum Principle (Corollary 4.16), we conclude that

$$\sup_{|w| \leq 1/R} |h_A(w)| \leq \sup_{|w|=1/R} |G_A(w^{-1})| + R \leq 5R.$$

Rescale and define $\tilde{h}_A(w) = (5R)^{-1} h_A(w/R)$. Then $\tilde{h}_A$ is holomorphic in $B(0,1)$, $\tilde{h}_A(0) = 0$ and $\tilde{h}_A(B(0,1)) \subseteq B(0,1)$. Hence, by Schwarz' lemma (Theorem 4.17), we get

$$\forall w \in B(0,1): \quad |\tilde{h}_A(w)| \leq |w|.$$

This completes the proof of (a).

As for (b), let $z \in \mathbb{C} \setminus B(0, R)$. Using (a), we observe the inequality $|G_A(z)| \geq |z| - |G_A(z) - z| \geq |z| - \frac{5R^2}{|z|}$. This estimate implies that $|G_A(z)| \geq 4R$ for all half-plane hulls $A \subseteq B(0, R)$ and z, with $|z| \geq 5R$. Therefore

$$F_A(B[0, 4R] \setminus J) \subseteq B[0, 5R]. \tag{19.26}$$

We proceed to argue, using the Weak Maximum Principle and Schwarz' lemma, as in the proof of (a). Let $h(w) = F_A(w^{-1}) - w^{-1}$, which extends to a holomorphic function with $h(0) = 0$, defined in $B(0, 1/(4R))$ and continuous on the boundary. Hence, using (19.26), we conclude that $|h(w)| \leq 5R + 4R = 9R$, for all $w \in B(0, 1/(4R))$. Applying Schwarz' lemma to $\tilde{h}(w) = (9R)^{-1} h(w/(4R))$, we conclude that $h(w) \leq 36R^2 |w|$, for all $w \in B(0, 1/(4R))$. Inserting the defining expression for h, we arrive at the remaining claimed bound for $w \in \setminus B(0, 4R)$. $\qquad \square$

Theorem 19.44. *Let $\{A_n\}_{n \in \mathbb{N}}$ be a sequence of half-plane hulls converging in Carathéodory sense to a half-plane hull A. Abbreviate*

$$C_n = \overline{A_n \cup A_n^*}, \qquad\qquad J_n = \mathbb{C} \setminus G_{A_n}(\mathbb{C} \setminus C_n),$$
$$C = \overline{A \cup A^*}, \qquad\qquad J = \mathbb{C} \setminus G_A(\mathbb{C} \setminus C),$$
$$\widetilde{C} = \mathbb{C}_\infty \setminus \ker(\{\mathbb{C}_\infty \setminus C_n\}_{n \in \mathbb{N}}, \infty) \quad and \quad \widetilde{J} = \mathbb{C} \setminus G_A(\mathbb{C} \setminus \widetilde{C}).$$

(a) *We have $C \subseteq \widetilde{C}$ and $J \subseteq \widetilde{J}$.*

(b) *$G_{A_n} \to G_A$ locally uniformly in $\mathbb{C} \setminus \widetilde{C}$.*

(c) *$\ker(\{\mathbb{C}_\infty \setminus J_n\}_{n \in \mathbb{N}}, \infty) = \mathbb{C}_\infty \setminus \widetilde{J}$.*

(d) *$F_{A_n} \to F_A$ locally uniformly in $\mathbb{C} \setminus \widetilde{J}$.*

Proof. We begin with (a). Let $R > 0$ be such that $A_n, A \subseteq B[0, R]$, and hence also $C_n, \widetilde{C} \subseteq B[0, R]$.

Let $w \in \mathbb{C} \setminus \widetilde{C}$. We have to show that $w \notin C$. By reflection symmetry in the real axis, we may assume that $w \in \overline{\mathbb{H}}$ and therefore it suffices to show that $w \notin \overline{A}$. Using that $\mathbb{C} \setminus \widetilde{C}$ is connected, there exists a curve $\gamma \colon [0, 1] \to \mathbb{C} \setminus \widetilde{C}$ with $\gamma(0) = w$ and $\gamma(1) = 2Ri$. Abbreviate $d(z, \gamma([0,1])) = \inf\{|z - \gamma(t)| \mid t \in [0,1]\}$ for $z \in \mathbb{C}$. Since $\widetilde{C}$ and $\gamma([0,1])$ are compact, there exists $r > 0$ such that $K_r = \{z \in \mathbb{C} \mid d(z, \gamma([0,1])) \leq r\} \cap \widetilde{C} = \emptyset$.

Since K_r is compact in $\mathbb{C}_\infty \setminus \widetilde{C}$, we must – by Remark 18.7 – have an N such that $K_r \cap C_n = \emptyset$ for $n \geq N$. Put $U = \mathbb{H} \cap (V_r \cup V_r^*)$, where $V_r = \{z \in \mathbb{C}_\infty \mid d(z, \gamma([0,1])) < r\} \subseteq K_r$. Then U is open, connected and contains $2Ri$. With this construction, we have $U \cap A_n = U \cap C_n = \emptyset$ for $n \geq N$. Appealing to Definition 18.6 and (18.16), we conclude that $U \in \mathcal{K}$ and, hence; $U \subseteq \ker(\{\mathbb{H} \setminus A_n\}_{n \in \mathbb{N}}, 2Ri) = \mathbb{H} \setminus A$. Since $B(w, r) \cap \mathbb{H} \subseteq U$, we may conclude that $w \notin \overline{A}$, which wraps up the proof of the first inclusion in (a). Note that J is (now) well-defined and the inclusion $J \subseteq \widetilde{J}$ is obvious.

We now turn to the proof of (b). It suffices to show that for any $w \in \mathbb{C} \setminus \widetilde{C}$, there exists $r > 0$, such that $B[w, r] \cap \widetilde{C} = \emptyset$ and $G_{A_n} \to G_A$ locally uniformly in $B(w, r)$.

If $w \in \mathbb{H}$, we choose $r > 0$ such that $B[w, r] \subseteq \mathbb{H}$ and $B[w, r] \cap A = \emptyset$. Locally uniform convergence $G_{A_n} \to G_A$ on $B(a, r)$ now follows from Theorem 19.38.

If $w \in \mathbb{H}^*$ convergence follows from what has just been proved together with the relation $G_{A_n}(\bar{z}) = \overline{G_{A_n}(z)}$ and $G_A(\bar{z}) = \overline{G_A(z)}$.

Finally, we are reduced to consider the case $w \in \mathbb{R} \setminus \widetilde{C}$. Choose $r > 0$ such that $B[w, r] \cap \widetilde{C} = \emptyset$ and subsequently N such that $B[w, r] \cap C_n = \emptyset$ for $n \geq N$. We claim that this choice of r works.

Pick an $R > 0$ such that $A_n, B[w, r] \subseteq B[0, R]$ for all $n \in \mathbb{N}$. By Lemma 19.43(a), we have

$$\sup_{n \geq N, |z - w| \leq r} |G_{A_n}(z)| \leq 4R. \tag{19.27}$$

Since $G_{A_n} \to G_A$ in $\mathbb{H} \cap B(w, r)$, it follows from (19.27) and Vitali-Porter's theorem (Exercise 8.10) that $G_{A_n} \to G_A$ locally uniformly in $B(w, r)$.

We now turn to (c). Put $V := \ker(\{\mathbb{C}_\infty \setminus J_n\}, \infty)$. We need to show that $V = \mathbb{C}_\infty \setminus \widetilde{J}$.

We first argue that $\mathbb{C}_\infty \setminus \widetilde{J} \subseteq V$, which we do by showing that $U := \mathbb{C}_\infty \setminus \widetilde{J} \in \mathcal{K}$, where we refer the reader to (18.16) and Definition 18.6 for the notation associated with Carathéodory kernels. Note that U is open, connected and contains the base point ∞. To conclude that $U \in \mathcal{K}$, it suffices to show that for any $\xi \in U$ (different from ∞), there exists $r > 0$ such that $B[\xi, r] \cap J_n = \emptyset$ eventually.

Let $\xi \in U$ and choose $s > 0$ such that $B[\xi, s] \cap \tilde{J} = \emptyset$. Abbreviate $K = F_A(B[\xi, s])$. Since K is compact and $K \cap C = \emptyset$, there exists N such that $K \cap C_n = \emptyset$ for $n \geq N$. This means that G_{A_n} is defined on K for $n \geq N$. Then, by (b), $G_{A_n} \to G_A$ uniformly in K. By Lemma 18.11, applied with $\Omega = F_A(B(\xi, s))$ and $a = F_A(\xi)$, there exists $0 < s' < s$ and $N' \geq N$ such that $B(\xi, s') \subseteq G_{A_n}(K)$ for all $n \geq N'$. Hence $B[\xi, s'/2] \cap J_n = \emptyset$ for $n \geq N'$. This show that $U \subseteq V$.

To argue that $\mathbb{C}_\infty \setminus \tilde{J} \supseteq V$, we fix an $x \in V$, which we may assume to be real since $\mathbb{H}, \mathbb{H}^* \subseteq \mathbb{C}_\infty \setminus \tilde{J}$. Since V is open we get an $r > 0$ such that $B[x, r] \subseteq V$. Since V is a Carathéodory kernel, and $B[x, r]$ is compact, we get an N such that $B[x, r] \cap J_n = \emptyset$ for $n \geq N$.

Since $F_{A_n} \to F_A$ locally uniformly in $B(x, r) \cap \mathbb{H}$ by Theorem 19.38, we obtain from Lemma 19.43(b) and Vitali-Porter's theorem (Exercise 8.10) that $F_{A_n} \to F$ locally uniformly in $B(x, r)$ where F is a holomorphic extension of F_A from $\mathbb{H} \cup \mathbb{H}^*$ across $]x - r, x + r[$. Since F is not constant, it must be univalent by Exercise 9.18.

Let $\Omega =]x - r, x + r[\times]-10Ri, 10Ri[$ and $U' = F(\Omega) \cup \{z \in \mathbb{C} \mid |z| > R\} \cup \{\infty\}$, which is open and connected. To see that U' is connected it suffices to argue that there exists $z \in F(\Omega)$ with $|z| > R$. But this follows from Lemma 19.43(b) ($z = F(x + 8Ri)$ will do).

We claim that $U' \subseteq \mathbb{C}_\infty \setminus \tilde{C} = \ker(\{\mathbb{C}_\infty \setminus C_n\}, \infty)$. To see this, it suffices to show that for any $w \in U'$ (not equal to ∞) there exist $s > 0$ and M such that $B[w, s] \cap C_n = \emptyset$ for $n \geq M$. Let $w \in U'$ (with $w \neq \infty$). Then there exists $a \in \Omega$ such that $F(a) = w$. By Lemma 18.11, there exists $\mathcal{O} \ni w$ open and M such that $\mathcal{O} \subseteq F_{A_n}(\Omega)$ for all $n \geq M$. Choose $s > 0$ such that $B[w, s] \subseteq \mathcal{O}$. Then $B[w, s] \cap C_n = \emptyset$ for $n \geq M$, which concludes the proof of (c).

The last item (d), was actually observed above to be a consequence of Vitali-Porter. Note that (a) and (c) ensure that the claim in (d) is meaningful. $\square$

Exercises

19.1. Show that $f(z) = e^{i\theta}z$, for $\theta \in \mathbb{R}$, are the only biholomorphic maps from $\mathbb{C}_\infty \setminus B[0,1]$ onto $\mathbb{C}_\infty \setminus B[0,1]$, satisfying $f(\infty) = \infty$.

19.2. Let K be a compact hull. Prove that K consists of uncountably many points.

19.3. Let $K = [a,b] \subseteq \mathbb{R}$ be a compact hull. Compute the normalized Riemann map $F_{[a,b]}$.

19.4. Let $F_K(z) = Rz + (Rz)^{-1}$, for $R > 1$. Argue that F_K is univalent on $\mathbb{C} \setminus B[0,1]$, and show that $K = \mathbb{C} \setminus F_K(\mathbb{C} \setminus B[0,1])$ is an ellipse of the form $\{x + iy \mid ax^2 + by^2 \le 1\}$.

Determine a and b as a function of $R \ge 1$.

What is $\mathrm{Cap}(K)$?

19.5. Let $f(z) = z + 1/z$ and $R, R' \ge 1$.

(α) Show that $f(\mathbb{C} \setminus (B[0,1] \cup [-R', -1] \cup [1, R])) = [-R' - 1/R', R + 1/R]$.

(β) Let $L = \{(s + is) \mid -1 - \sqrt{2} \le s \le 1 + \sqrt{2}\}$ and $K = B[0,1] \cup L$. Compute $\mathrm{Cap}(K)$.

19.6. Let $f(z) = z + 1/z$ and $R > 1$.

(α) Compute $f([i, iR]))$ and $f([-iR, -i])$.

(β) Let $a, b > 0$, $c, d \ge 0$ and put $K = [-a, b] \cup [-ic, id]$. Compute $\mathrm{Cap}(K)$.

19.7. Let K be a compact hull. Argue that $K^* = \{z \in \mathbb{C} \mid \bar{z} \in K\}$ is also a compact hull. Compute F_{K^*} in terms of F_K, and prove that $\mathrm{Cap}(K) = \mathrm{Cap}(K^*)$.

19.8. Let $\gamma \colon [a, b] \to \mathbb{C}$ be a simple (continuous) curve. For $s \in \,]a, b]$, define $K_s = \gamma([a, s])$. Argue that $]a, b] \ni s \to \mathrm{Cap}(K_s)$ is a continuous and strictly increasing function with $\lim_{s \to a} \mathrm{Cap}(K_s) = -\infty$.

Hint: Show that $s \to \mathrm{Cap}(K_s)$ is sequentially continuous.

19.9. Let K be a compact hull and put $\Omega = \mathbb{C} \setminus K$.

(α) Argue that there exists exactly one real-valued function $\mathcal{G}(\,\cdot\,; \infty)$ harmonic in Ω, vanishing on $\partial\Omega$ and with $\mathcal{G}(z; a) - \mathrm{Log}(|z|)$ bounded in a neighborhood of infinity.

(β) Show that the limit $\lim_{z \to \infty}(\mathcal{G}(z; \infty) - \mathrm{Log}(|z|))$ exists and equals $\mathrm{Cap}(K)$.

19.10. Let $\{K_n\}$ be a sequence of compact hulls (or singletons) satisfying that there exists a $z_0 \in \mathbb{C} \setminus B[0,1]$ for which $\lim_{n\to\infty} |F_{K_n}(z_0)| = \infty$. Show that $K_n \to \mathbb{C}$ in Carathéodory sense.

19.11. Let A be a half-plane hull and $z_0 \in \mathbb{H} \setminus A$. Show that the set K from (19.10) is a compact hull.

19.12. Let $r > 0$ and G be holomorphic in $B(1,r)$ with $G'(1) \neq 0$. Suppose furthermore that $|G(z)| = 1$ for $z \in B(1,r)$ with $|z| = 1$.

(α) Prove that $G'(1)/G(1) \in \mathbb{R}$ and

$$\frac{G'(1)}{G(1)} + \operatorname{Re}\frac{G''(1)}{G(1)} = |G'(1)|^2.$$

(β) Conclude that $\operatorname{Re} G''(1)/G'(1) = -1 + G'(1)/G(1)$.

19.13. Let $A \neq \emptyset$ be a half-plane hull, $I \subseteq \mathbb{R}$ and put $K = A \cup A^* \cup I$.

(α) Show that that K is a compact hull if and only if I is a compact interval (or singleton) containing $\inf(\overline{A} \cap \mathbb{R})$ and $\sup(\overline{A} \cap \mathbb{R})$.

(β) In the rest of this exercise, I is fixed such that K is a compact hull. Argue that $\overline{A \cup A^*} \subseteq K$.

(γ) Put $J = \mathbb{C} \setminus G_A(\mathbb{C} \setminus K)$, where G_A has been Schwartz reflected across the real line. Prove that $J \subseteq \mathbb{R}$ is a compact interval.

(δ) Write $J = [\alpha, \beta]$. Show that $\operatorname{Cap}(K) = \operatorname{Log}((\beta - \alpha)/4)$.

19.14. Let $a > 0$ and $0 < \varepsilon < a$. Denote by $C_\pm$ the circles passing through the points $-a, a$ and $\pm\varepsilon$. Let $J_\pm$ be the closed subarcs of $C_\pm$ with endpoints $-a$ and a, containing $\pm\varepsilon$. Denote by K_ε the compact hull, with boundary $J_- \cup J_+$.

(α) Prove, using Example 19.29, that

$$\operatorname{Cap}(K_\varepsilon) = \frac{a\pi}{2\pi - 2\arctan(\frac{2\varepsilon a}{a^2 - \varepsilon^2})}.$$

(β) Compute $K = \bigcap_{0 < \varepsilon < a} K_\varepsilon$ and $\operatorname{Cap}(K)$. Observe that $\lim_{\varepsilon \to 0} \operatorname{Cap}(K_\varepsilon) = \operatorname{Cap}(K)$. Compare with Theorem 19.17.

19.15. Let $A_t = \{e^{is} \mid 0 < s \le t\}$ for $t \in [0, \pi[$.

(α) Argue that the A_t's are half-plane hulls.

(β) Find a Möbius transformation M_1 taking $\Omega_t = \mathbb{H} \setminus A_t$ onto $\mathbb{H} \setminus]0, iR_t]$, where $R_t = \tan(t/2)$.

(γ) Find a biholomorphic map M_2 taking $\mathbb{H} \setminus]0, iR_t]$ onto $\mathbb{C} \setminus [0, \infty[$.

(δ) Compose with a biholomorphic map M_3 taking $\mathbb{C} \setminus [0, \infty[$ onto $\mathbb{H}$.

(ε) Define a Möbius transformation $M_4(z) = (a_t + z)/(a_t - z)$, where $a_t = \cos(t/2)^{-1}$. Prove that M_4 is an automorphism of the upper half-plane with $\lim_{z \to \infty, z \in \Omega_t} M_4 \circ M_3 \circ M_2 \circ M_1(z) = \infty$.

(ζ) As in Example 19.29, expand the M_j's to identify the correct normalization of $M_4 \circ M_3 \circ M_2 \circ M_1$ that gives G_{A_t} and verify that $\text{hcap}(A_t) = 1 - \cos(t/2)^4$.

(η) Identify a half-plane hull A such that $A_{\pi - 1/n} \to A$ in Carathéodory sense, and verify by inspection that $\text{hcap}(A_{\pi - 1/n}) \to \text{hcap}(A)$.

(θ) Finally, for $t \in [0, \pi[$ let $K_t = \{e^{is} \mid |s| \le t\}$. Observe that the K_t's are compact hulls and verify that $\text{Cap}(K_t) = \text{Log}(\sin(t/2))$.

Chapter 20

Loewner theory

In this the last chapter we derive the Loewner equation both for the disc, the classical Loewner equation, and for the half-plane. The latter forms the complex function theoretic backbone of the stochastic Schramm-Loewner equation, which played a key role in the works for which W. Werner (2006) and S. Smirnov (2010) were awarded the Fields medal. Our focus is on the half-plane setup, and the more standard disc case is presented here in a manner parallel with the half-plane setup in order to ease the comparison of the two equations. We refer the reader to Lawler (2005) for Schramm-Loewner theory.

20.1 Loewner's equation in the exterior of the disc

Let $T \in \,]0, +\infty]$ and $\gamma \colon [0, T[\to \mathbb{C}$ be a simple curve with $\gamma(0) \in \partial B(0, 1)$ and $|\gamma(t)| > 1$ for $t > 0$. For $t \in [0, T[$, define

$$K_t = \gamma(]0, t]) \cup B[0, 1], \quad \Omega_t := \mathbb{C} \setminus K_t.$$

Clearly, $K_t \subseteq \mathbb{C}$ is a compact hull,[1] and we write $F_t = F_{K_t}$ and $G_t = G_{K_t}$ for the corresponding biholomorphic mappings introduced in Section 19.1.

Note that, by Proposition 17.22, $\gamma(t)$ is a simple accessible boundary point of $\mathbb{C} \setminus K_t$, so $\lambda(t) = G_t(\gamma(t)) \in \partial B[0, 1]$ – the so-called *driving function* is well-defined through the continuous extension of G_t, which exists by Corollary 17.20. Define furthermore $b(t) = \operatorname{Cap}(K_t)$.

---

[1] After a Cayley transform, we find one of the examples treated in Example 19.34. Therefore, $\mathbb{C}_\infty \setminus K_t$ is connected and simply connected. This is the only non-obvious thing to check.

For $0 \le s_1 \le s_2 < T$, we define a new curve by $\gamma_{s_1,s_2}(t) = G_{s_1}(\gamma(t+s_1))$ for $t \in [0, s_2 - s_1]$. This defines a new compact hull $K_{s_1,s_2} = B[0,1] \cup \gamma_{s_1,s_2}^*$ with associated functions $F_{s_1,s_2} = F_{K_{s_1,s_2}}$ and $G_{s_1,s_2} = G_{K_{s_1,s_2}}$. By the uniqueness part of Proposition 19.4, $G_{s_1,s_2} = G_{s_2} \circ F_{s_1}$, where F_{s_1} is restricted to $\mathbb{C} \setminus K_{s_1,s_2}$. Notice that F_t extends by continuity to $\partial B[0,1]$ by Corollary 17.20 for any t. We denote also $A_{s_1,s_2} = \gamma_{s_1,s_2}^*$ and $B_{s_1,s_2} = F_{s_2}^{-1}(\gamma([s_1,s_2])) \subseteq \partial B[0,1]$ (preimage).

Lemma 20.1. *The functions $b(t)$ and $\lambda(t)$ depend continuously on t. Furthermore, $t \mapsto b(t)$ is strictly increasing.*

Proof. The continuity and monotonicity of $b(t)$ follows immediately from Theorem 19.17 and Theorem 19.10. So we only need to consider continuity of λ.

Suppose that $t_n \searrow t$. Let $\varepsilon > 0$. Since γ_{t,t_n} is a continuous curve, we have that $\Theta(\gamma_{t,t_n}^*) \cup \gamma_{t,t_n}^* \subseteq B(\lambda(t), \varepsilon/2)$ for all n sufficiently large, where $\Theta(z) = 1/\bar{z}$ is inversion in the unit circle. Now, G_{t,t_n} extends by continuity to the simple boundary points $\partial B(0,1) \setminus B[\lambda(t), \varepsilon/2]$ by Corollary 17.20, where it takes values in $\partial B(0,1)$. So we can extend further by Schwarz reflection in the circle to $\mathbb{C} \setminus B[\lambda(t), \varepsilon/2]$. Let $\tilde{G}_n$ be the extended function, i.e.,

$$\tilde{G}_n(z) = \begin{cases} G_{t,t_n}(z), & |z| > 1, \\ 1/\overline{G_{t,t_n}(1/\bar{z})}, & |z| < 1. \end{cases}$$

For given $\varepsilon > 0$ we have seen that $\tilde{G}_n(z)$ is defined on $\mathbb{C} \setminus B[\lambda(t), \varepsilon/2]$ for all sufficiently large n. Since K_{t,t_n} converges to $B[0,1]$ in the sense of Carathéodory, we get that $\tilde{G}_n(z) \to z$ locally uniformly in $\mathbb{C} \setminus \{|z| = 1\}$, by Theorem 19.17. We will prove that actually $\tilde{G}_n(z) \to z$ locally uniformly in $\mathbb{C} \setminus \{\lambda(t)\}$. Notice first that $\tilde{G}_n$ is uniformly bounded on $|z| \le 2$ by Lemma 19.13. Let $z_0 \in B(0,2) \setminus \{\lambda(t)\}$ and $r > 0$ such that $B[z_0, r] \subseteq B(0,2) \setminus \{\lambda(t)\}$. Choose n so large that $\tilde{G}_n$ is defined on $B[z_0, r]$ for all $n \ge N$. Now, $\tilde{G}_n \to z$ locally uniformly on $(B(z_0, r) \cap \{|z| < 1\}) \cup (B(z_0, r) \cap \{|z| > 1\})$. By Vitali-Porter's theorem (Exercise 8.10) we therefore get that $\tilde{G}_n \to z$ locally uniformly on $B(z_0, r)$.

So $\tilde{G}_n(\partial B(\lambda(t), \varepsilon)) \subseteq B[\lambda(t), 2\varepsilon]$ for all sufficiently large n. Clearly, $\tilde{G}_n(\partial B(\lambda(t), \varepsilon))$ is a Jordan curve which must have $\lambda(t)$ in its interior. Therefore, $\lambda(t_n) \in B[\lambda(t), 2\varepsilon]$ for all sufficiently large n.

Suppose next that $t_n \nearrow t$. Notice that $\lambda(t) \in B_{t_n,t}$ and $\lambda(t_n) \in \gamma_{t_n,t}^*$. By the continuity of F_t extended to $\partial B[0,1]$ and the continuity of G_t at $\gamma(t)$, we get that for all $\varepsilon > 0$ exists N such that $B_{t_n,t} \subseteq B(\lambda(t), \varepsilon)$ for all $n \ge N$.

Clearly, $K_{t_n} \to K_t$ in Carathéodory sense as $n \to \infty$. So $F_{t_n} \to F_t$ and $G_{t_n} \to G_t$ locally uniformly. Therefore, also $F_{t_n,t} \to z$ locally uniformly on $\mathbb{C} \setminus B[0,1]$ as $n \to \infty$.

The function $F_{t_n,t}$ extends by continuity to $\partial B(0,1) \setminus B_{t_n,t}$ and by Schwarz reflection to $B(0,1)$. By the same arguments as in the previous part of the proof $F_{t_n,t}(z) \to z$ locally uniformly on $\mathbb{C} \setminus \{\lambda(t)\}$. Let $\varepsilon > 0$. The local uniform convergence yields that $B_{t_n,t} \subseteq F_{t_n,t}(\{\partial B(\lambda(t),\varepsilon)\}) \subseteq B(\lambda(t),2\varepsilon)$ for n sufficiently large. Therefore $|\lambda(t_n) - \lambda(t)| < 2\varepsilon$ for all large values of n. $\qquad\square$

After possibly reparametrizing γ, we may assume that $b(t) = t$, i.e., that $\mathrm{Cap}(K_t) = t$. We say that γ is *parametrized by logarithmic capacity*.

Theorem 20.2. *Let* $\gamma \colon [0, T[\to \mathbb{C} \setminus B(0,1)$ *be a simple curve with* $|\gamma(0)| = 1$ *and* $|\gamma(t)| > 1$ *for all* $t > 0$. *Assume that* γ *is parametrized by logarithmic capacity and let* G_t *be the associated normalized Riemann map. Then, for any* $t \in [0, T[$ *and* $w \in \Omega_t$, *the function* $t \mapsto G_t(w)$ *is differentiable and satisfies the equation*

$$\frac{dG_t(w)}{dt} = G_t(w)\frac{\lambda(t) + G_t(w)}{\lambda(t) - G_t(w)}. \tag{20.1}$$

Proof. Define, for $z \in B(0,1)$, $0 \le s \le t < T$,

$$\phi_{s,t}(z) = \frac{1/z}{G_s \circ F_t(1/z)},$$

Clearly, $\phi_{s,t}$ has a removable singularity at $z = 0$ and extends by continuity with $\phi_{s,t}(0) = e^{s-t}$. Furthermore, by Theorem 17.19, $\phi_{s,t}$ extends by continuity to $\partial B(0,1)$ with

$$|\phi_{s,t}(z)| \begin{cases} \le 1, & \text{for } z^{-1} \in B_{s,t}, \\ = 1, & \text{for } z^{-1} \in \partial B(0,1) \setminus B_{s,t}. \end{cases} \tag{20.2}$$

Since $\phi_{s,t}$ has no zeroes, we can define a unique holomorphic logarithm $\Phi_{s,t}$ on $B(0,1)$, continuous on $B[0,1]$, such that

$$e^{\Phi_{s,t}(z)} = \phi_{s,t}(z), \quad \Phi_{s,t}(0) = s - t.$$

By the Schwarz Formula (Theorem 14.6), we have

$$\Phi_{s,t}(z) = \frac{1}{2\pi} \int_0^{2\pi} \mathrm{Re}[\Phi_{s,t}(e^{i\theta})]\frac{e^{i\theta} + z}{e^{i\theta} - z}\, d\theta. \tag{20.3}$$

By changing variable in the integral to $-\theta$ and parametrising $B_{s,t}$ as $\exp(i[\alpha_{s,t}, \beta_{s,t}])$ with $\beta_{s,t} - \alpha_{s,t} < 2\pi$, we get

$$\Phi_{s,t}(z) = \frac{1}{2\pi} \int_{\alpha_{s,t}}^{\beta_{s,t}} \mathrm{Re}[\Phi_{s,t}(e^{-i\theta})] \frac{e^{-i\theta} + z}{e^{-i\theta} - z} \, d\theta. \tag{20.4}$$

Evaluating at $z = 0$ yields

$$s - t = \Phi_{s,t}(0) = \frac{1}{2\pi} \int_{\alpha_{s,t}}^{\beta_{s,t}} \mathrm{Re}[\Phi_{s,t}(e^{-i\theta})] \, d\theta. \tag{20.5}$$

So the Mean-Value Theorem for Integrals (applied to the real and imaginary parts) yields the existence of $\sigma_{s,t,z}, \tau_{s,t,z} \in [\alpha_{s,t}, \beta_{s,t}]$ such that

$$\frac{\Phi_{s,t}(z)}{s-t} = \mathrm{Re}\left(\frac{e^{-i\sigma_{s,t,z}} + z}{e^{-i\sigma_{s,t,z}} - z}\right) + i\,\mathrm{Im}\left(\frac{e^{-i\tau_{s,t,z}} + z}{e^{-i\tau_{s,t,z}} - z}\right). \tag{20.6}$$

Let now $t_0 \in \,]0, T[$, $\varepsilon > 0$ and set $z_\varepsilon = 1/G_{t_0+\varepsilon}(w)$, $w \in \Omega_{t_0}$ (this makes sense for ε sufficiently small). Then

$$\frac{G_{t_0+\varepsilon}(w) - G_{t_0}(w)}{\varepsilon} = G_{t_0}(w)\varepsilon^{-1}\left(\frac{G_{t_0+\varepsilon}(w)}{G_{t_0}(w)} - 1\right)$$

$$= G_{t_0}(w)\varepsilon^{-1}(\exp(\Phi_{t_0,t_0+\varepsilon}(z_\varepsilon)) - 1).$$

So we see that, since $\exp(i[\alpha_{s,t}, \beta_{s,t}])$ shrinks to $\lambda(s)$ as t decreases to s, and using (20.6),

$$\lim_{\varepsilon \to 0_+} \frac{G_{t_0+\varepsilon}(w) - G_{t_0}(w)}{\varepsilon} = -G_{t_0}(w)\frac{1/\lambda(t_0) + 1/G_{t_0}(w)}{1/\lambda(t_0) - 1/G_{t_0}(w)}.$$

A similar argument works for the left-hand derivative, thereby finishing the proof. □

20.2   Loewner's equation in the slit plane

In this section we will deduce the best known version of the Loewner equation. For this we need to introduce some notation.

Definition 20.3. Let v be a biholomorphic map from $B(0,1)$ to $\Omega = \mathbb{C} \setminus \Gamma^*$, where $\Gamma \colon [0, \infty[\,\to \mathbb{C}$ is injective and $\lim_{t \to \infty} \Gamma(t) = \infty$. Let furthermore $0 \notin \Gamma^*$, $v(0) = 0$ and $v'(0) = 1$. In that case we call v a single slit mapping.

For $t_0 \geq 0$, we denote by Γ_{t_0} the restricted curve $\Gamma|_{[t_0,\infty[}$. We let $v_{t_0}: B(0,1) \to \mathbb{C} \setminus \Gamma_{t_0}^*$ be the unique biholomorphic map with $v_{t_0}(0) = 0$, $v_{t_0}'(0) > 0$. Furthermore, we write $u_{t_0}: \mathbb{C} \setminus \Gamma_{t_0}^* \to B(0,1)$ for the inverse of v_{t_0}.

Remark 20.4. As we have seen before, by Proposition 17.22, $\Gamma(t_0)$ is a simple accessible boundary point of $\mathbb{C} \setminus \Gamma_{t_0}^*$, so the *driving function*

$$\tilde{\lambda}(t_0) = u_{t_0}(\gamma(t_0)) \in \partial B[0,1] \tag{20.7}$$

is well-defined through the continuous extension of u_{t_0}, which exists by Corollary 17.20.

Proposition 20.5 (Standard Parametrization). *Let* $\mathbb{C} \setminus \Gamma$ *be a slit plane. There exists a parametrization of* Γ *called the standard parametrization such that* $v_t'(0) = e^t$.

Proof. It is an easy consequence of Schwarz' lemma (Theorem 4.17), that $t \mapsto v_t'(0)$ is an increasing function. Also, $t \mapsto v_t'(0)$ is continuous. This follows from Theorem 18.12, since if $t_n \to t$, then $\mathbb{C} \setminus \Gamma_{t_n}^* \to \mathbb{C} \setminus \Gamma_t^*$ in the sense of Carathéodory (with base point 0).

It follows that we can reparametrize Γ by the interval $[0, T[$ in order to get $v_t'(0) = e^t$. It only remains to check that $T = \infty$ in this parametrization.

Let $M > 0$ be given. For all t sufficiently large, $B(0, M) \subseteq \mathbb{C} \setminus \Gamma_t^*$, so $z \mapsto u_t(Mz)$ defines a holomorphic map from $B(0,1)$ to itself. By Schwarz' lemma (Theorem 4.17),

$$1 \geq M u_t'(0) = M e^{-t},$$

which shows that $t \geq \log M$. □

Theorem 20.6. *Let* $v: B(0,1) \to \mathbb{C} \setminus \Gamma^*$ *be a single slit mapping. Assume that* Γ *is given the standard parametrization and that* u_t *is the inverse of* v_t *as defined above. Then* $t \mapsto u_t$ *satisfies the differential equation,*

$$\frac{\partial u_t(z)}{\partial t} = -u_t(z) \frac{\tilde{\lambda}(t) + u_t(z)}{\tilde{\lambda}(t) - u_t(z)}, \tag{20.8}$$

with $\tilde{\lambda}(t)$ *the continuous function from* (20.7). *Furthermore,*

$$e^t u_t(z) \to z, \tag{20.9}$$

locally uniformly on compact subsets of $\mathbb{C}$.

This is the, which originated in the celebrated paper Löwner (1923). For alternative treatments we refer the reader to Duren (1983), Goluzin (1969) and Lawler (2005).

Proof. We start by proving (20.8). We will deduce it from (20.1). Pick $T_0 > 0$ (large). Define, for $0 \leq t \leq T_0$,

$$\gamma_{T_0}(t) = 1/u_{T_0}(\Gamma(T_0 - t)).$$

Using Remark 20.4, $|\gamma_{T_0}(0)| = 1$, $|\gamma_{T_0}(s)| > 1$, for $0 < s \leq T_0$. So, $K_t = B[0,1] \cup \gamma_{T_0}([0,t])$ defines a compact hull. Notice that $z \mapsto 1/u_{T_0-t}(v_{T_0}(1/z))$ maps $\mathbb{C} \setminus K_t$ biholomorphically to $\mathbb{C} \setminus B[0,1]$. Furthermore, $u_{T_0-t}(v_{T_0}(0)) = 0$, $u_{T_0-t}(v'_{T_0}(0)) = e^t$, so γ_{T_0} is parametrized by logarithmic capacity. Therefore, $1/u_{T_0-t}(v_{T_0}(1/z)) = G_t(z)$ is the normalized Riemann map associated to K_t and satisfies the differential equation (20.1) with

$$\lambda(t) = G_t(\gamma(t)) = 1/u_{T_0-t}(\Gamma(T_0 - t)). \qquad (20.10)$$

Applying (20.1) at the point $w = 1/u_{T_0}(z)$ yields,

$$\frac{\partial}{\partial t}\left(\frac{1}{u_{T_0-t}(z)}\right) = \frac{1}{u_{T_0-t}(z)}\frac{\lambda(t) + 1/u_{T_0-t}(z)}{\lambda(t) - 1/u_{T_0-t}(z)},$$

i.e.,

$$\frac{\partial}{\partial t}u_{T_0-t}(z) = -u_{T_0-t}(z)\frac{\lambda(t)u_{T_0-t}(z) + 1}{\lambda(t)u_{T_0-t}(z) - 1}.$$

Now, for $r < T_0$, we can write $\frac{\partial}{\partial r}u_r(z) = -\frac{\partial}{\partial t}u_{T_0-t}(z)|_{t=T_0-r}$. Therefore, we get (using also (20.10))

$$\frac{\partial}{\partial r}u_r(z) = u_r(z)\frac{\lambda(T_0 - r)u_r(z) + 1}{\lambda(T_0 - r)u_r(z) - 1} = -u_r(z)\frac{\tilde{\lambda}(r) + u_r(z)}{\tilde{\lambda}(r) - u_r(z)}, \qquad (20.11)$$

with $\tilde{\lambda}(r) = u_r(\Gamma(r))$.

We proceed to prove (20.9). Notice that $(e^{-t}v_t)'(0) = 1$, so by (18.8) of Theorem 18.5,

$$\frac{|z|}{(1 + |z|)^2} \leq e^{-t}|v_t(z)| \leq \frac{|z|}{(1 - |z|)^2} \qquad (20.12)$$

Consider K compact, say $K \subseteq B(0, M)$. Then u_t is defined on K for all sufficiently large t. Inserting $z = u_t(w)$, with $w \in K$, in (20.12) one finds

$$(1 - |u_t(w)|)^2 \leq \frac{e^t|u_t(w)|}{|w|} \leq (1 + |u_t(w)|)^2. \qquad (20.13)$$

In particular, we see that $|u_t(w)| \leq 4e^{-t}M$. This implies that $u_t \to 0$ locally uniformly on $\mathbb{C}$. Reinserting this information in (20.13) we see that

$$\left| \frac{e^t u_t(w)}{w} \right| \to 1,$$

locally uniformly.

Let $\{t_n\}_n \subseteq \mathbb{R}_+$ with $t_n \to \infty$. By Montel's theorem (Theorem 8.15), there exists a holomorphic function G on $\mathbb{C}$ and a subsequence $\{t_{n_j}\}_j$ such that

$$\frac{e^{t_{n_j}} u_{t_{n_j}}(w)}{w} \to G(w),$$

locally uniformly. Since $|G(w)| = 1$ for all w by (20.13), we get from the Open Mapping Theorem, Theorem 4.14, that G is constant. Since $G(0) = 1$, we get $G(w) = 1$ for all w. Since the sequence $\{t_n\}_n$ was arbitrary, this proves that $e^t u_t(w)/w \to 1$ locally uniformly, i.e., (20.9). $\qquad\square$

20.3 Loewner's equation in the upper half-plane

In this our final section, we derive and analyze the half-plane Loewner equation, originally due to Kufarev *et al.* (1968). The equation was derived an analyzed in Lawler (2005) with the aid of stochastic techniques.

Let $\gamma\colon [0, T[\to \mathbb{C}$ be a simple curve with $\gamma(0) \in \mathbb{R}$ and $\operatorname{Im}\gamma(t) > 0$ for $t \in \,]0, T[$. Here $0 < T \leq \infty$. For $t \in [0, T[$, write $A_t = \{\gamma(s) \mid 0 < s \leq t\}$, $\Omega_t = \mathbb{H} \setminus A_t$ and put $b(t) = \operatorname{hcap}(A_t)$. Note that $A_0 = \emptyset$. We know from Example 19.34 that A_t is a half-plane hull for all $t \in [0, T[$. We abbreviate throughout this section $\Omega_t = \Omega_{A_t}$, $G_t = G_{A_t}$ and $F_t = F_{A_t}$. Put $C_t = \overline{A_t \cup A_t^*} = A_t \cup A_t^* \cup \{\gamma(0)\}$ and recall that G_t can be extended by Schwarz reflection to a univalent map defined in $\mathbb{C} \setminus C_t$ (Proposition 19.30).

Note that $\partial\Omega_t$ is a closed curve (in the Riemann sphere). Hence, by Corollary 17.20, F_t extends by continuity to $\mathbb{R}$. Put

$$J_t = \{x \in \mathbb{R} \mid F_t(x) \in \gamma([0, t])\} = \mathbb{C} \setminus G_t(\mathbb{C} \setminus C_t), \qquad (20.14)$$

which is a compact interval (Proposition 19.30). We leave it to the reader, cf. Exercise 20.3, to argue for the identity in (20.14) and that $F_t(J_t) = \gamma([0, t])$.

It follows from Corollary 17.23 that $\gamma(t)$ is a simple boundary point for $\mathbb{H} \setminus \gamma([0, t])$. Hence, by Theorem 17.17 (and an application of the Cayley transform), the function G_t extends by continuity to the tip $\gamma(t)$ and we put $\lambda(t) = G_t(\gamma(t)) \in J_t \subseteq \mathbb{R}$. We define $\lambda(0) = \gamma(0)$. The function $\lambda \colon [0, T[\to \mathbb{R}$ is called the *half-plane driving function*.

Theorem 20.7. *Let $b(t) = \mathrm{hcap}(A_t)$ and $\lambda(t) = G_t(\gamma(t))$ for $t \in [0, T[$ be as above. Then $b, \lambda \colon [0, T[\to \mathbb{R}$ are both continuous.*

Proof. That b is (sequentially) continuous follows from Theorem 19.38(d) after observing that if $t_n \to t$ with $t_n, t \in [0, T[$, then $A_{t_n} \to A_t$ in Carathéodory sense, cf. Definition 19.37. It remains to prove that the Loewner driving function λ is continuous.

Let $0 \leq s < t < T$ and define a new curve $\gamma_{s,t} \colon [0, t - s] \to \overline{\mathbb{H}}$ by setting $\gamma_{t,s}(r) = G_s(\gamma(s + r))$. Then $A_{s,t} = \gamma_{s,t}^* \setminus \{\lambda(s)\}$ is again a half-plane hull (of the same type as the A_t's). That $\gamma_{s,t}$ is continuous at $r = 0$ follows from the discussion preceding the theorem, i.e., that G_s extends by continuity to $\gamma(s)$, where it takes the value $\lambda(s)$. We write $G_{s,t}$ and $F_{s,t}$ for the associated biholomorphic functions and note that

$$G_t = G_{t,s} \circ G_s,$$

which follows from the uniqueness part of Proposition 19.24.

Denote by $\sigma \colon \,]-T, T[$ the extension of γ by reflection, given by $\sigma(t) = \gamma(t)$ and $\sigma(-t) = \overline{\gamma(t)}$ for $t \in [0, T[$. Similarly, we extend $\gamma_{s,t}$ by reflection to $\sigma_{s,t} \colon [s - t, t - s] \to \mathbb{C}$ by setting $\sigma_{s,t}(r) = \gamma_{s,t}(r)$ and $\sigma_{s,t}(-r) = \overline{\gamma_{s,t}(r)}$ for $r \in [0, t - s]$.

We may now identify the sets $C_{t,s}$ and $J_{s,t}$ from Proposition 19.30 to be

$$C_{t,s} = \sigma_{t,s}^* \quad \text{and} \quad J_{t,s} = F_{s,t}^{-1}(\gamma_{s,t}^*) = F_t^{-1}(\gamma([s, t])).$$

Note that $\lambda(t) \in J_{s,t}$ and that $F_{s,t}^{-1}(\cdot)$ and $F_t^{-1}(\cdot)$ denotes preimages, since the inverse is not defined on the sets in the argument. Finally, we write $B_{s,t} = [\alpha_{s,t}, \beta_{s,t}] \ni \lambda(t)$ for the compact interval $F_{s,t}^{-1}(\gamma_{s,t}^*) = F_t^{-1}(\gamma([s, t]))$.

We argue first that λ is (sequentially) right-continuous. Let $0 \leq s < T$ and suppose $s < t_n < T$ is a sequence of real numbers with $t_n \to s$. Since $A_{s,t} = G_s(\gamma(]s, t]))$, we observe that $A_{s,t_n} \to \emptyset$ in Carathéodory sense. The sets $C_{s,t}$ are in fact compact hulls in this setup and $C_{s,t_n} \to \{\lambda(s)\}$ in Carathéodory sense (as compact hulls).

By Proposition 19.30, we conclude that $G_{s,t_n}(z) \to z$ locally uniformly in $\mathbb{C} \setminus \{\lambda(s)\}$. Proposition 19.30 also implies that $\mathbb{C}_\infty \setminus J_{s,t_n} \to \mathbb{C}_\infty \setminus \{\lambda(s)\}$ in Carathéodory sense with base point ∞. Since $\lambda(t_n) \in J_{s,t_n}$ for all n, we conclude that $\lambda(t_n) \to \lambda(s)$. This proves right-continuity of λ.

As for left-continuity, let $0 < s_n < t < T$ with $s_n \to t$. Since G_t is continuous at $\gamma(t)$, we conclude that $J_{s_n,t} \to \{\lambda(t)\}$ in Carathéodory sense as compact hulls, cf. Definition 19.14. Indeed, the diameter of the intervals, which all contain $\lambda(t)$, goes to zero implying the Carathéodory convergence by Lemma 19.20. Proposition 19.30 now implies that $\widetilde{C} = \mathbb{C}_\infty \setminus \ker(\{\mathbb{C}_\infty \setminus C_{s_n,t}\}_{n\in\mathbb{N}}, \infty)$ is a singleton and therefore $A_{s_n,t} \to \emptyset$ in Carathéodory sense. Hence, $F_{A_n} \to z$ locally uniformly in $\mathbb{C} \setminus \{\lambda(t)\}$ and $\widetilde{C} = \{\lambda(t)\}$. Since $\lambda(s_n) \in C_{s_n,t}$ for all n, we conclude $\lambda(s_n) \to \lambda(t)$ establishing left-continuity of λ. $\qquad\square$

Note that by Proposition 19.36, the half-plane capacity $b(t)$ is a strictly increasing function of t with $b(0) = 0$. The range $b([0,T[)$ is a half-open interval $[0,\widetilde{T}[$, and we denote by $b^{-1} \colon [0,\widetilde{T}[\to [0,T[$ the continuous inverse of b. By passing to the reparametrized continuous curve $\tilde{\gamma} \colon [0,\widetilde{T}/2[\to \mathbb{C}$, defined by $\tilde{\gamma}(t) = \gamma(b^{-1}(2t))$, we may assume that $b(t) = 2t$. The driving function λ transforms similarly to the new continuous driving function $\tilde{\lambda} \colon [0,\widetilde{T}/2[\to \mathbb{C}$, defined by $\tilde{\lambda}(t) = \lambda(b^{-1}(2t))$. A curve γ with the property that $b(t) = 2t$ is said to be *parametrized by half-plane capacity*.

Theorem 20.8. *Let γ be a Jordan arc in the upper half-plane parametrized by half-plane capacity. Denote by G_t the associated normalized Riemann map. Then, for any $t \in [0,T[$ and $z \in \Omega_t$, the function $t \to G_t(z)$ is differentiable at t and satisfies Loewner's equation in the half-plane*

$$\frac{dG_t(z)}{dt} = \frac{2}{G_t(z) - \lambda(t)}. \tag{20.15}$$

Proof. Let $0 \le s < t < T$ and put

$$H_{t,s}(z) = G_s \circ F_t(z) - z.$$

The function $H_{t,s} \colon \mathbb{H} \to \mathbb{C}$ is holomorphic. Furthermore, it extends by continuity to the closed upper half-plane with

$$H_{s,t}(\mathbb{R} \setminus B_{s,t}) \subseteq \mathbb{R} \quad \text{and} \quad H_{s,t}(B_{s,t}) = J_{s,t}.$$

Hence, $\operatorname{Im} H_{s,t}(x) \ge 0$, for $x \in \mathbb{R}$, and equals zero for $x \in \mathbb{R} \setminus B_{s,t}$.

Applying Schwarz' formula (Theorem 14.7) in the upper half-plane, we get

$$H_{s,t}(z) = \frac{1}{\pi} \int_{B_{s,t}} \frac{\operatorname{Im} H_{s,t}(x)}{x - z}\, dx. \tag{20.16}$$

Here we used that $H_{s,t} = O(|z|^{-1})$. Multiplying both sides by z and taking the limit $z \to \infty$ we find

$$2(s - t) = -\frac{1}{\pi} \int_{-\infty}^{\infty} \operatorname{Im} H_{s,t}(x) \, dx, \qquad (20.17)$$

where we used that γ is parametrized by half-plane capacity.

Substituting $z = G_t(w)$ in (20.16), for $w \in A_t$, we get

$$G_s(w) - G_t(w) = \frac{1}{\pi} \int_{B_{s,t}} \frac{\operatorname{Im} H_{s,t}(x)}{x - G_t(w)} \, dx.$$

Using the Mean-Value Theorem for Integrals twice (for the real and imaginary part) we find that

$$
\begin{aligned}
G_s(w) &- G_t(w) \\
&= \left(\operatorname{Re}\{(\sigma - G_t(w))^{-1}\} + i \operatorname{Im}\{(\tau - G_t(w))^{-1}\} \right) \\
&\quad \times \frac{1}{\pi} \int_{B_{s,t}} \operatorname{Im} H_{s,t}(x) \, dx,
\end{aligned}
$$

where $\sigma = \sigma(t, s)$ and $\tau = \tau(t, s)$ are elements of $B_{s,t}$. Appealing to (20.17) and that $B_{s,t}$ shrinks to $\lambda(t)$, we find that

$$\frac{G_s(w) - G_t(w)}{s - t} = \frac{2}{G_t(w) - \lambda(t)} + o(1),$$

where $\lim_{|s-t| \to 0} o(1) = 0$. Here one should either fix t and take $s \to t$, or vice versa. In the latter case, w should actually be fixed in A_s, and t chosen a priori close enough to s to ensure that $w \in A_t$ as well. This completes the proof. $\qquad \square$

Example 20.9. Let $\gamma \colon [0, \infty[$ be given by $\gamma(t) = it$, such that $A_t = \,]0, it]$. We know from Example 19.27 that A_t is a half-plane hull for all t and that $\operatorname{hcap}(A_t) = t^2/2$. To obtain a curve parametrized by half-plane capacity, we replace γ by $\tilde{\gamma}(t) = i\sqrt{4t}$.

We have $G_t(z) = \sqrt{z^2 + 4t}$ and the driving function $\tilde{\lambda}(t) = G_t(\tilde{\gamma}(t)) = 0$ is particularly simple. We observe by explicit computation that $t \to G_t(z)$ satisfies the Loewner equation $\frac{d}{dt} G_t(z) = 2/G_t(z)$.

Example 20.10. Here we consider the simple curve $\gamma \colon [0, \pi[\to \overline{\mathbb{H}}$ from Exercise 19.15, given by $\gamma(t) = e^{it}$. In that exercise the reader was lead through the construction of $G_t = G_{A_t}$, and one may identify the driving function to be

$$\lambda(t) = G_t(e^{it}) = 2 - \cos(t/2)^2. \tag{20.18}$$

The reader is asked to do this computation in Exercise 20.7.

In Exercise 19.15 the reader was also asked to verify that the half-plane capacity $b(t) = \operatorname{hcap}(A_t) = 1 - \cos(t/2)^4$. We therefore find that $b^{-1}(s) = 2\arccos((1-s)^{1/4})$. Reparametrizing by half-plane capacity, we get $\tilde{\gamma}\colon [0,1/2[\to \overline{\mathbb{H}}$ defined by

$$\tilde{\gamma}(t) = e^{2i\arccos((1-2s)^{1/4})}$$

with driving function

$$\tilde{\lambda}(t) = 2 - \sqrt{1 - 2t}. \tag{20.19}$$

See also Exercise 20.6.

The following corollary, along with the proof we follow, appeared first in Lalley *et al.* (2009). The reader may find a different proof in Dubinin (2010).

Corollary 20.11. *Let A be a half-plane hull. Then $\operatorname{hcap}(A) \geq \frac{1}{2}\operatorname{Im}(A)^2$, where $\operatorname{Im}(A) = \sup_{z\in A} \operatorname{Im} z$.*

Proof. Assume that the corollary is false, that is; there exists $z \in A$ with $\operatorname{Im} z > \sqrt{2\operatorname{hcap}(A)}$. Put $\varepsilon = (\operatorname{Im} z)^2 - 2\operatorname{hcap}(A) > 0$. By Lemma 19.40, there exists a half-plane hull $A' \supseteq A$ with a closed Jordan curve as boundary and $\operatorname{hcap}(A) \leq \operatorname{hcap}(A') \leq \operatorname{hcap}(A) + \varepsilon/8$. Our point z is an interior point of A' and may be connected to $\overline{A}' \cap \mathbb{R}$ with a Jordan arc $\gamma\colon [0,1] \to \mathbb{C}$ with $\gamma(0) \in \mathbb{R}$ and $\gamma(t) \in A'$ for $t \in \,]0,1]$. Put $A'' = \gamma(]0,1])$. Then $A'' \subseteq A'$ is a half-plane hull with

$$\operatorname{hcap}(A'') \leq \operatorname{hcap}(A') \leq \operatorname{hcap}(A) + \varepsilon/8. \tag{20.20}$$

Let $b(t) = \operatorname{hcap}(\gamma(]0,t]))$ and reparametrize by half-plane capacity to obtain a new curve $\sigma =: [0,T''] \to \overline{\mathbb{H}}$, where $T'' = \operatorname{hcap}(A'')/2$ and $\sigma(t) = \gamma(b^{-1}(2t))$ for $t \in [0,T'']$. Let G_t, $t \in [0,T'']$ be the associated Riemann maps satisfying the Loewner equation (20.15).

Since $G_{T''}$ extends by continuity to $z = \sigma(T'')$ and $\operatorname{Im} G_{T''}(z) = 0$, we can pick $z'' \in \mathbb{H} \setminus A''$ with

$$\operatorname{Im} z'' \geq 2\operatorname{hcap}(A) + \varepsilon/2 \quad \text{and} \quad \operatorname{Im} G_{T''}(z'') < \sqrt{\varepsilon/8}. \tag{20.21}$$

We compute using Theorem 20.8 and estimate for $t \in [0, T''[$:

$$\frac{d\operatorname{Im} G_t(z'')}{dt} = \frac{-2\operatorname{Im} G_t(z'')}{|G_t(z'') - \lambda(t)|^2} \geq \frac{-2\operatorname{Im} G_t(z'')}{(\operatorname{Im} G_t(z''))^2} = \frac{-2}{\operatorname{Im} G_t(z'')},$$

where $\lambda \colon [0, T''] \to \mathbb{R}$ is the driving function associated with γ. Hence,

$$\frac{d(\operatorname{Im} G_t(z''))^2}{dt} \geq -4$$

and we may conclude that $(\operatorname{Im} G_t(z''))^2 \geq (\operatorname{Im} z'')^2 - 4t$ for all $0 \leq t < T''$. Taking the limit $t \to T'' = \operatorname{hcap}(A'')/2$, we find that

$$\begin{aligned}\tfrac{\varepsilon}{8} &\geq (\operatorname{Im} G_{T''}(z''))^2 \geq (\operatorname{Im} z'')^2 - 4T'' \\ &\geq 2\operatorname{hcap}(A) + \tfrac{\varepsilon}{2} - 2\operatorname{hcap}(A'') \geq \tfrac{\varepsilon}{4},\end{aligned}$$

where we made use of (20.20) and (20.21). Having arrived at a contradiction, we may conclude the corollary. $\qquad\square$

Note that the above result is optimal since we have equality for half-plane hulls of the form $A = {]}0, Ri]$, cf. Example 19.27.

We now turn to the inverse problem. Let $T \in {]}0, \infty]$ and $\lambda \colon [0, T[\to \mathbb{R}$ be a continuous function. A function $\zeta \colon [0, \tau[\to \mathbb{H}$ is a solution of Loewner's equation (20.15) if $\tau \in {]}0, T]$, ζ is differentiable in $[0, \tau[$, and $\frac{d\zeta_t}{dt} = 2/(\zeta_t - \lambda(t))$ for all $t \in [0, \tau[$. Furthermore, for the solution to be maximal ζ may not be the restriction of a solution defined with a strictly larger τ. For $z \in \mathbb{C} \setminus \{\lambda(0)\}$, we write $t \to G_t(z)$ for the unique maximal solution with $G_0(z) = z$ and write $\tau(z) \in [0, T]$ for the endpoint of the interval on which $G_t(z)$ is defined.

Note that if $x \in \mathbb{R} \setminus \{\lambda(0)\}$, then $\zeta_t(x) \in \mathbb{R} \setminus \{\lambda(t)\}$ for all $t \in [0, \tau(x)[$. Hence, by uniqueness of solutions, if $z \in \mathbb{H}$ then $\zeta_t(z) \in \mathbb{H}$ for all $t \in [0, \tau(z)[$.

We define a collection of subsets $\{A_t\}_{t \in [0,T[}$ of the upper half-plane

$$A_t = \mathbb{H} \setminus \Omega_t, \qquad \text{where } \Omega_t = \{z \in \mathbb{H} \mid \tau(z) > t\}. \tag{20.22}$$

Note that $A_0 = \emptyset$ and $A_s \subseteq A_t$ for all $0 \leq s \leq t < T$.

Theorem 20.12. *Let $T \in {]}0, \infty]$ and $\lambda \colon [0, T[\to \mathbb{R}$ be a continuous function. Then the A_t's defined in (20.22) are half-plane hulls with $\operatorname{hcap}(A_t) = 2t$ and $A_s \subsetneq A_t$ for $0 \leq s < t < T$. The functions $G_t \colon \Omega_t \to \mathbb{H}$ are the associated normalized Riemann maps.*

Proof. That Ω_t is an open set for all t follows from lower semi-continuity of $\tau \colon \mathbb{H} \to \,]0, T]$ (Hartman, 1964, Theorem 2.1). Hence, $\overline{A}_t \cap \mathbb{H} = A_t$. Extend λ to $]-\infty, 0]$ by $\lambda(s) = \lambda(0)$ for $s < 0$.

We need some extra notation. For $w \in \mathbb{H}$ and $t_0 < T$, write $t \to \zeta_{t,t_0}(w)$ for the unique maximal solution of Loewner's equation with $\zeta_{t_0,t_0}(w) = w$ defined on (an automatically open) interval $]\tau_-, \tau_+[$. With this notation $G_t(z) = \zeta_{t,0}(z)$ for $t \geq 0$. Note that $\Omega_t \ni z \to G_t(z)$ is invertible with inverse $F_t(w) = \zeta_{0;t}(w)$. To see that F_t is defined in $\mathbb{H}$, we argue that $G_t(\Omega_t) = \mathbb{H}$ as follows: Let $w \in \mathbb{H}$ and $t_0 \in \,]0, T]$. Since $s \to \operatorname{Im} \zeta_{s,t_0}(w)$ is a decreasing function of s (its derivative is $-2\operatorname{Im} \zeta_{s,t_0}(w)/|\zeta_{s,t_0}(w) - \lambda(s)|^2 < 0$), we observe that $\tau' = -\infty$, since $|\zeta_{s,t_0}(w)|$ cannot diverge to ∞ in finite time as $s \to \tau'_-$ if τ' is finite. Put $z = \zeta_{0,t_0}(w)$. By uniqueness of solutions of ODE's, $\tau(z) > t_0$ and $G_t(z) = w$. This proves the claim.

Since $\mathbb{H} \ni \zeta \to 2/(\zeta - \lambda(t))$ is holomorphic for each t, we find that $\mathbb{H} \ni w \to F_t(w)$ is holomorphic, and consequently that its inverse G_t is holomorphic in Ω_t.

It is now easy to see that Ω_t is connected and simply connected. Let $z_1, z_2 \in \Omega_t$ and put $w_j = G_t(z_j)$, $j = 1, 2$. Connect w_1 and w_2 be a curve γ in $\mathbb{H}$. Then $F_t \circ \gamma$ is a curve in Ω_t connecting z_1 and z_2. If σ is a closed curve in Ω_t, then $\gamma = G_t \circ \sigma$ is a closed curve in $\mathbb{H}$. Let $\Gamma \colon [0,1] \times [0,1] \to \mathbb{H}$ be the continuous map $\Gamma(r, s) = (1 - s)\gamma(r) + si$. Then $F_t \circ \Gamma$ is a homotopy in Ω_t contracting σ to the point $F_t(i)$.

In the next step we establish in one fell swoop that the A_t's are bounded, that the G_t's are the properly normalized Riemann maps and that $\operatorname{hcap}(A_t) = 2t$. By Proposition 19.36, we may then wrap up the proof concluding that the half-plane hulls are strictly increasing with t. Let

$$h_t(z) = G_t(z) - z - \frac{2t}{z},$$

defined for all $z \in \mathbb{H}$ and $t \in [0, \tau(z)[$. Fix a $0 < T' < T$. Compute for $z \in \mathbb{H}$ and $0 \leq t < \tau(z)$

$$\frac{dh_t(z)}{dt} = \frac{2}{G_t(z) - \lambda(t)} - \frac{2}{z} = \frac{-2h_t(z) + 2\lambda(t) - \frac{4t}{z}}{z(h_t(z) + z + \frac{2t}{z} - \lambda(t))}.$$

Put $\Lambda = \sup_{0 \leq s \leq T'} |\lambda(s)|$ and $C_1 = 1 + 2T' + \Lambda > 1$. Pick $R > 2C_1$ and estimate for $|z| > R$ and $t < \min\{T', \tau(z)\}$ with $|h_t(z)| < C_1$

$$\left| \frac{dh_t(z)}{dt} \right| \leq \frac{2|h_t(z)|}{|z|(|z| - 2C_1)} + \frac{2C_1}{|z|(|z| - 2C_1)}$$

By Gronwall, using that $h_0(z) = 0$, this implies that for the z's and t's considered

$$|h_t(z)| \leq \frac{2C_1 T'}{|z|(|z| - 2C_1)} e^{\frac{2T'}{R(R-2C_1)}}.$$

Since $h_t(z) = O(R^{-2})$, we conclude that for R large enough $\tau(z) > T'$, implying that $A_t \subseteq B[0, R]$ for $t \in [0, T']$, and that

$$\forall t \in [0, T'], |z| > R : \quad |h_t(z)| \leq \frac{C}{R^2}$$

for some constant $C > 0$. This shows that G_t has the right normalization at infinity and that $\mathrm{hcap}(A_t) = 2t$.   □

Remark 20.13. It is a natural question to ask under what conditions on the driving function λ, the associated family of half-plane hulls A_t are of the form $\gamma(]0, t])$ for some simple curve γ with λ as its driving function. It turns out that this question is related to the degree of regularity of λ.

In Example 20.10, we saw that the half-plane hull A_t (reparametrized by half-plane capacity) ceases to be the range of a simple curve when $t = 1/2$. At $t = 1/2$, the driving function (20.19) goes from being differentiable to being Hölder continuous of degree $1/2$. The role of Hölder continuity is explored in Exercise 20.8.

Kager *et al.* (2004) studied the family of driving functions $\lambda_\kappa(t) = \kappa\sqrt{1-t}$ for $t \in [0, 1]$ and $\kappa > 0$. They solved the inverse problem and identified curves γ_κ. Here one can explicitly see the role of the critical value $\kappa = 4$ discovered by Lind (2005). The curve does not return to the real line if $\kappa < 4$ and it runs back down into the real line if $\kappa \geq 4$, tangentially if $\kappa = 4$. This particular inverse problem was revisited in Lind *et al.* (2010).

The illustration on the cover of this book is a plot of curves γ_κ, which solve the inverse problem with driving function $\lambda_\kappa(t) = \kappa\sqrt{1-t}$, for $\kappa = 1$, 2.5, 3.5, 4, 4.5, 6, 8. The choice of κ's are as in Kager *et al.* (2004).

Remark 20.14. We note that there is a multiple slit version of the theory developed in this section, where one builds half-plane hulls using a finite number of simple curves growing up into the upper half-plane from distinct points on the real axis. We refer the reader to Schleissinger (2012) and Roth and Schleissinger (2014) for details on this topic. See also Exercise 20.9.

Exercises

20.1. Let $\Gamma(s) = -s$ for $s \in [0, \infty[$ be a parametrization of the negative half-axis and let $\Gamma_t^* = \Gamma([t, \infty[)$ for $t \geq 0$.

(α) For $t > 0$ find the biholomorphic map f_t from $B(0,1)$ to $\mathbb{C} \setminus \Gamma_t^*$ with $f_t(0) = 0$, $f_t'(0) > 0$ and the inverse function $g_t = f_t^{-1}$.

(β) Find a change of parametrization $t = t(s)$ of Γ such that $v_s = f_{t(s)}$ has $v_s'(0) = e^s$.

(γ) Verify that $u_s = v_s^{-1}$ satisfies (20.8) with $\tilde{\lambda}(s) = -1$ and (20.9).

20.2. In this exercise the reader is asked to investigate the relation between transformations of a slit plane and its driving function. Let $\Gamma : [0, \infty[\to \mathbb{C} \setminus \{0\}$ be injective and $\lim_{t \to \infty} \Gamma(t) = \infty$ (but not necessarily given the standard parametrization). Let $\tilde{\lambda} : [0, \infty[\to \partial B(0,1)$ be the driving function as introduced in (20.7).

(α) *Translations*: Let $\Sigma = \Gamma + c$, where $c \in \mathbb{C}$ and assume that $0 \notin \Sigma^*$. Show that the driving function of Σ is also $\tilde{\lambda}$.

(β) *Scaling*: Let $\Sigma = a\Gamma$, where $a = re^{i\theta}$ with $r > 0$, $\theta \in \mathbb{R}$. Show that the driving function of Σ is $e^{i\theta} \tilde{\lambda}$.

(γ) *Reflection*: Let $\Sigma = -\bar{\Gamma}$. Show that the driving function of Σ is $\bar{\tilde{\lambda}}$ (complex conjugation).

(δ) *Reparametrization*: Let $\Sigma = \Gamma \circ \tau : [0, S[\to \mathbb{C}$, where $\tau : [0, S[\to [0, \infty[$ is continuous and strictly increasing with $\tau(0) = 0$. Here $0 < S \leq \infty$. Show that the driving function of Σ is $\tilde{\lambda} \circ \tau$.

20.3. Let $A_t = \gamma(]0, t])$ for a continuous function $\gamma : [0, t] \to \mathbb{H}$ with $\gamma(0) \in \mathbb{R}$ and $\operatorname{Im} \gamma(s) > 0$ for $s > 0$. Denote by $F_t : \mathbb{H} \to \mathbb{C}$ the associated Riemann map extended by continuity to $\mathbb{R}$. See the discussion around (20.14).

(α) Argue that $F_t(\overline{\mathbb{H}}) = \overline{\mathbb{H}}$ and $F_t(\mathbb{R}) = \mathbb{R} \cup \gamma([0, t])$.

(β) Show that the two sets in (20.14) are identical as claimed. Note that G_A has been extended by Schwarz reflection to $\mathbb{C} \setminus C_t$ as usual.

(γ) Prove that $F_t(J_t) = \gamma([0, t])$ with J_t given by (20.14).

20.4. Let $T \in]0, \infty]$ and let $\lambda : [0, T[\to \mathbb{R}$ be a continuous function. Denote by $\{A_t\}_{t \in [0, T[}$ the associated family of half-plane hulls.

(α) Let $0 < T' < T$. Argue that there exists $R > 0$ such that $A_t \subseteq B(0, R)$ for all $t \in [0, T']$.

   Hint: See the proof of Theorem 20.12.

(β) Show that $t \to A_t$ is continuous in the sense that: for any $t \in [0, T[$ and sequence $\{t_n\}_{n \in \mathbb{N}} \subseteq [0, T[$ with $t_n \to t$, we have $A_{t_n} \to A_t$ in Carathéodory sense.

   Hint: Prove that $F_{t_n} \to F_t$ locally uniformly in $\mathbb{H}$ using continuity of solutions of ODE's w.r.t. initial conditions.

20.5. Let $A_1 \subseteq A_2$ be two half-plane hulls and assume that there exists a Jordan arc $\gamma : [0, 1] \to \mathbb{C}$ with $\gamma(0) < \gamma(1) \in \mathbb{R}$ and $\partial A_2 \cap \mathbb{H} = \gamma(]0, 1[)$.

(α) Define a holomorphic map $H : \mathbb{H} \to \mathbb{C}$ by setting

$$H(z) = G_{A_1}(F_{A_2}(z)) - z.$$

Argue that $H(\mathbb{H}) \subseteq \mathbb{H}$.

(β) Put $U = \mathbb{R} \setminus [\gamma(0), \gamma(1)]$. Observe that G_{A_1} and G_{A_2} both extend by continuity to U and $G_{A_j}(U) \subseteq \mathbb{R}$. Argue that $B_2 = \mathbb{R} \setminus G_{A_2}(U)$ is a compact interval and that H extends by continuity to $\mathbb{R} \setminus B_2$ with $\operatorname{Im} H(x) = 0$, for $x \in \mathbb{R} \setminus B_2$.

(γ) Show that $\operatorname{Im} H$ extends by continuity to $\overline{\mathbb{H}}$ and that for any z in $\mathbb{H} \setminus A_2$, we have

$$G_{A_1}(z) - G_{A_2}(z) = \frac{1}{\pi} \int_{B_2} \frac{\operatorname{Im} H(x)}{x - G_{A_2}(z)} \, dx.$$

   Hint: Apply Theorem 14.7 with $f = -iH$.

(δ) Conclude that $G_{A_2}(x) \geq G_{A_1}(x)$, for any real $x > \gamma(1)$.

(ε) Show that $A_1 = A_2$ if and only if there exists $x \in U$ such that $G_{A_1}(x) = G_{A_2}(x)$.

(ζ) Finally, let $A \subseteq B[0, 1]$ be a half-plane hull and $x > 1$. Show that $x \leq G_A(x) \leq x + 1/x$.

20.6. In this exercise the reader is asked to investigate the relation between transformations of a growing slit and its driving function. Let $\gamma : [0, T[\to \overline{\mathbb{H}}$ be a simple curve[2] with $\gamma(0) \in \mathbb{R}$ and $\operatorname{Im} \gamma(t) > 0$ for $t \in]0, T[$. Here $0 < T \leq \infty$. Denote by $A_t = \gamma(]0, t])$ the associated half-plane hulls, G_t the Riemann maps and $\lambda : [0, T[\to \mathbb{R}$ the driving function.

---

[2] Not necessarily parametrized by half-plane capacity.

(α) *Translations:* Let $\sigma = \gamma + c$, where $c \in \mathbb{R}$. Show that σ's driving function is $\lambda + c$.

(β) *Scaling:* Let $\sigma = a\gamma$, where $a > 0$. Show that σ's driving function is $a\lambda$.

(γ) *Reflection:* Let $\sigma = -\bar{\gamma}$. Show that σ's driving function is $-\lambda$.

(δ) *Reparametrization:* Let $\sigma = \gamma \circ \tau \colon [0, S[\to \overline{\mathbb{H}}$, where $\tau \colon [0, S[\to [0, T[$ is continuous and strictly increasing with $\tau(0) = 0$. Here $0 < S \leq \infty$. Show that σ's driving function is $\lambda \circ \tau$.

20.7. Verify the claimed expression for the driving function in Example 20.10.

Hint: See Exercise 19.15.

20.8. Let $\lambda \colon [0, \infty[\to \mathbb{R}$ be a continuous function. Suppose there exists $c > 0$ and $0 < p \leq 1$ such that $|\lambda(t) - \lambda(s)| \leq c|t - s|^p$ for all $s, t \geq 0$. Suppose $x \colon [0, T[\to \mathbb{R}$ is a real solution of the Loewner equation $\dot{x} = 2/(x - \lambda(t))$ with $T < \infty$. In particular $x(t) \neq \lambda(t)$ for all $t \in [0, T[$.

(α) Argue that the solution x is strictly monotone.

(β) Argue that the solution x is maximal if and only if $x(T) := \lim_{t \to T_-} = \lambda(T)$.

(γ) From now on we suppose that the solution is maximal and $x(0) > \lambda(0)$. Show that $0 < x(t) - \lambda(t) \leq c(T - t)^p$.

(δ) Conclude that $\lambda(T) - x(t) \geq 2(1 - p)^{-1}c^{-1}(T - t)^{1-p}$ for all $t \in [0, T[$ and show that this implies that we must have $p \leq 1/2$.

(ε) Observe that $\lambda(T) - x(0) \leq cT^p$ and arrive at the estimate $cT^p \leq 2(1 - p)^{-1}c^{-1}T^{1-p}$.

(ζ) Assuming that $p = 1/2$, show that we must have $c > 2$. (If $p < 1/2$, we get no restrictions on c.)

When $p = 1/2$, one may iterate the above argument to improve the constraint on c. Doing this will yield $c > 4$. See Lind (2005). The mechanism explored here goes to back to Marshall and Rohde (2005), who studied the classical Loewner equation in the slit plane.

20.9. Let $\gamma_1, \ldots, \gamma_n \colon [0, T[\to \overline{\mathbb{H}}$ with $0 < T \leq \infty$ be simple curves satisfying, $\gamma_i^* \cap \gamma_j^* = \emptyset$ for $i \neq j$, $\gamma_1(0) < \gamma_2(0) < \cdots < \gamma_n(0)$ real numbers and $\gamma_j(]0, T[) \subseteq \mathbb{H}$ for $j = 1, 2, \ldots, n$.

(α) Show that $A_t = \cup_{j=1}^n \gamma_j(]0, t[)$ is a half-plane hull for each $t \in [0, T[$ and that $t \to \mathrm{hcap}(A_t)$ is continuous. (Here $A_0 = \emptyset$.)

(β) Argue that G_{A_t} extends by continuity to the tips $\gamma_1(t), \ldots, \gamma_n(t)$.

(γ) Define $\lambda_j(t) = G_{A_t}(\gamma_j(t))$ for $j = 1, 2, \ldots, n$. Prove that the driving functions $\lambda_1, \lambda_2, \ldots, \lambda_n : [0, T[\to \mathbb{R}$ are continuous.

Hint: Consult the proof of Theorem 20.7.

Bibliography

Åkerberg, B. (1961). Proof of Poisson's formula, *Proc. Cambridge Philos. Soc.* **57**, p. 186.

Anonyme (1889). Sur l'intégrale $\int_0^\infty e^{-x^2} dx$, *Bull. Sci. Math., II. Sér.* **13**, p. 84.

Apostol, T. M. (1976). *Introduction to analytic number theory* (Springer-Verlag, New York), undergraduate Texts in Mathematics.

Apostol, T. M. (1983). A proof that Euler missed: evaluating $\zeta(2)$ the easy way, *Math. Intelligencer* **5**, 3, pp. 59–60.

Arora, A. K., Goel, S. K. and Rodriguez, D. M. (1988). Special integration techniques for trigonometric integrals, *Amer. Math. Monthly* **95**, 2, pp. 126–130.

Artmann, B. and Gerecke, W. (1990). Didaktische Beobachtungen zu einem Divergenzbeweis für $\sum 1/p$ von Paul Erdős, *Math. Semesterber.* **37**, 1, pp. 46–56.

Axler, S. (1986). Harmonic functions from a complex analysis viewpoint, *Amer. Math. Monthly* **93**, 4, pp. 246–258.

Ayoub, R. (1974). Euler and the zeta function, *Amer. Math. Monthly* **81**, pp. 1067–1086.

Baker, J. A. (1999). The Dirichlet problem for ellipsoids, *Amer. Math. Monthly* **106**, 9, pp. 829–834.

Beck, A. (1986). Notes: The broken spiral theorem, *Amer. Math. Monthly* **93**, 4, p. 293.

Berndt, B. C. (1975). Elementary evaluation of $\zeta(2n)$, *Math. Mag.* **48**, pp. 148–154.

Bieberbach, L. (1916). Über die Koeffizienten derjenigen Potenzreihen, welche eine schlichte Abbildung des Einheitskreises vermitteln, *Berl. Ber.* **1916**, pp. 940–955.

Boas, H. P. and Boas, R. P. (1988). Short proofs of three theorems on harmonic functions, *Proc. Amer. Math. Soc.* **102**, 4, pp. 906–908.

Burckel, R. B. (1979). *An introduction to classical complex analysis. Vol. 1*, *Pure and Applied Mathematics*, Vol. 82 (Academic Press Inc. [Harcourt Brace Jovanovich Publishers], New York).

Campbell, D. M. (1985). Beauty and the beast: the strange case of Andre Bloch, *Math. Intelligencer* **7**, 4, pp. 36–38.

Carathéodory, C. (1954). *Theory of functions of a complex variable. Vol. II*, second english edn. (Chelsea Publishing Company, New York).

Cartan, H. and Ferrand, J. (1988). The case of André Bloch, *Math. Intelligencer* **10**, 1, pp. 23–26.

Chernoff, P. R. (1972). Liouville's theorem for harmonic functions, *Amer. Math. Monthly* **79**, pp. 310–311.

Chernoff, P. R. and Waterhouse, W. C. (1972). Problems and solutions: Solutions of advanced problems: 5781, *Amer. Math. Monthly* **79**, 3, pp. 310–311.

Conway, J. B. (1995). *Functions of one complex variable. II*, *Graduate Texts in Mathematics*, Vol. 159 (Springer-Verlag, New York).

Crowe, J. and Samperi, D. (1989). On open maps, *Amer. Math. Monthly* **96**, 3, pp. 242–243.

Davidson, K. R. (1983). Pointwise limits of analytic functions, *Amer. Math. Monthly* **90**, 6, pp. 391–394.

Davis, P. J. (1959). Leonhard Euler's integral: A historical profile of the gamma function. *Amer. Math. Monthly* **66**, pp. 849–869.

de Branges, L. (1985). A proof of the Bieberbach conjecture, *Acta Math.* **154**, 1-2, pp. 137–152.

Dewan, K. K. and Govil, N. K. (1984). On the Eneström-Kakeya theorem, *J. Approx. Theory* **42**, 3, pp. 239–244.

Dixon, J. D. (1971). A brief proof of Cauchy's integral theorem, *Proc. Amer. Math. Soc.* **29**, pp. 625–626.

Dubinin, V. N. (2010). Lower bounds for the half-plane capacity of compact sets and symmetrization. *Sb. Math.* **201**, 11, pp. 1635–1646.

Dugundji, J. (1966). *Topology* (Allyn and Bacon Inc., Boston, Mass.).

Dunford, N. and Schwartz, J. T. (1958). *Linear Operators. I. General Theory*, With the assistance of W. G. Bade and R. G. Bartle. Pure and Applied Mathematics, Vol. 7 (Interscience Publishers, Inc., New York).

Duren, P. (1983). *Univalent functions, Grundlehren Math. Wiss.*, Vol. 259 (Springer Verlag, New York).

Edwards, H. M. (1974). *Riemann's zeta function* (Academic Press [A subsidiary of Harcourt Brace Jovanovich, Publishers], New York-London), pure and Applied Mathematics, Vol. 58.

Folland, G. B. (1995). *Introduction to partial differential equations*, 2nd edn. (Princeton University Press, New Jersey).

García Arenas, F. and Luz Puertas, M. (1998). Jordan's curve theorem, *Divulg. Mat.* **6**, 1, pp. 43–60.

Gårding, L. (1979). The Dirichlet problem, *Math. Intelligencer* **2**, pp. 43–53.

Gauthier, P. M. (1985). Remarks on a theorem of Keldysh-Lavrent'ev, *Russ. Math. Surv.* **40**, 4, pp. 179–180.

Gilbarg, D. and Trudinger, N. S. (2001). *Elliptic partial differential equations of second order*, Classics in Mathematics (Springer).

Globevnik, J. (1990). Zero integrals on circles and characterizations of harmonic and analytic functions, *Trans. Amer. Math. Soc.* **317**, 1, pp. 313–330.

Goluzin, G. M. (1969). *Geometric theory of functions of a complex variable*, Translations of Mathematical Monographs, Vol. 26 (American Mathematical Society, Providence, R.I.).

Gray, J. (1994). On the history of the Riemann mapping theorem, *Rend. Circ. Mat. Palermo (2) Suppl.*, No. 34, pp. 47–94.

Gray, J. D. and Morris, S. A. (1978). When is a function that satisfies the Cauchy-Riemann equations analytic? *Amer. Math. Monthly* **85**, 4, pp. 246–256.

Gronwall, T. H. (1914). Some remarks on conformal representation, *Ann. Math. (2)* **16**, pp. 72–76.

Halmos, P. R. (1980/81). Does mathematics have elements? *Math. Intelligencer* **3**, 4, pp. 147–153.

Hartman, P. (1964). *Ordinary differential equations* (John Wiley & Sons, Inc.).

Hayman, W. K. and Kennedy, P. B. (1976). *Subharmonic functions. Vol. I* (Academic Press [Harcourt Brace Jovanovich Publishers], London), london Mathematical Society Monographs, No. 9.

Hille, E. (1959). *Analytic function theory. Vol. 1*, Introduction to Higher Mathematics (Ginn and Company, Boston).

Hille, E. (1962). *Analytic function theory. Vol. II*, Introductions to Higher Mathematics (Ginn and Co., Boston, Mass.-New York-Toronto, Ont.).

Hörmander, L. (1963). *Linear partial differential operators*, Die Grundlehren der mathematischen Wissenschaften, Bd. 116 (Academic Press, Inc., Publishers, New York; Springer-Verlag, Berlin-Göttingen-Heidelberg).

Hörmander, L. (1973). *An introduction to complex analysis in several variables*, revised edn. (North-Holland Publishing Co., Amsterdam), north-Holland Mathematical Library, Vol. 7.

Hörmander, L. (1990). *The analysis of linear partial differential operators. I*, 2nd edn., Springer Study Edition (Springer-Verlag, Berlin), distribution theory and Fourier analysis.

Hurwitz, A. (1888). Ueber die Nullstellen der Bessel'schen Function, *Math. Ann.* **33**, 2, pp. 246–266.

Kager, W., Nienhuis, B. and Kadanoff, L. P. (2004). Exact solutions for Loewner evolutions, *J. Statist. Phys.* **115**, 3-4, pp. 805–822.

Knopp, K. (1951). *Theory and Application of Infinite Series*, 2nd edn. (Blackie & Son Ltd., Glasgow).

Kufarev, P. P., Sobolev, V. V. and Sporyševa, L. V. (1968). A certain method of investigation of extremal problems for functions that are univalent in the half-plane, *Trudy Tomsk. Gos. Univ. Ser. Meh.-Mat.* **200**, pp. 142–164.

Lalley, S., Lawler, G. and Narayanan, H. (2009). A geometric interpretation of half-plane capacity, *Electron. Commun. Probab.* **14**, pp. 566—-571.

Lawler, G. F. (2005). *Conformally invariant processes in the plane*, Math. Surveys Monogr., Vol. 114 (American Mathematical Society, Providence, RI).

Lenard, A. (1986). A note on Liouville's theorem, *Amer. Math. Monthly* **93**, 3, pp. 200–201.

Lichtenstein, L. (1916–17). Ueber die erste Randwertaufgabe der Potentialtheorie, *Sitzungsberichte, Berliner Mathematischer Gesellschaft; Archiv Für Mathematik und Physik* **25**, pp. 92–96.

Lind, J. R. (2005). A sharp condition for the Loewner equation to generate slits, *Ann. Acad. Sci. Fenn. Math.* **30**, 1, pp. 143–158.

Lind, J. R., Marshall, D. E. and Rohde, S. (2010). Collisions and spirals of Loewner traces, *Duke Math. J.* **154**, 3, pp. 527–573.

Löwner, K. (1923). Untersuchingen über slichte konforme Abbildungen des Einheitskreises I, *Math. Ann.* **89**, pp. 103–121.

MacGregor, T. H. (1972). Geometric problems in complex analysis, *Amer. Math. Monthly* **79**, pp. 447–468.

Maehara, R. (1984). The jordan curve theorem via the brouwer fixed point theorem, *Am. Math. Mon.* **91**, pp. 641–643.

Marshall, D. E. and Rohde, S. (2005). The Loewner differential equation and slit mappings, *J. Amer. Math. Soc.* **18**, 4, pp. 763–778.

Meyerson, M. D. (1981). Every power series is a Taylor series, *Amer. Math. Monthly* **88**, 1, pp. 51–52.

Minda, D. (1990). The Dirichlet problem for a disk, *Amer. Math. Monthly* **97**, 3, pp. 220–223.

Minda, D. and Schober, G. (1983). Another elementary approach to the theorems of Landau, Montel, Picard and Schottky, *Complex Variables Theory Appl.* **2**, 2, pp. 157–164.

Moise, E. E. (1977). *Geometric topology in dimensions 2 and 3*, no. 47 in Graduate Texts in Mathematics (Springer Verlag, New York).

Narasimhan, R. (1985). *Complex Analysis in One Variable* (Birkhäuser Boston Inc., Boston, MA).

Nelson, E. (1961). A proof of Liouville's theorem, *Proc. Amer. Math. Soc.* **12**, p. 995.

Neuenschwander, E. (1978). Studies in the history of complex function theory. I. The Casorati-Weierstrass theorem, *Historia Math.* **5**, 2, pp. 139–166.

Newman, D. J. (1974). Fourier uniqueness via complex variables, *Amer. Math. Monthly* **81**, pp. 379–380.

Newman, M. H. A. (1954). *Topology of plane sets*, 2nd edn. (Cambridge Univeristy Press).

Novinger, W. P. (1975). Some theorems from geometric function theory: applications, *Amer. Math. Monthly* **82**, pp. 507–510.

Oppenheim, A. V. and Schafer, R. W. (1975). *Digital Signal Processing* (Prentice-Hall Inc., Inglewodd Cliffs, New Jersey).

Osgood, W. F. (1900). On the existence of the green's function for the most general simply connected plane region, *Trans. Amer. Math. Soc.* **1**, pp. 310–314.

Patin, J. M. (1989). A very short proof of Stirling's formula, *Amer. Math. Monthly* **96**, 1, pp. 41–42.

Petrowski, I. G. (1955). *Vorlesungen über partielle Differentialgleichungen* (B. G. Teubner Verlagsgesellschaft, Leipzig).

Protter, M. H. and Weinberger, H. F. (1967). *Maximum principles in differential equations* (Prentice-Hall Inc., Englewood Cliffs, N.J.).

Rao, K. N. S. (1972). A contour for the Poisson integral, *Elem. Math.* **27**, pp. 88–90.

Rosay, J.-P. (1982). Injective holomorphic mappings, *Amer. Math. Monthly* **89**, 8, pp. 587–588.

Rosay, J.-P. and Rudin, W. (1989). Arakelian's approximation theorem, *Amer. Math. Monthly* **96**, 5, pp. 432–434.

Roth, O. and Schleissinger, S. (2014). The Schramm-Loewner equation for multiple slits, *arXiv:1311.0672v2* .

Rudin, W. (1974). *Real and complex analysis*, 2nd edn. (McGraw-Hill Book Co., New York), mcGraw-Hill Series in Higher Mathematics.

Sansone, G. and Gerretsen, J. (1960). *Lectures on the theory of functions of a complex variable. I. Holomorphic functions* (P. Noordhoff, Groningen).

Sansone, G. and Gerretsen, J. (1969). *Lectures on the theory of functions of a complex variable. II: Geometric theory* (Wolters-Noordhoff Publishing, Groningen).

Schleissinger, S. (2012). The multiple-slit version of Loewner's differential equation and pointwise Hölder continuity of driving functions, *Ann. Acad. Sci. Fenn. Math.* **37**, 1, pp. 191–201.

Selberg, A. (1949). An elementary proof of the prime-number theorem, *Ann. of Math. (2)* **50**, pp. 305–313.

Spanier, E. H. (1966). *Algebraic topology* (McGraw-Hill Book Co., New York).

Thomassen, C. (1992). The Jordan-Schönflies theorem and the classification of surfaces, *Amer. Math. Monthly* **99**, 2, pp. 116–130.

Tideman, M. (1954). Elementary proof of a uniqueness theorem for positive harmonic functions, *Nordisk Mat. Tidskr.* **2**, pp. 95–96.

Trèves, F. (1975). *Basic linear partial differential equations* (Academic Press [A subsidiary of Harcourt Brace Jovanovich, Publishers], New York-London), pure and Applied Mathematics, Vol. 62.

Tsarpalias, A. (1989). A version of Rouché's theorem for continuous functions, *Amer. Math. Monthly* **96**, 10, pp. 911–913.

van de Lune, J. and te Riele, H. J. J. (1983). On the zeros of the Riemann zeta function in the critical strip. III, *Math. Comp.* **41**, 164, pp. 759–767.

van der Poorten, A. (1978/79). A proof that Euler missed...Apéry's proof of the irrationality of $\zeta(3)$, *Math. Intelligencer* **1**, 4, pp. 195–203, an informal report.

Vitushkin, A. G. (1987). Uniform approximation of functions by holomorphic functions, *Trudy Mat. Inst. Steklov.* **176**, pp. 300–307, 328, translated in Proc. Steklov Inst. Math. 1988, no. 3, 301–308, Mathematical physics and complex analysis (Russian).

Wagon, S. (1986). Where are the zeros of zeta of *s*? *Math. Intelligencer* **8**, 4, pp. 57–62.

Walter, W. (1982). Old and new approaches to Euler's trigonometric expansions, *Amer. Math. Monthly* **89**, 4, pp. 225–230.

Young, R. M. (1986). Notes: On Jensen's formula and $\int_0^{2\pi} \log|1 - e^{i\theta}|d\theta$, *Amer. Math. Monthly* **93**, 1, pp. 44–45.

Zalcman, L. (1974). Real proofs of complex theorems (and vice versa), *Amer. Math. Monthly* **81**, pp. 115–137.

Glossary

Page numbers refer to relevant definitions.

$\mathrm{Arg}\,z$	principal branch of $\arg z$, 29
$\arg z$	argument of z, 27
$\mathrm{Cap}(K)$	logarithmic capacity of K, 344
$\mathbb{C}_\infty$	extended complex plane, 123
$\partial/\partial \bar{z}$	Cauchy-Riemann operator, 14, 153
Δ	Laplace operator, 203
$\partial_A G$	set of equivalence classes of accessible boundary points, 277
E_γ	limit set of γ, 304
$\exp$	exponential function, 25
Γ	Gamma function, 173
γ^*	image of the curve γ, 18
$\mathrm{hcap}(A)$	half-plan capacity of A, 358
H^p	Hardy space, 237
$\int_\gamma F(z) \cdot dz$	integration of a vector field F along a path γ, 295, 296
$\int_\gamma f$	line integral of f along γ, 19
$l(\gamma)$	length of the path γ, 18
Log	principal branch of $\log z$, 6, 29
$\log$	any logarithm, 27
$\prod_{k=1}^\infty c_k$	infinite product, 162
P_R	Poisson kernel, 208
ρ	radius of convergence, 2

$\mathrm{Res}(f;a)$	residue of f at a, 91
$\rho_A(B)$	distance of the set B from the set A, 280
$\sqrt{z}$	principal square root of z, 40
usc	upper semi-continuous, 225
$\mathrm{Ind}_\gamma(z)$	winding number of γ relative to z, 34
ζ	ζ-function, 187

Index

411

Printed in the United States
By Bookmasters